George Angelos Papadopoulos and Costin Badica (Eds.)

Intelligent Distributed Computing III

AF786027

Studies in Computational Intelligence, Volume 237

Editor-in-Chief
Prof. Janusz Kacprzyk
Systems Research Institute
Polish Academy of Sciences
ul. Newelska 6
01-447 Warsaw
Poland
E-mail: kacprzyk@ibspan.waw.pl

Further volumes of this series can be found on our homepage: springer.com

Vol. 216. Matthew Taylor
Transfer in Reinforcement Learning Domains, 2009
ISBN 978-3-642-01881-7

Vol. 217. Horia-Nicolai Teodorescu, Junzo Watada, and Lakhmi C. Jain (Eds.)
Intelligent Systems and Technologies, 2009
ISBN 978-3-642-01884-8

Vol. 218. Maria do Carmo Nicoletti and Lakhmi C. Jain (Eds.)
Computational Intelligence Techniques for Bioprocess Modelling, Supervision and Control, 2009
ISBN 978-3-642-01887-9

Vol. 219. Maja Hadzic, Elizabeth Chang, Pornpit Wongthongtham, and Tharam Dillon
Ontology-Based Multi-Agent Systems, 2009
ISBN 978-3-642-01903-6

Vol. 220. Bettina Berendt, Dunja Mladenic, Marco de de Gemmis, Giovanni Semeraro, Myra Spiliopoulou, Gerd Stumme, Vojtech Svatek, and Filip Zelezny (Eds.)
Knowledge Discovery Enhanced with Semantic and Social Information, 2009
ISBN 978-3-642-01890-9

Vol. 221. Tassilo Pellegrini, Sören Auer, Klaus Tochtermann, and Sebastian Schaffert (Eds.)
Networked Knowledge - Networked Media, 2009
ISBN 978-3-642-02183-1

Vol. 222. Elisabeth Rakus-Andersson, Ronald R. Yager, Nikhil Ichalkaranje, and Lakhmi C. Jain (Eds.)
Recent Advances in Decision Making, 2009
ISBN 978-3-642-02186-2

Vol. 223. Zbigniew W. Ras and Agnieszka Dardzinska (Eds.)
Advances in Data Management, 2009
ISBN 978-3-642-02189-3

Vol. 224. Amandeep S. Sidhu and Tharam S. Dillon (Eds.)
Biomedical Data and Applications, 2009
ISBN 978-3-642-02192-3

Vol. 225. Danuta Zakrzewska, Ernestina Menasalvas, and Liliana Byczkowska-Lipinska (Eds.)
Methods and Supporting Technologies for Data Analysis, 2009
ISBN 978-3-642-02195-4

Vol. 226. Ernesto Damiani, Jechang Jeong, Robert J. Howlett, and Lakhmi C. Jain (Eds.)
New Directions in Intelligent Interactive Multimedia Systems and Services - 2, 2009
ISBN 978-3-642-02936-3

Vol. 227. Jeng-Shyang Pan, Hsiang-Cheh Huang, and Lakhmi C. Jain (Eds.)
Information Hiding and Applications, 2009
ISBN 978-3-642-02334-7

Vol. 228. Lidia Ogiela and Marek R. Ogiela
Cognitive Techniques in Visual Data Interpretation, 2009
ISBN 978-3-642-02692-8

Vol. 229. Giovanna Castellano, Lakhmi C. Jain, and Anna Maria Fanelli (Eds.)
Web Personalization in Intelligent Environments, 2009
ISBN 978-3-642-02793-2

Vol. 230. Uday K. Chakraborty (Ed.)
Computational Intelligence in Flow Shop and Job Shop Scheduling, 2009
ISBN 978-3-642-02835-9

Vol. 231. Mislav Grgic, Kresimir Delac, and Mohammed Ghanbari (Eds.)
Recent Advances in Multimedia Signal Processing and Communications, 2009
ISBN 978-3-642-02899-1

Vol. 232. Feng-Hsing Wang, Jeng-Shyang Pan, and Lakhmi C. Jain
Innovations in Digital Watermarking Techniques, 2009
ISBN 978-3-642-03186-1

Vol. 233. Takayuki Ito, Minjie Zhang, Valentin Robu, Shaheen Fatima, and Tokuro Matsuo (Eds.)
Advances in Agent-Based Complex Automated Negotiations, 2009
ISBN 978-3-642-03189-2

Vol. 234. Aruna Chakraborty and Amit Konar
Emotional Intelligence, 2009
ISBN 978-3-540-68606-4

Vol. 235. Reiner Onken and Axel Schulte
System-Ergonomic Design of Cognitive Automation, 2009
ISBN 978-3-642-03134-2

Vol. 236. Natalio Krasnogor, Belén Melián-Batista, José A. Moreno-Pérez, J. Marcos Moreno-Vega, and David Pelta
Nature Inspired Cooperative Strategies for Optimization, 2009
ISBN 978-3-642-03134-2

Vol. 237. George Angelos Papadopoulos and Costin Badica (Eds.)
Intelligent Distributed Computing III, 2009
ISBN 978-3-642-03213-4

George Angelos Papadopoulos and Costin Badica (Eds.)

Intelligent Distributed Computing III

Proceedings of the 3rd International Symposium on Intelligent Distributed Computing – IDC 2009, Ayia Napa, Cyprus, October 2009

Springer

George Angelos Papadopoulos
Department of Computer Science
University of Cyprus
75 Kallipoleos Str.
P.O. Box 20537, CY-1678
Nicosia, Cyprus
E-mail: george@cs.ucy.ac.cy

Costin Badica
Software Engineering Department
Faculty of Automatics
Computers and Electronics
University of Craiova
Bvd. Decebal, Nr. 107
Craiova, RO-200440, Romania
E-mail: badica_costin@software.ucv.ro

ISBN 978-3-642-26930-1 e-ISBN 978-3-642-03214-1

DOI 10.1007/978-3-642-03214-1

Studies in Computational Intelligence ISSN 1860-949X

© 2009 Springer-Verlag Berlin Heidelberg
Softcover reprint of the hardcover 1st edition 2009

This work is subject to copyright. All rights are reserved, whether the whole or part of the material is concerned, specifically the rights of translation, reprinting, reuse of illustrations, recitation, broadcasting, reproduction on microfilm or in any other way, and storage in data banks. Duplication of this publication or parts thereof is permitted only under the provisions of the German Copyright Law of September 9, 1965, in its current version, and permission for use must always be obtained from Springer. Violations are liable to prosecution under the German Copyright Law.

The use of general descriptive names, registered names, trademarks, etc. in this publication does not imply, even in the absence of a specific statement, that such names are exempt from the relevant protective laws and regulations and therefore free for general use.

Typeset & Cover Design: Scientific Publishing Services Pvt. Ltd., Chennai, India.

Printed in acid-free paper

9 8 7 6 5 4 3 2 1

springer.com

Preface

Intelligent computing covers a hybrid palette of methods and techniques derived from classical artificial intelligence, computational intelligence, multi-agent systems a.o. Distributed computing studies systems that contain loosely-coupled components running on networked computers and that communicate and coordinate their actions by exchange of messages. The emergent field of intelligent distributed computing is expected to pose special challenges of adaptation and fruitful combination of results of both areas with a great impact on the development of new generation intelligent distributed information systems.

Intelligent Distributed Computing – IDC Symposium Series was started as an initiative of research groups from: (i) *Systems Research Institute, Polish Academy of Sciences in Warsaw, Poland* and (ii) *Software Engineering Department of the University of Craiova, Craiova, Romania*. IDC aims at bringing together researchers and practitioners involved in all aspects of intelligent distributed computing. IDC 2009 was the third event in this series and was hosted by *Department of Computer Science, University of Cyprus* in *Ayia Napa, Cyprus* during October 13-14, 2009.

This book represents the peer-reviewed proceedings of the IDC 2009. We received 47 submissions from 24 countries. Each submission was carefully reviewed by at least 3 members of the Program Committee. Acceptance and publication were judged based on the relevance to the symposium themes, clarity of presentation, originality and accuracy of results and proposed solutions. Finally 16 regular papers and 18 short papers were selected for presentation and were included in this volume, resulting in acceptance rates of 34.04 % for regular papers and 72.34 % for regular and short papers. The book contains also 2 invited papers authored by well-known professors in the field.

The 36 contributions in this book address many topics related to theory and applications of intelligent distributed computing, including: actor-agent systems, agent-based simulation, autonomic computing, computational service economies, defeasible reasoning, distributed data mining, distributed logic programming, e-learning, emergent properties in complex systems, formal methods of intelligent distributed systems, genetic and evolutionary algorithms, information retrieval, knowledge fusion, multi-sensor networks, mobile ad hoc networks, mobile computing, ontologies and metadata, peer-to-peer networks, process modeling and integration, remote sensing distributed systems, secure e-payment systems, social networks, surveillance and disaster management applications, swarm computing, Web services and systems.

We would like to thank to Prof. Janusz Kacprzyk, editor of *Studies in Computational Intelligence* series and member of the Steering Committee for their kind support and encouragement in starting and continuing the IDC Symposium Series. We would like to thank to the Program Committee members for their work in promoting the event and refereeing submissions and also to all colleagues who submitted papers to the IDC 2009. We deeply appreciate the efforts of our invited speakers (in alphabetical order): Prof. Nicholas R. Jennings and Prof. Dana Petcu and thank them for their interesting lectures. Special thanks also go to members of the Department of Computer Science, University of Cyprus for their help with organizing the IDC 2009 event.

July 2009
Nicosia, Craiova

George A. Papadopoulos
Costin Bădică

Organization

Organizers

Department of Computer Science, University of Cyprus, Cyprus
Software Engineering Department, Faculty of Automation, Computers and Electronics, University of Craiova, Romania

Conference Chairs

George A. Papadopoulos	University of Cyprus, Cyprus
Costin Bădică	University of Craiova, Romania

Steering Committee

Costin Bădică	University of Craiova, Romania
Janusz Kacprzyk	Polish Academy of Sciences, Poland
Michele Malgeri	Università di Catania, Italia
George A. Papadopoulos	University of Cyprus, Cyprus
Marcin Paprzycki	Polish Academy of Sciences, Poland

Organizing Committee

George A. Papadopoulos	University of Cyprus, Cyprus
Konstantinos Kakousis	University of Cyprus, Cyprus
Dumitru Dan Burdescu	University of Craiova, Romania
Mihai Mocanu	University of Craiova, Romania
Elvira Popescu	University of Craiova, Romania

Invited Speakers

Nicholas R. Jennings	Southampton University, UK
Dana Petcu	Western University of Timişoara and Institute e-Austria Timişoara, Romania

Program Committee

Razvan Andonie	Central Washigton University, USA
Galia Angelova	Bulgarian Academy of Sciences, Bulgaria
Nick Bassiliades	Aristotle University of Thessaloniki, Greece
Doina Bein, ARL	Pennsylvania State University, USA
Frances Brazier	Vrije Universiteit, Amsterdam, The Netherlands
Dumitru Dan Burdescu	University of Craiova, Romania
Giacomo Cabri	University of Modena and Reggio Emilia, Italy
David Camacho	Universidad Autonoma de Madrid, Spain
Vincenza Carchiolo	University of Catania, Italy
Jen-Yao Chung	IBM T.J. Watson Research Center, USA
Gabriel Ciobanu	"A.I.Cuza" University of Iaşi, Romania
Valentin Cristea	"Politehnica" University of Bucharest, Romania
Paul Davidsson	Blekinge Institute of Technology, Sweden
Beniamino Di Martino	Second University of Naples, Italy
Vadim A. Ermolayev	Zaporozhye National University, Ukraine
Adina Magda Florea	"Politehnica" University of Bucharest, Romania
Chris Fox	University of Essex, UK
Maria Ganzha	Elblag University of Humanities and Economics, Poland
Adrian Giurca	Brandenburg University of Technology at Cottbus, Germany
Nathan Griffiths	University of Warwick, UK
De-Shuang Huang	Chinese Academy of Sciences, China
Mirjana Ivanović	University of Novi Sad, Serbia
Ivan Jelinek	Czech Technical University, Czech Republic
Igor Kotenko	Russian Academy of Sciences, Russia
Halina Kwasnicka	Wroclaw University of Technology, Poland
Ioan Alfred Leţia	Technical University of Cluj-Napoca, Romania
Alessandro Longheu	University of Catania, Italy
Heitor Silverio Lopes	Federal University of Technology - Parana, Brazil
José Machado	University of Minho, Portugal

Giuseppe Mangioni	University of Catania, Italy
Yannis Manolopoulos	Aristotle University of Thessaloniki, Greece
Ronaldo Menezes	Florida Institute of Technology, USA
Mihai Mocanu	University of Craiova, Romania
Alberto Montresor	University of Trento, Italy
Alexandros Nanopoulos	Aristotle University of Thessaloniki, Greece
Viorel Negru	Western University of Timişoara, Romania
José Neves	University of Minho, Portugal
Peter Noerr	MuseGlobal, Inc., USA
Andrea Omicini	University of Bologna, Italy
Gregor Pavlin	D-CIS Lab / Thales Research & Technology, The Netherlands
Dana Petcu	Western University of Timişoara, Romania
Joël Quinqueton	CNRS & Montpellier University, France
Shahram Rahimi	Southern Illinois University, USA
Yucel Saygin	Sabanci University, Turkey
Rainer Unland	University of Duisburg-Essen, Germany
Laurenţiu Vasiliu	National University of Ireland, Ireland
Laurent Vercouter	Ecole des Mines de St-Etienne, France
Lucian Vinţan	Academy of Technical Sciences, Romania
Nikos Vlassis	Technical University of Crete, Greece
Peter Vojtáš	Charles University, Czech Republic
Gerd Wagner	Brandenburg University of Technology at Cottbus, Germany
Niek Wijngaards	D-CIS Lab / Thales Research & Technology, The Netherlands
Franz Wotawa	Graz University of Technology, Austria
Franco Zambonelli	University of Modena and Reggio Emilia, Italy
Filip Zavoral	Charles University, Czech Republic

Contents

Computational Service Economies: Design and Applications

Nicholas R. Jennings and Alex Rogers

Many modern computing systems have to operate in environments that are highly interconnected, highly unpredictable, in a constant state of flux, have no centralized point of control, and have constituent components owned by a variety of stakeholders that each have their own aims and objectives. Relevant exemplars include the Web, Grid Computing, Peer-to-Peer systems, Sensor Networks, Cloud Computing, Pervasive Computing and many eCommerce applications. Now, although these areas have somewhat different areas of emphasis, we believe they can all be viewed as operating under the same conceptual model. Namely, one in which: (i) entities offer a variety of services in some form of institutional setting; (ii) other entities connect to these services (covering issues such as service discovery, service composition and service procurement); and (iii) entities enact services, subject to service agreements, in a flexible and context sensitive manner. Moreover, we believe agent-based computing, with its emphasis on autonomous, flexible action and interaction, is an appropriate computational model for conceptualizing, designing and implementing such systems [12, 13]. In particular, such agents are a natural way of viewing flexible service providers and consumers that need both to respond to environmental change, while being able to work towards their individual and collective aims. Moreover, the interactions between these agents need to take place in some form of electronic institution that structures the interactions and can provide an effective matching of the appropriate producers and consumers. Everyday examples of such institutions might include: eBay, Second Life, Betfair exchanges or World of Warcraft. Nevertheless, in the work described here, we will focus specifically on trading institutions and in particular computational service economies that mediate the exchanges between software agents in the area of software services.

In more detail, when thinking about the design of such computational service economies there are two major perspectives that need to be considered [4]. First,

Nicholas R. Jennings · Alex Rogers
School of Electronics and Computer Science, University of Southampton,
Southampton SO17 1BJ, UK
e-mail: nrj@ecs.soton.ac.uk, acr@ecs.soton.ac.uk

G.A. Papadopoulos and C. Badica (Eds.): Intelligent Distributed Computing III, SCI 237, pp. 1–7.
springerlink.com © Springer-Verlag Berlin Heidelberg 2009

there is the design of the institution itself; this can essentially be thought of as the rules of the game. This covers issues such as who are the valid participants (e.g. buyers, sellers, intermediaries), what are the interaction states (e.g. accepting bids, negotiation closed), what are the events causing state transitions (e.g. bid, time out, proposal accepted), what are the valid actions of the participants (e.g. propose, accept, reject, counter-propose) and what are the reward structures (e.g. who pays and who gets paid for what). Second, there is the design of the strategies for the agents that participate in the institution; this can essentially be thought of as how to succeed in the game. Such strategies are very much determined by the institutional rules and are essentially the decision making employed to achieve the agents trading objectives. They can range from the very simple, to the very complex and they can be targeted at maximising the benefit to the individual agent (self interest) or to the wider group (social welfare).

With the team here at Southampton, we have focused on both the techniques for constructing such computational service economies and on developing applications using such techniques. In the former case, we have made advances in the areas of auctions [5, 6, 28, 32, 11], coalition formation [3, 7, 19, 1, 20], automated negotiation [8, 14, 22, 7], trust and reputation [21, 23, 24], flexible reasoning strategies for workflows [29] and decentralized coordination [26, 10]. In the latter case, we have built applications using these techniques in areas such as: virtual organizations [16], emergency response [2], sensor networks [17, 15, 25], mobile sensors [31], computational grids [30] and personalized recommendations [35, 18].

In the context of this work, it is possible to identify three broad types of application that we have been involved with. First, are those in which we, as system designers, have control over the institution, but not the strategies of the participating agents. Second, are those in which we have control over the agents strategies, but not the institution. Finally, are those in which we have control over both the strategy and the institution. We will now briefly provide details of exemplar applications that we have recently developed from each class.

First, we consider an application of computational service economies for environmental sensor networks in general [27] and glacial monitoring in particular (see figure 1). In this work we developed an energy-aware self-organized routing algorithm for the networking of simple battery-powered wireless micro-sensors that might be owned by different organisations or individuals. In these networks, the battery life of individual sensors is typically limited by the power required to transmit their data to a receiver or sink. Thus effective network routing algorithms allow us to reduce this power and extend both the lifetime and the coverage of the sensor network as a whole. In particular, if agents offer the service of forwarding information toward the basestation for one another, then the overall system can sense the environment for longer. However, implementing such routing algorithms with a centralized controller is undesirable due to the physical distribution of the sensors, their limited localization ability and the dynamic nature of such networks (given that sensors may fail, move or be added at any time and the communication links between sensors are subject to noise and interference). Against this background, we devised a distributed mechanism that enabled individual sensors to follow locally

Fig. 1 Deploying the environmental sensors at the Briksdalsbreen glacier in Norway. Left part is the sensors being blasted into the ice, right part is an individual sensor node.

selfish strategies, which, in turn, result in the self-organization of a routing network with desirable global properties. The mechanism uses a locally calculated payment scheme that allows individual sensors to make local decisions as to whether their own limited battery resources would be better used to transmit their own information, or used to forward the information of other sensors in the network. We showed that our mechanism performs close to the optimal solution (as computed by a centralized optimizer), it deals adaptively with changing sensor numbers and topology, and it extends the useful life of the network by a factor of three over the traditional approach.

Second, we consider work on developing strategies for auctions that allocate aircraft engines that require maintenance to repair and overhaul bases [34]. In this context, the structure of the electronic institution is fixed. In particular, it is a variant of the Continuous Double Auction (CDA) in which there are multiple buyers (aircraft that require maintenance or repair services for their engines) and multiple sellers (workshops that are able to offer the service of undertaking such work) that need to be matched on an ongoing basis (see figure 2). The challenge in this case is to design an effective strategy for the participating agents. In particular, we developed a novel strategy that is based on both short and long-term learning that allows such agents to adapt their bidding behaviour to be efficient in a wide variety of environments [33]. For the short-term learning, the agent updates the aggressiveness of its bidding

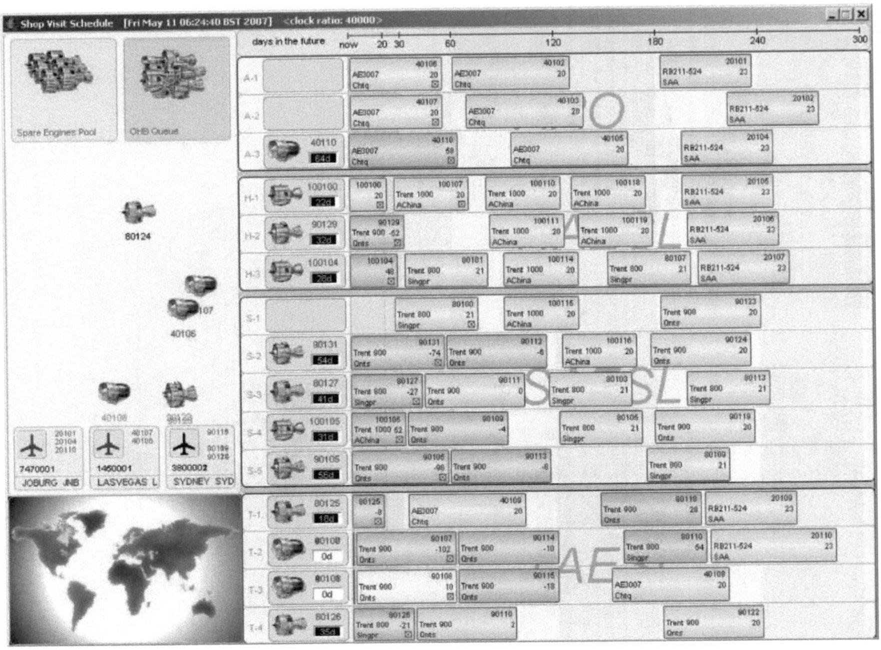

Fig. 2 Aero-Engine Repair and Overhaul Interface. The multiple service lines of four overhaul bases (A, H, S and T) are shown on the right of the figure with the current engine schedule. The left of the figure shows the current state of the engine pool and indicates which aircraft are currently awaiting engine removal.

behaviour (more aggressive means it will trade off profit to improve its chance of transacting, less aggressive that it targets more profitable transactions and is willing to trade of its chance of transacting to achieve them) based on market information observed after any bid or ask appears in the market. The long-term learning then determines how this aggressiveness factor influences an agents choice of which bids or asks to submit in the market, and is based on market information observed after every transaction (successfully matched bid and ask). The principal motivation for the short-term learning is to enable the agent to immediately respond to market fluctuations, while for the long-term learning it is to adapt to broader trends in the way in which the market demand and supply changes over time. We benchmark our strategy against the current state of the art and show that it outperforms these benchmarks in both static and dynamic environments.

Finally, we consider the case where we, as system designers, have control over both the agents strategies and the institution design. Specifically we consider the case where teams of sensors need to coordinate with one another in order to focus the system resources on high priority targets or locations (see figure 3). Now, in most realistic settings, this needs to be carried out in a decentralised fashion because of the scale of the operation and the inherent dynamism of the domain. In this setting,

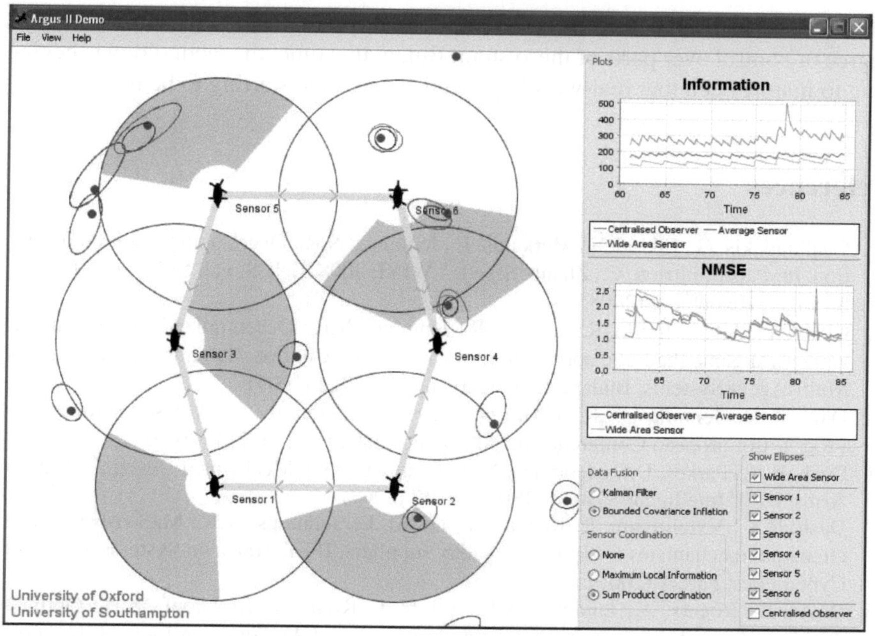

Fig. 3 Coordinated tracking with teams of sensors (represented as helicopters). Blue shaded region represents the agents direction of observation, yellow lines indicate inter-agent communication, and red dots are targets of interests.

the design of the institution involved determining the type and content of messages that can be exchanged between the interacting agents, the solution concept that is required (in this case, the agents adopted a cooperative strategy of maximising the total amount of information within the system, and were thus attempting to maximise the social welfare of the system), and finally, the computational algorithm that is used to compute this solution. The particular algorithm developed within this work is a derivative of the max-sum message passing algorithm [10] and this proved to be very effective. In particular, it finds good solutions to this global optimisation problem in a decentralised fashion, it is communication efficient, there is no aggregation of calculations at a single agent, it operates with asynchronous communications and calculations, it degrades gracefully with lossy communication and it continuously adapts solution within dynamic setting.

In summary, we believe that computational service economies are a good basis for designing and building many complex systems. They provide a good set of conceptual structures and there is a strong toolset available from the fields of decision theory, game theory, mechanism design, information theory and machine learning. Nevertheless, determining system behaviour and/or the effectiveness of the individual participants in such systems is always going to be a challenging task, because of their decentralised nature, the presence of multiple stakeholders, and the limited degree of control over parts of the system. But, at this time, the technology is beginning to mature such that real-world applications are now starting to be practicable.

References

1. Chalkiadakis, G., Elkind, E., Markakis, E., Jennings, N.R.: Overlapping coalition formation. In: Papadimitriou, C., Zhang, S. (eds.) WINE 2008. LNCS, vol. 5385, pp. 307–321. Springer, Heidelberg (2008)
2. Chapman, A.C., Micillo, R.A., Kota, R., Jennings, N.R.: Decentralized task allocation: a practical game theoretic approach. In: Proc. 8th Int. Conf on Autonomous Agents and Multi-Agent Systems, Budapest, Hungary, pp. 915–922 (2009)
3. Dang, V.D., Jennings, N.R.: Coalition structure generation in task-based settings. In: Proc. 17th European Conference on AI, Trento, Italy, pp. 210–214 (2006)
4. Dash, R.K., Parkes, D.C., Jennings, N.R.: Computational Mechanism Design: A Call to Arms. IEEE Intelligent Systems 18(6), 40–47 (2003)
5. Dash, R.K., Vytelingum, P., Rogers, A., David, E., Jennings, N.R.: Market-based task allocation mechanisms for limited capacity suppliers. IEEE Trans. on Systems, Man and Cybernetics, Part A (2007)
6. David, E., Rogers, A., Jennings, N.R., Schiff, J., Kraus, S., Rothkopf, M.H.: Optimal design of English auctions with discrete bid levels. ACM Trans. on Internet Technology 7(2), 34, article 12 (2007)
7. Fatima, S.S., Wooldridge, M., Jennings, N.R.: An agenda based framework for multi-issues negotiation. Artificial Intelligence Journal 152(1), 1–45 (2004)
8. Fatima, S.S., Wooldridge, M., Jennings, N.R.: Multi-issue negotiation with deadlines. Journal of AI Research 27, 381–417 (2006)
9. Fatima, S.S., Wooldridge, M., Jennings, N.R.: A linear approximation method for the Shapley value. Artificial Intelligence Journal 172(14), 1673–1699 (2008)
10. Farinelli, A., Petcu, A., Rogers, A., Jennings, N.R.: Decentralised coordination of low-power embedded devices using the max-sum algorithm. In: Proc. 7th Int Conf. on Autonomous Agents and Multi-Agent Systems, Estoril, Portugal, pp. 639–646 (2008)
11. Gerding, E.H., Dash, R.K., Byde, A., Jennings, N.R.: Optimal strategies for bidding agents participating in simultaneous Vickrey auctions with perfect substitutes. Journal of AI Research 32, 939–982 (2008)
12. Jennings, N.R.: On Agent-Based Software Engineering. Artificial Intelligence Journal 117(2), 277–296 (2000)
13. Jennings, N.R.: An agent-based approach for building complex software systems. Comms. of the ACM 44(4), 35–41 (2001)
14. Karunatillake, N.C., Jennings, N.R., Rahwan, I., McBurney, P.: Dialogue games that agents play within a society. Artificial Intelligence Journal 173(9-10), 935–981 (2009)
15. Kho, J., Rogers, A., Jennings, N.R.: Decentralised control of adaptive sampling in wireless sensor networks. ACM Trans. on Sensor Networks 5(3), article 19, 35 (2009)
16. Norman, T.J., Preece, A., Chalmers, S., Jennings, N.R., Luck, M., Dang, V.D., Nguyen, T.D., Deora, V., Shao, J., Gray, A., Fiddian, N.: Agent-based formation of virtual organizations. Int. J. Knowledge Based Systems 17(2-4), 103–111 (2004)
17. Padhy, P., Dash, R.K., Martinez, K., Jennings, N.R.: A utility-based sensing and communication model for a glacial sensor network. In: Proc. 5th Int. Conf. on Autonomous Agents and Multi-Agent Systems, Hakodate, Japan, pp. 1353–1360 (2006)

18. Payne, T.R., David, E., Jennings, N.R., Sharifi, M.: Auction mechanisms for efficient advertisement selection on public displays. In: Proc. 17th European Conference on AI, Trento, Italy, pp. 285–289 (2006)
19. Rahwan, T., Jennings, N.R.: An algorithm for distributing coalitional value calculations among cooperating agents. Artificial Intelligence Journal 171(8-9), 535–567 (2007)
20. Rahwan, T., Ramchurn, S.D., Giovannucci, A., Jennings, N.R.: An anytime algorithm for optimal coalition structure generation. Journal of AI Research 34, 521–567 (2009)
21. Ramchurn, S.D., Dash, R.K., Giovannucci, A., Rodriguez-Aguilar, J., Mezzetti, C., Jennings, N.R.: Trust-based mechanisms for robust and efficient task allocation in the presence of execution uncertainty. Journal of AI Research 35, 119–159 (2009)
22. Ramchurn, S.D., Sierra, C., Godo, L., Jennings, N.R.: Negotiating using rewards. Artificial Intelligence Journal 171(10-15), 805–837 (2007)
23. Reece, S., Rogers, A., Roberts, S., Jennings, N.R.: Rumours: evaluating multi-dimensional trust within a decentralized reputation system. In: Proc. 6th Int. J. Conference on Autonomous Agents and Multi-agent Systems, Hawaii, USA, pp. 1063–1070 (2007)
24. Reece, S., Rogers, A., Roberts, S., Jennings, N.R.: A multi-dimensional trust model for heterogeneous contract observations. In: Proc. 22nd Conference on Artificial Intelligence (AAAI), Vancouver, Canada, pp. 128–135 (2007)
25. Rogers, A., Corkill, D.D., Jennings, N.R.: Agent technologies for sensor networks. IEEE Intelligent Systems 24(2), 13–17 (2009)
26. Rogers, A., Dash, R.K., Ramchurn, S.D., Vytelingum, P., Jennings, N.R.: Coordinating team players within a noisy iterated Prisoners Dilemma tournament. Theoretical Computer Science 377(1-3), 243–259 (2007)
27. Rogers, A., David, E., Jennings, N.R.: Self organized routing for wireless micro-sensor networks. IEEE Trans. on Systems, Man and Cybernetics (Part A) 35(3), 349–359 (2005)
28. Rogers, A., David, E., Schiff, J., Jennings, N.R.: The Effects of Proxy Bidding and Minimum Bid Increments within eBay Auctions. ACM Transactions on the Web 1(2), article 9, 28 pages (2007)
29. Stein, S., Payne, T.R., Jennings, N.R.: Flexible provisioning of web service workflows. ACM Trans. on Internet Technology 9(1), article 2, 45 pages (2009)
30. Stein, S., Payne, T.R., Jennings, N.R.: Flexible selection of heterogeneous and unreliable services in large-scale grids. Philosophical Trans. of the Royal Society A: Mathematical, Physical and Engineering Sciences 367(1897), 2483–2494 (2009)
31. Stranders, R., Farinelli, A., Rogers, A., Jennings, N.R.: Decentralised coordination of mobile sensors using the max-sum algorithm. In: Proc. 21st Int. Joint Conf. on AI (IJCAI), Pasadena, USA (2009)
32. Vetsikas, I., Jennings, N.R., Selman, B.: Generating Bayes-Nash equilibria to design autonomous trading agents. In: Proc. 20th Int. Joint Conf. on AI (IJCAI), Hyderabad, India, pp. 1543–1550 (2007)
33. Vytelingum, P., Cliff, D., Jennings, N.R.: Strategic bidding in continuous double auctions. Artificial Intelligence Journal 172(14), 1700–1729 (2008)
34. Vytelingum, P., Macbeth, D.K., Dutta, P., Stranjak, A., Rogers, A., Jennings, N.R.: A market-based approach to multiple-factory scheduling. In: Proc. 1st Int. Conf. on Auctions, Market Mechanisms and their Applications, Boston, USA (2009)
35. Wei, Y.Z., Moreau, L., Jennings, N.R.: A market-based approach to recommender systems. ACM Trans. on Information Systems 23(3), 227–266 (2005)

Challenges of Data Processing for Earth Observation in Distributed Environments

Dana Petcu

Abstract. Remote sensing systems have a continuous growth in the capabilities that can be handled nowadays only using distributed systems. In this context, the challenges for the distributed systems coming from Earth observation field are reviewed in this paper. Moreover, the technological solutions used to built a platform for Earth observation data processing are exposed as proof of concept of current distributed system capabilities.

1 Introduction

Earth observation systems are gathering daily large amounts of information about our planet and are nowadays intensively used to monitor and assess the status of the natural and built environments. Earth observation (EO) is most often referring to satellite imagery or satellite remote sensing, the result of sensing process being an image or a map. Remote sensing refers to receiving and measuring reflected or emitted radiation from different parts of the electromagnetic spectrum (in ultraviolet, visible, reflected infrared, thermal infrared, or microwave). Remote sensing systems systems involve not only the collection of the data, but also their processing and distribution. The rate of increase in the remote sensing data volume is continuously growing. Moreover, the number users and applications is also increasing and the data and resource sharing became a key issue in remote sensing systems. Furthermore, EO scientists are often hindered by difficulties locating and accessing the data and services. These needs lead to a shift in the design of remote sensing systems from centralized environments towards wide-area distributed environments.

The current paper is a short survey of the current challenges imposed on distributed systems and coming from remote sensing application field. It is based on

Dana Petcu
Computer Science Department, West University of Timişoara, Romania
e-mail: petcu@info.uvt.ro

G.A. Papadopoulos and C. Badica (Eds.): Intelligent Distributed Computing III, SCI 237, pp. 9–19.
springerlink.com © Springer-Verlag Berlin Heidelberg 2009

several recent research reports of EO and distributed systems communities. The paper is organized as follows. The next section is presenting the identified challenges. The third section is dedicated to a review of the Grid usage benefits for EO. The fourth section is describing a case study on building a distributed environment for training in EO. A short list of conclusions is provided in the last section.

2 Challenges of Earth Observation on Distributed Environments

Satellite image processing is usually a computational and storage consuming task and special techniques are required for both data storage and processing in distributed environments. In what follows we point some main topics. This section is a survey of the ideas exposed in the recent scientific reports, like [2, 4, 5].

Data Management. The management of the distribution of data, from storing to long-term archiving, is currently an important topic in EO systems. The first issue is the data format that is varying from image files, databases, or structured file. Usually an EO data contain metadata describing the data, such as the dimensionality or reference coordinates. Another issue is related to the user need to access remotely the EO data. Due to the size of the EO data, a distributed file system is needed. For more than three decades there are several distributed file systems enabling multiple, distributed servers to be federated under the same file namespace. Another issue is the data discovery and this is currently done usually exploiting metadata catalogs, Moreover, for EO, replica management services are essential, allowing to determine an optimal physical location for data access based on data destination aiming to reducing the network traffic and the response time. Data transfers secure protocols were developed to extends the traditional file transfer protocol.

While for the basic needs mentioned above there are several stable and standardized solutions, the current key issue in EO data management is to make the data reachable and useful for any application through interoperability. Interoperability is achieved through the usage of standard interfaces and protocols. Interoperable interfaces are attractive to users allowing the fast design of distributed application based on multiple components. Achieving interoperability include also building adapter interfaces providing different front end to basic services and bridging protocols (namely interoperate). There are at least two layers for interoperability: for resource format and domain encoding, and semantic interoperability.

Interoperability solutions for resources structures and content are often application-field dependent. The solutions are related to different levels, like device, communication, middleware and deployment ones. At device level, the solutions are mostly standardized and are referring to the interfaces to the storage devices. At communication level, there are standardized data transfer protocols (as HTTP, HTTPS, or GridFTP), standardized protocols for Web services, and less standardized data movers for heterogeneous computing environments. At middleware level there are fewer standard solutions. For example, for data storage it is necessary a single consistent interface to different storage systems – a solution is coming from Grid community through the open standard storage resource manager, a control protocol for accessing mass storage. In what concerns file catalogs there are no current standards, but several implementations are available in Grid environments that are using special file catalogs allowing data replications. The same situation is valid also for metadata catalogs; fortunately, in the particular case of EO this issue is pursued by Open Geospatial Consortium (http://www.opengeospatial.org). In what concerns the interoperability of federated databases, a standard again proposed by Grid community is the Open Grid Services Architecture Data Movement Interface (OGSA-DMI, http://forge.gridforum.org/sf/projects/ogsa-dmi-wg). At deployment level, interoperability degradation is related to the event of new deployments – currently there are no automated tools or standard interfaces allowing the propagation of updates.

While resource-level interoperability is ensuring the compatibility of implementations at hardware and software levels, the semantic interoperability is enabling data and information flows to be understood at a conceptual level. Research efforts are currently devoted to the definition of generic data models for specific structured linguistic data types with the intention to represent a wide class of documents without loosing the essential characteristics of the linguistic data type.

EO data particularities. Data provision services in EO are not satisfying the today's user needs due to current application and infrastructure limitations. The process of identifying and accessing data takes up the a lot of time, according [4], due to: physical discontinuity of data, diversity of metadata formats, large volume of data, unavailability of historic data, and many different actors involved.

In this context, there is a clear need for an efficient data infrastructure able to provide reliable long-term access to EO data via the Internet, and to allow the users to easily and quickly derive information and share knowledge. Recognizing these needs, the European INSPIRE Directive (http://inspire.jrc.ec.europa.eu) requires all public authorities holding spatial data to provide access to that data through common metadata, data and network service standards. OPeNDAP (http://opendap.org/) is a data transport architecture and protocol widely used in EO; it is based on HTTP and includes standards for encapsulating structured data, annotating the data with attributes, and adding semantics that describe the data. Moreover, it is widely used by governmental agencies to EO data [4]. The Committee on EO Satellites (www.ceos.org) maintains a Working Group on Information Systems and Services with the responsibility to promote the development of interoperable systems for the management of EO data internationally. This group plans to build in the next decade the Global EO System of Systems (GEOSS) targeting the development of a global, interoperable geospatial services architecture [8].

Data Processing. To address the computational requirements introduced by time-critical satellite image applications, several research efforts have been oriented towards parallel processing strategies. According to the Top500 list of supercomputer sites, NASA, for example, is maintaining two massively parallel clusters for remote sensing applications. The recent book [14] presents the latest achievements in the field of high performance computing (HPC).

Currently ongoing research efforts are aiming also the efficient distributed processing of remote sensing data. Recent reports are related to the use of new versions of data processing algorithms developed for heterogeneous clusters as [13]. Moreover, distributed application framework specifically have been developed for remote sensed data processing, like JDAF [18]. EO applications are also good candidates for building architectures based on components encapsulating complex data processing algorithms and being exposed through standard interfaces like in [7].

Web services technology emerged as standard for integrating applications using open standards. In EO Web services play a key role. A concrete example is the Web mapping implementation specification proposed by OpenGIS (`http://www.opengis.org`). Web technologies are allowing also the distribution of scientific data in a decentralized approach and are exposing catalogue services of dataset metadata.

The promise of a Grid for EO community is to be a shared environment that provide access to an wide range of resources: instrumentation, data, HPC resources, and software tools. There are at least three reasons for using Grids for EO: (a) the required computing performance is not available locally, the solution being the remote computing; (b) the required computing performance is not available in one location, the solution being cooperative computing; (c) the required services are only available in specialized centres, the solution being application specific computing.

3 Large Distributed Environments for Earth Observation

Realizing the potential of the Grid computing for EO, several projects were launched to make the Grid usage idea a reality. We review the most important ones.

Grid-based EO Initiatives. Within the DataGrid project funded by the European Commission, an experiment aiming to demonstrate the use of Grid technology for remote sensing applications has been carried out; the results can be found for example in the paper [9]. Several other international Grid projects were focused on EO, like SpaceGrid (`http://www.spacegrid.org`), Earth Observation Grid (`http://www.e-science.clrc.ac.uk/web/projects/earthobservation`), or Genesis [22]. The MediGrid project (`http://www.eu-medigrid.org`) aimed to integrate and homogenize data and techniques for managing multiple natural hazards. The authors of paper [1] present an overview of SARA Digital Puglia, a remote sensing environment that shows how Grid and HPC technologies can be efficiently used to build dynamic EO systems for the management of space mission data and for their on-demand processing and delivering to final users. A frequent approach is to use the Grid as a HPC facility for processor-intensive operations. The paper [17], for example, focuses on the Grid-enabled parallelization of the computation-intensive satellite image geo-rectification problem. The aim of the proposed classification middleware on Grid from [19] is to divide jobs into several assignments and submit them to a computing pool. The parallel remote-sensing image processing software PRIPS was encapsulated into a Grid service in [20]. In the paper [21] is discussed the architecture of a spatial information Grid computing environment, based on Globus Toolkit, OpenPBS, and Condor-G; a model of the image division is proposed, which can compute the most appropriate image pieces and make the processing time shorter.

CrossGrid (`http://www.crossgrid.org`) aimed at developing techniques for real-time, large-scale grid-enabled simulations and visualizations, and the issues addressed included distribution of source data and the usefulness of Grid in crisis scenarios. DEGREE (`http://www.eu-degree.eu`) delivered a study on the challenges that the Earth Sciences are imposing on Grid infrastructure. D4Science (`http://www.d4science.org`) studied the data management of satellite images on Grid infrastructures. G-POD (`http://eogrid.esrin.esa.int/`) aims to offer a Grid-based platform for remote processing the satellite images provided by European Space Agency. The GlobAEROSOL service of BEinGRID [15] is processing data gathered from satellite sensors and generates an multi-year global aerosol information in near real time. The GEOGrid project [16] provides an e-Science infrastructure for Earth sciences community and integrates a wide varieties of existing data sets including satellite imagery, geological data, and ground sensed data, through Grid technology, and is accessible as a set of services. LEAD (`https://portal.leadproject.org/`) is creating an integrated, scalable infrastructure for meteorology research; its applications are characterized by large amounts of streaming data from sensors. The Landsat Data Continuity Mission Grid Prototype (LGP) offers a specific example of distributed processing of remotely sensed data [5] generating single, cloud and shadow scenes from the composite of multiple input scenes. GENESI-DR (`http://genesi-dr.eu`) intends to prove reliable long-term access to Earth Science data allowing scientists to locate, access, combine and integrate data from space, airborne and in-situ sensors archived in large distributed repositories; its discovery service allows to query information about data existing in heterogeneous catalogues, and can be accessed by users via a Web portal, or by external applications via open standardized interfaces (OpenSearch-based) exposed by the system [4]. Several other smaller projects, like MedioGrid [11], were also initiated to provide Grid-based services at national levels.

Remote Sensing Grid. A RSG is defined in [5] as a highly distributed system that includes resources that support the collection, processing, and utilization of the remote sensing data. The resources are not under a single central control. Nowadays it is possible to construct a RSG using standard, open, protocols and interfaces. In the vision of [5] a RSG is made up of resources from a variety of organizations provide specific capabilities, like observing elements, data management elements, data processing and utilization elements, communications, command, and control elements, and core infrastructure. If a service oriented architecture is used, modular services can be discovered and used to build complex applications by clients. The services should have the following characteristics [5]: composition, communication, workflow, interaction, and advertise. These requirements are mapped into the definition of specific services for workflow management, data management and processing, resource management, infrastructure core functions, policy specification, and performance monitoring. The services proposed in [5] are distributed in four categories: workflow management services, data management services, applications in the form of services, and core Grid services. In the next section we describe a case study of a recent Grid-based satellite imagery system that follows the RSG concepts.

4 Case Study: GiSHEO

The rapid evolution of the remote sensing technology is not followed at the same developing rate by the training and high education in this field. Currently there is only a few number of resources involved in educational activities in EO. The CEOS Working Group of Education, Training and Capacity Building for example is collecting an index of free EO educational materials (`http://oislab.eumetsat.org/CEOS/webapps/`). Recognizing the gap between research activities and the educational ones, we have developed recently a platform, namely GiSHEO (On Demand Grid Services for Training and High Education in EO `http://gisheo.info.uvt.ro`) addressing the issue of specialized services for training in EO. Contrary to the existing platforms providing tutorials and training materials, GiSHEO intends to be a living platform where experimentation and extensibility are the key words. Moreover, special solutions were proposed for data management, image processing service deployment, workflow-based service composition, and user interaction. A particular attention is given to the basic services for image processing that are reusing free image processing tools, like GDAL. A special features of the platform is the connection with the GENESI-DR catalog mentioned in the previous section.

While the Grid is usually employed to respond to the researcher requirements to consume resources for computational-intensive or data-intensive tasks, we aim to use it for near-real time applications for short-time data-intensive tasks. The data sets that are used for each application are rather big (at least of several tens of GBs), and the tasks are specific for image processing (most of them very simple). In this particular case a scheme of instantiating a service where the data are located is required in order to obtain a response in near-real time. Grid services are a quite convenient solution in this case: a fabric service is available at the server of the platform that serves the user interface and this service instantiates the processing service where the pointed data reside. In our platform the Web services serve as interfaces for the processing algorithms. These interfaces allow the remote access and application execution on a Grid using different different strategies for fast response.

The platform design concepts were shortly presented in [10] and the details about the e-learning component can be found in [6]. The EO services were described in [12] and the data management is detailed in [11]. In this section we describe the approaches that were undertaken to solve the previously mentioned problems.

Conceptual View of the Platform's Architecture. The GiSHEO architecture is a Grid-enabled platform for satellite image processing. Figure 1 presents the conceptual view of the implemented service-oriented architecture. The WMS is the standard Web Mapping Service ensuring the access to the distributed database. WAS

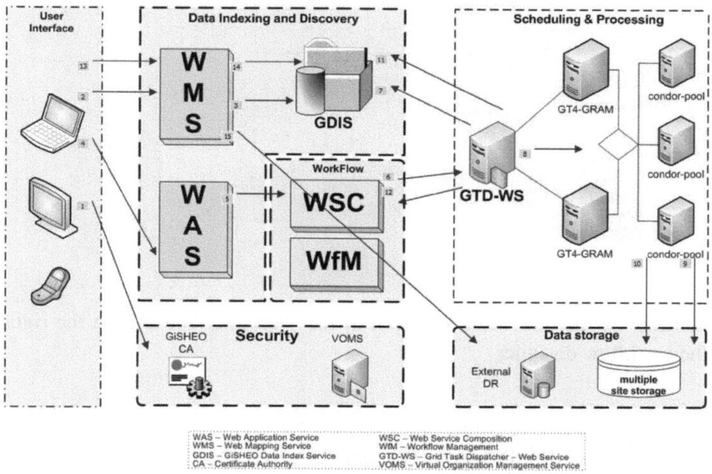

Fig. 1 GiSHEO platform components. Numbers are representing the stages in its usage process: authenticate and authorize; select area of interest and available datasets; select processing unit (list of applications) and submit it to WSC; submit a task to GTD-WS; discover datasets, schedule the task, retrieve datasets for processing; upload output to GDB, register output to GDIS, notify WSC; UI requests WMS to display the output, query GDIS, select output datasets from GDB, display

(Web Application Service) is invocated by user interface at run-time and allows workflows description. GDIS is a data index service, a Web service providing information about the available data to its clients; it intermediates access to data repositories, stores the processing results, ensures role based access control to the data, retrieves data from various information sources, queries external data sources and has a simple interface that is usable by various data consumers. The platform has distributed data repositories – it uses PostGIS for storing raster extent information and in some cases vector data. The data search is based on PostGIS spatial operators. The Workflow Service Composition (WSC) and Workflow Manager (WfM) are the engines behind WAS and are connected with the tasks manager. Each basic image processing operation is viewed as a task; several tasks can be linked together to form a workflow in an order that is decided at client side. The GTD-WS (Grid Task Dispatcher Web Service) is a service-enabled interface for interoperability with the Grid environment. Digital certificate are required to access the full facilities of the platform. A particular component of WAS is eGLE that uses templates to allow teachers specialized in EO to develop new lessons that uses EO data – details are given in [6]. The physical platform is based on four clusters that are geographically distributed. Due to the low security restriction between the four institutions, data distribution between the clusters is done using Apache Hadoop distributed file system. The data transfer from and to external databases is done using GridFTP - this is for example the case of the connection with GENESI-DR database.

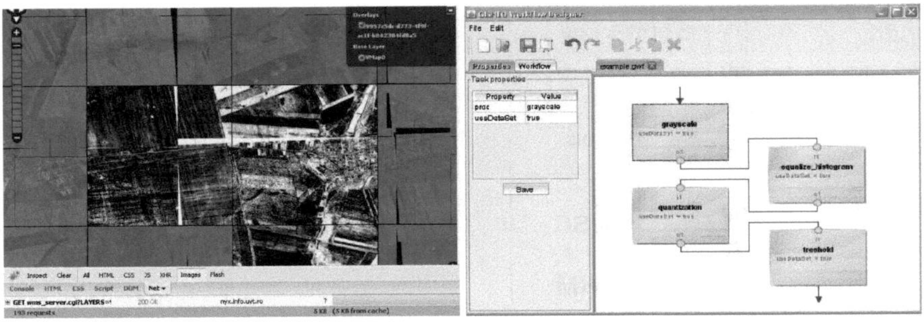

Fig. 2 GiSHEO interface: on the left side a snapshot of the Web interface and on the right side a snapshot of the workflow designer

Platform's Levels. The platform architecture has several levels including user, service, security, processing and a data level. The *user level* is in charge with the access to the Web user interface (built by using DHTML technologies). A workflow language was developed together with a set of tools for users not familiar with programming which can be used both for visually creating a workflow (details in [3]). After defining the workflow the user can then select a region containing one or more images on which the workflow is to be applied (Figure 2). The *service level* exposes internal mechanisms belonging to the platform and consists in: EO services – processing applications; workflow service – the internal workflow engine accessible through a special Web service; data indexing and discovery services – allowing the access to the platform's data management mechanisms. The *security level* provides security context for both users and services. Each user is identified by either using a username-password pair or a canonical name provided by a digital certificate. The services use a digital certificate for authentication, authorization, and trust delegation. A VOMS service is used for authorization.

At *processing level* the platform enables two models for data processing: direct job submission trough Condor's specific Web services, or through WS-GRAM tool of Globus Toolkit 4 (GT4). The Grid service interface GTD (Grid Task Dispatcher) is responsible for the interaction with other internal services as the Workflow Composition Engine in order to facilitate access to the processing platform. It receives tasks from the workflow engine or directly from user interface. A task description language (the ClassAd meta language for example in case of Condor HTC) is used in order to describe a job unit, to submit and check the status of jobs inside the workload management system and to retrieve job logs for debugging purposes. GTD with Condor is used mainly for development purposes (application development and testing). For production GTD with GT4 is used because it offers a complete job management-tracking system.

At *data level* two different types of data are involved: datasets database which contain the satellite imagery repository and processing application datasets used by applications to manipulate satellite images. The GiSHEO Data Indexing and Storage Service (GDIS) provides features for data storage, indexing data, finding data by various conditions, querying external services and for keeping track of

Fig. 3 An example of GiSHEO simple service: multi-image transformation using a binary decision tree to detect areas with water, clouds, forest, non-forest and scrub; the top two images in gray scale are the input data (infrared and red bands), while the bottom image is the output color image

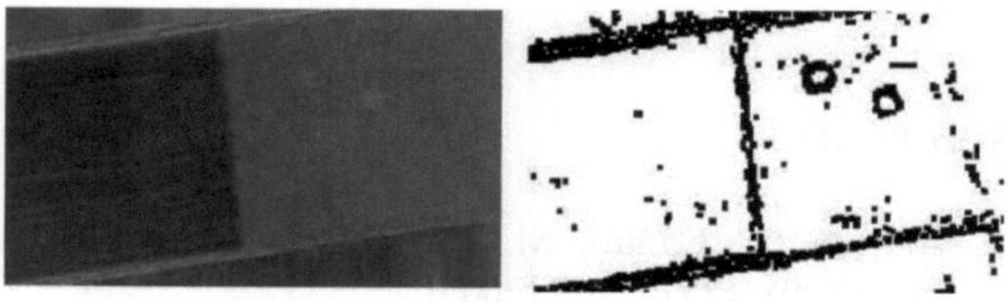

Fig. 4 An example of GiSHEO complex service: identification of soil marks of burial mounds: on the left the original image and in the right the result of a sequence of operations allowing the detection of round shapes

temporary data generated by other components. GDIS is available to other components or external parties using a specialized Grid service. This service is also responsible for enforcing data access rules based on specific grid credentials (VO attributes, etc.). The data storage component part of GDIS is responsible for storing the data by using available storage back-ends. The data distributed across various storage domains should be exposed through a unique interface – this is achieved by implementing a front-end GridFTP service that has native access to the Hadoop distributed file system (HDFS) offering access to data stored inside the internal HDFS and providing the required access control facilities. The data indexing service is performed by PostGIS to index the metadata and location of the geographical data available in the storage layer. The PostGIS layer provides advanced geographical operations allowing searching the data by using various criteria. An advanced and highly flexible interface for searching the project's geographical repository was also designed and built around a custom query language designed to provide fine grained access to the data in the repository and to query external services (TerraServer, GENESI-DR, etc) – details can be found in [3].

In what concerns the platform's EO services, we divided the remote sensing processing operations into basic and complex types. Basic operations represent image processing algorithms that can be applied on a satellite image (obtaining the negative, gray level conversion, histogram equalization, quantization, thresholding, band extraction, embossing, equalization, layers subtraction etc.) – see an example in Figure 3. Complex operations are represented by the complex image processing algorithms (i.e. topographic effect regression) or by a composition of two or more basic operations – see an example in Figure 4. Applications for training in archaeology are presented in details in [12].

5 Conclusions

The paper reviewed the challenges imposed to current distributed environments serving the EO community. The data processing and management are the key issues. A special emphasis has been put in the last decade on using wide-area distributed systems, namely Grids. Their usage benefits were underlined in this paper. The current standards and specific services for EO are allowing the design of new platforms focusing on the user needs. Such a distributed platform aiming to serve the training needs in EO was presented as proof of the concepts discussed in this paper.

Acknowledgements. This research is supported by ESA PECS Contract no. 98061 GiSHEO – On Demand Grid Services for High Education and Training in Earth Observation.

References

1. Aloisio, G., Cafaro, M.: A dynamic Earth observation system. Parallel Computing 29(10), 1357–1362 (2003)
2. Coghlan, B., et al.: e-IRG Report on Interoperability Issues in Data Management (2009)
3. Frincu, M.E., Panica, S., Neagul, M., Petcu, D.: Gisheo: On demand Grid service based platform for EO data processing. In: Procs. HiperGrid 2009, pp. 415–422 (2009)
4. Fusco, L., Cossu, R., Retscher, C.: Open Grid services for Envisat and Earth observation applications. In: High Performance Computing in Remote Sensing, pp. 237–280 (2008)
5. Gasster, S.D., Lee, C.A., Palko, J.W.: Remote sensing Grids: architecture and implementation. In: High Performance Computing in Remote Sensing, pp. 203–236 (2008)
6. Gorgan, D., Stefanut, T., Bacu, V.: Grid based training environment for Earth observation. LNCS, vol. 5529, pp. 98–109 (2009)
7. Larson, J.W., et al.: Components, the common component architecture, and the climate/weather/ocean community. In: Procs. 84th AMS Annual Meeting (2004)
8. Lee, C.A.: An introduction to Grids for remote sensing applications. In: Plaza, A., Chang, C. (eds.) High Performance Computing in Remote Sensing, pp. 183–202 (2008)
9. Nico, G., Fusco, L., Linford, J.: Grid technology for the storage and processing of remote sensing data: description of an application. SPIE, vol. 4881, pp. 677–685 (2003)
10. Panica, S., Neagul, M., Petcu, D., Stefanut, T., Gorgan, D.: Desiging a Grid-based training platform for Earth observation. In: Procs. SYNASC 2008, pp. 394–397 (2009)
11. Petcu, D., Gorgan, D., Pop, F., Tudor, D., Zaharie, D.: Satellite image processing on a Grid-based platform. International Scientific Journal of Computing 7(2), 51–58 (2008)
12. Petcu, D., Zaharie, D., Neagul, M., Panica, S., Frincu, M., Gorgan, D., Stefanut, T., Bacu, V.: Remote sensed image processing on Grids for training in Earth observation. In: Kordic, V. (ed.) Image Processing, In-Tech, Vienna (2009)
13. Plaza, A., Plaza, J., Valencia, D.: Ameepar: Parallel morphological algorithm for hyperspectral image classification in heterogeneous NoW. LNCS, vol. 3391, pp. 888–891 (2006)
14. Plaza, A., Chang, C. (eds.): High Performance Computing in Remote Sensing. Chapman & Hall/CRC, Taylor & Francis Group, Boca Raton (2008)
15. Portela, O., Tabasco, A., Brito, F., Goncalves, P.: A Grid enabled infrastructure for Earth observation. Geophysical Research Abstracts 10 (2008)

16. Sekiguchi, et al.: Design principles and IT overviews of the GEOGrid. IEEE Systems Journal 2(3), 374–389 (2008)
17. Teo, Y.M., Tay, S.C., Gozali, J.P.: Distributed geo-rectification of satellite images using Grid computing. In: Procs. IPDPS 2003, pp. 152–157 (2003)
18. Votava, P., Nemani, R., Golden, K., Cooke, D., Hernandez, H.: Parallel distributed application framework for Earth science data processing. In: Procs. IGARSS 2002, pp. 717–719 (2002)
19. Wang, J., Sun, X., Xue, Y., et al.: Preliminary study on unsupervised classification of remotely sensed images on the Grid. In: Bubak, M., van Albada, G.D., Sloot, P.M.A., Dongarra, J. (eds.) ICCS 2004. LNCS, vol. 3039, pp. 981–988. Springer, Heidelberg (2004)
20. Yang, X.J., Chang, Z.M., Zhou, H., Qu, X., Li, C.J.: Services for parallel remote-sensing image processing based on computational Grid. In: Jin, H., Pan, Y., Xiao, N., Sun, J. (eds.) GCC 2004. LNCS, vol. 3252, pp. 689–696. Springer, Heidelberg (2004)
21. Yang, C., Guo, D., Ren, Y., Luo, X., Men, J.: The architecture of SIG computing environment and its application to image processing. In: Zhuge, H., Fox, G.C. (eds.) GCC 2005. LNCS, vol. 3795, pp. 566–572. Springer, Heidelberg (2005)
22. Yunck, T., Wilson, B., Braverman, A., Dobinson, E., Fetzer, E.: GENESIS: the general Earth science investigation suite. In: Procs. 4th annual NASAs Earth Technology Conference (2008)

A Protocol for Execution of Distributed Logic Programs

László Aszalós and Andreas Herzig

Abstract. In the last fifteen years many work in logic programming has focused on parallel implementations both in shared memory and in distributed memory multi-processor systems. In both cases it is supposed that each processor knows the entire logic program. In this paper we shall study a case where each agent only knows part of the program. We give a multi-modal logical language and its semantics which defines a protocol that is able to acquire the information needed to run the logic program.

1 Introduction

Suppose you want to plan a journey where you need to travel both by train and by plane. Up to now such a task cannot be done by a single information system, and you have to use several ones to do that, e.g. several train time tables and several plane timetables. In general we can treat such information systems as logic databases (or logic programs), and putting questions as querying them by logical formulas.

In general there are numerous solutions, and their number makes it difficult to find the best ones by hand. If you want to find them automatically, you need some mechanism making the different information systems cooperate.

This might be implemented in a centralized client-server system. The difficulty in such solutions is first that they get very complex and are difficult to maintain, and second that they are sensitive to failure of the central server. A solution that is more robust is to integrate the whole information in a single database, and copy it

László Aszalós
Universitity of Debrecen, H4010, Debrecen PoBox 12, Hungary
e-mail: laszalos@unideb.hu

Andreas Herzig
IRIT, Université Paul Sabatier, 118 route de Narbonne F-31062 Toulouse Cedex 4, France
e-mail: herzig@irit.fr

G.A. Papadopoulos and C. Badica (Eds.): Intelligent Distributed Computing III, SCI 237, pp. 21–30.
springerlink.com © Springer-Verlag Berlin Heidelberg 2009

to several servers. The problem here is to maintain and update these databases in a coherent way. We can construct more robust and adaptive systems if all the nodes of the network can act both as server and client. These are the peer-to-peer systems. They are interesting from a technical point of view because they have characteristics previously unseen.

Our hypotheses about such multi-agent systems (MAS) are the following:

- Agents receive a huge mass of information, but they can communicate only a little fraction of it. Nowadays memory capacity is progressing more rapidly than communication capacity, so this assumption is realistic. (Hence we have very clever but short-spoken agents.) Different agent might receive different information from different information sources, so in this case they know different things. It can occur that none of them have enough information about something, but the sum of their knowledge is sufficient. This what is called their distributed knowledge in the sense of [8], and it is in this sense that we use it here.
- The agents are cooperative: if they know some fact they are asked they answer immediately, and if they do not know the answer then they forward the question to their neighbors. This information sharing method similar to Gnutella's i.e. if some agent asks something then he asks all his neighbors and if some agent answers to some question, he tells the answer to the neighbor(s) who asked him.
- Agents and communication are error-free, so each agent can trust every statement he received.
- For the sake of simplicity we assume that the environment does not change. So we exclude that some agent observes A and A becomes false later while the agent still believes A.

Our article is organized as follows: in the next section we describe why we do not break questions into parts. Next we show the propositional case and example in detail. Finally we present the case of Horn clauses.

2 Motivation

Sometimes one question can be divided into several subquestions. We shall show in this section that asking the entire question is in some cases more economical than asking each subquestion in parallel.

Suppose questions are about truth of logical formulas, and suppose agent 1, 2 and 3 knows p, q and r, respectively (Fig. 2). The agents that are connected listen to each other. If agent 4 asks $p \vee q \vee r$ then each of the three other agents can answer this question, and they do not forward this question. If agent 4 asks p and asks q and asks r, then agent 3 can answer r, but does not know the answer to p and q, so he will forward these questions according to our information sharing method (protocol) Similar statements hold for the other two agents, too.

If agent 4 asks $p \wedge q \wedge r$ then its neighbors together have enough information to answer the question. If they utter what they know then agent 4 can assemble the final answer from their answers. The neighbors of agents 4 are not neighbors of each other, so they do not know the other agents' knowledge. They can assume that

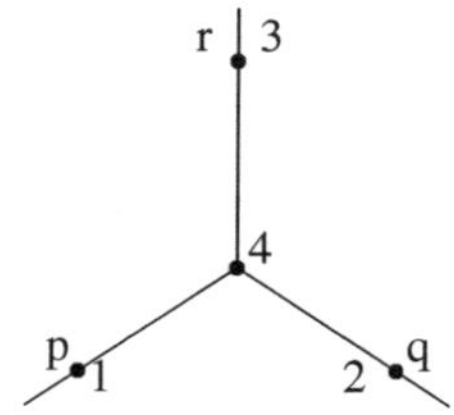

Fig. 1 Agent 4 asks $p \vee q \vee r$.

the agent behind them (not depicted in Fig. 2) may know the missing information. For example if agent 3 knows r, and he knows that the agents behind him has the distributed knowledge $p \wedge q$ (or its negation), then he can find out the answer. So he does not need to ask the original question, just the unknown part of it. Moreover if he knows q, too, then he only needs to ask p.

This demonstrates that it might be more efficient not to split questions into sub-questions.

3 Propositional Case

We describe with propositional letters the facts of the world of the agents. Let $AGT = \{i, j, \ldots\}$ be the set of agents, and $Atom = \{p, q, \ldots\}$ the set of propositional letters. We associate modal operators S_i, K_i and Q_i to every $i \in AGT$. The formulae S_iA, K_iA and Q_iA are read "agent i said A", "agent i knows A", and "agent i asks if A", respectively. In general the modal operators S_i and Q_i are non-normal [6], and are hence neither closed under logical truth, logical consequence, conjunction, nor material implication [3]. In the first part of this article for the sake of simplicity we restrict S_i operators to propositional letters or their negations, as well as the modal operator K_i.

If agent i gets to know some fact that he knows some other agent j is interested in then he communicates it to j, unless he had already said it before. To handle this exception we need to work with time. Therefore we shall use lineral temporal logic [9] and the *next-time* (X), *always in the future* (G) and *always in the past* (H) modal operators.

Now we are ready to define the set of formulae (FRM) and the set of propositional formulae (FRM_0):

- FRM_0: $a ::= p \mid \bot \mid \neg a \mid a \wedge a$
- FRM: $A ::= a \mid K_i p \mid K_i \neg p \mid S_i p \mid S_i \neg p \mid Qa \mid \neg A \mid A \wedge A \mid XA \mid GA \mid HA$

where p ranges over ATM and i over AGT. The agents system has some topology, i.e. they are connected in some sense. For this we introduce a reflexive, non-transitive relation *Listen*. $Listen(i, j)$ means that agent i listens agent j's reports.

For this language we choose the following type of models $\vartheta = \langle \vartheta_A, \vartheta_Q, \vartheta_S, \vartheta_K \rangle$, where $\vartheta_A \subseteq Atom$ is the set of atoms that are true in the real world, $\vartheta_S, \vartheta_K \subseteq AGT \times Atom \times \mathbb{N}$ associate mental attitudes with agents, atoms, and time points. $\vartheta_Q \subseteq AGT \times FRM_0 \times \mathbb{N}$ is a similar construction, but in questions we can use not only atoms but propositional formulae, too. To satisfy our assumptions we need to require the following conditions hold in this model for all $i \in AGT$, $p \in Atom$, $A \in FRM_0$ and $l \in \mathbb{N}$:

- $\vartheta_S \subseteq \vartheta_K$. (Our agents are sincere, they only say things they know. The opposite direction ($K_iA \to S_iA$) does not hold, agents do not say everything they know.)
- If $\langle i,p,l\rangle \in \vartheta_K$ then $\langle i,p,n\rangle \in \vartheta_K$ for all $n > l$. (Our agents don't forget anything.)
- If $Listen(i,j)$ and $\langle j,p,l\rangle \in \vartheta_S$ then $\langle i,p,l+1\rangle \in \vartheta_K$. (If some agent utters something then at the next time point all the adjacent agents will know it.)
- If $Listen(i,j)$, $\langle j,A,n\rangle \in \vartheta_Q$ for some $n < l$, $\langle i,p,l\rangle \in \vartheta_K$, p subformula of A, and $\langle i,p,m\rangle \notin \vartheta_S$ for all $m < l$ then $\langle i,p,l\rangle \in \vartheta_S$. (Our agents are cooperative, so if some neighbor has asked something the agent knows, then he will utter it, unless he had uttered it before.)
- If $Listen(i,j)$, $\langle j,A,n\rangle \in \vartheta_Q$ for some $n < l$, and A or $\neg A$ is not a logical consequence (in propositional logic) of the set

  $$\{p|\langle i,p,l\rangle \in \vartheta_K \text{ and } p \in \vartheta_A\} \cup \{\neg p|\langle i,p,l\rangle \in \vartheta_K \text{ and } p \notin \vartheta_A\} \cup$$

  $$\{B|\langle i,B,m\rangle \in \vartheta_Q \text{ for some } m < l\}$$

  then $\langle i,A^\star,l\rangle \in \vartheta_Q$, where

  $$A^\star = A(p_1,\ldots,p_n\|\top,\ldots,\top)(q_1,\ldots,q_m\|\bot,\ldots,\bot),$$

  $\langle i,p_j,l\rangle \in \vartheta_K$ and $p \in \vartheta_A$, or $\langle i,q_k,l\rangle \in \vartheta_K$ ans $p \notin \vartheta_A$. (If the agent does not know some part of the answer he forwards it, unless he had asked it before.)[1]

Now truth of a formula in the model ϑ at time point l can defined as follows:

1. $\vartheta,l \models p$, iff $p \in \vartheta_A$
2. $\vartheta,l \models K_ip$ iff $\langle i,p,l\rangle \in \vartheta_K$ and $p \in \vartheta_A$
3. $\vartheta,l \models K_i\neg p$ iff $\langle i,p,l\rangle \in \vartheta_K$ and $p \notin \vartheta_A$
4. $\vartheta,l \models S_ip$ iff $\langle i,p,l\rangle \in \vartheta_S$ and $p \in \vartheta_A$
5. $\vartheta,l \models S_i\neg p$ iff $\langle i,p,l\rangle \in \vartheta_S$ and $p \notin \vartheta_A$
6. $\vartheta,l \models Q_iA$ iff $\langle i,A,l\rangle \in \vartheta_Q$
7. $\vartheta,l \models HA$ iff for all $n < l$, $\vartheta,n \models A$
8. $\vartheta,l \models GA$ iff for all $n > l$, $\vartheta,n \models A$
9. $\vartheta,l \models XA$ iff $\vartheta,l+1 \models A$

The cases left out are the traditional ones. According to the points above $S_iA \to A$, $K_iA \to A$ and $S_iA \to K_iA$ are valid, so our agents know and say only true things, and they know things that they say, which was our assumption.

Let us return to the example on Figure 2! If the agent i gets the question A at time point l, then he constructs a new question $A^\star$ as follows. If the propositional letter p is a sub-formula of A, and he knows at this moment the truth value of p then he replaces all occurrences of p in A by $\top$ or $\bot$, depending on whether p is true or not. In this example, A was $p \wedge q \wedge r$ and agent 3 knew that r. Hence he replaces r with $\top$, so we get $p \wedge q \wedge \top$, and after simplification $p \wedge q$.

[1] $A(p_1,\ldots,p_n\|B_1,\ldots,B_n)$ denotes the substitution of every p_i by B_i in A.

4 Example in Detail

In Figure 2 there are five agents. We shall refer to the agents with numbers besides them, as before. In the figure we display the agents' knowledge at time point 0. For example agent 2 knows at this moment propositional letters p and t. Agent 1 wants to know whether the propositional formula $p \vee q \rightarrow r \wedge s$ is true or not.

Fig. 2 Example in detail

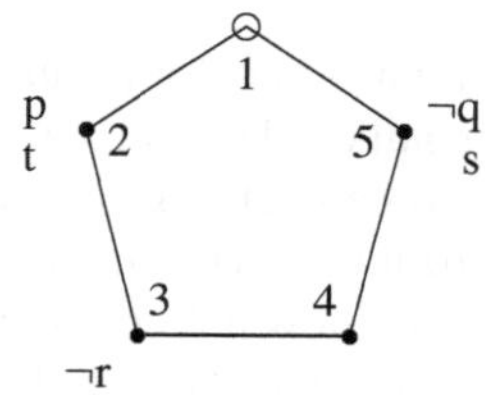

In this MAS the following happens:

Time point 0. Agent 1 asks his question: $Q_1(p \vee q \rightarrow r \wedge s)$.
Time point 1. Now agent 2 and 5 received agent 1's question.

- Propositional letter p occurs in $p \vee q \rightarrow r \wedge s$, so agent 2 utters information about p: $S_2 p$. Propositional letter t does not occur in $p \vee q \rightarrow r \wedge s$, so agent 2 does not distribute information about t Neither of $(p \wedge t) \rightarrow (p \vee q \rightarrow r \wedge s)$ and $(p \wedge t) \rightarrow \neg(p \vee q \rightarrow r \wedge s)$ is a logical law, so agent 2 must forward some question. If we replace $\top$ for p, then we get $\top \vee q \rightarrow r \wedge s$ which simplifies to $r \wedge s$. Hence agent 2 shall ask this: $Q_2(r \wedge s)$.
- Propositional letters q and s occur in $p \vee q \rightarrow r \wedge s$, so $S_5 \neg q$ and $S_5 s$. Neither of $(\neg q \wedge s) \rightarrow (p \vee q \rightarrow r \wedge s)$ and $(\neg q \wedge s) \rightarrow \neg(p \vee q \rightarrow r \wedge s)$ is a logical law, so agent 5 must forward some question. If we replace $\bot$ and $\top$ for q and s, then we get $p \vee \bot \rightarrow r \wedge \top$, and finally $p \rightarrow r$. Hence agent 5 shall ask this: $Q_2(p \rightarrow r)$.

Time point 2. Now agent 1 and 3 received agent 2's answer and question, and agent 1 and 4 received agent 5's answers and question.

- Agent 1 received messages p, $\neg q$ and s, so now these are parts of agent 1's knowledge. All of them occur in $(p \vee q \rightarrow r \wedge s)$, so agent 1 will utter them: $S_1 p$, $S_1 \neg q$ and $S_1 s$ both of $p \wedge \neg q \wedge s \wedge (p \vee q \rightarrow r \wedge s) \rightarrow r \wedge s$ and $p \wedge \neg q \wedge s \wedge (p \vee q \rightarrow r \wedge s) \rightarrow (p \rightarrow r)$ are logical laws, and agent 1 did not receive any other question, so he did not to need forward any questions.
- Agent 3 received the question $Q_2(r \wedge s)$. Propositional letter r occurs in $r \wedge s$, so $S_3 \neg r$. The formula $\neg r \rightarrow \neg(r \wedge s)$ is a logical law, he did not receive any other question, so agent 3 does not forwards any question.
- Agent 4 received the question $Q_5(p \rightarrow r)$ and answers $S_5 \neg q$ and $S_5 s$. Neither of $(\neg q \wedge s) \rightarrow (p \rightarrow r)$ and $(\neg q \wedge s) \rightarrow \neg(p \rightarrow r)$ are logical law, agent 4 has is not able to reduce the question, so he forwards it without modification: $Q_4(p \rightarrow r)$.

Time point 3. From now we only add the messages: $S_2\neg r$ (agent 1 has all the information to construct the answer for the original question), $S_2\neg q$, $S_2 s$, $S_3 p$, $S_4\neg r$, and $S_5 p$.
Time point 4. $S_1\neg r$ and $S_5\neg r$.

5 The Case of Horn Clauses

First-order Horn clauses have more expressive power than propositional logic, so we shall use Horn clauses instead of propositional variables when some agent knows or says something, and when some agent asks something we shall express this by call-formulae (conjunctions of atoms) instead of propositional formulae. Hence agents can know not only facts but rules, too.

Let us take a Prolog program on Table 1 which describes two families. Figure 3 shows the phases of the execution this Prolog program for the

```
?-father(Grandfather,Mother), mother(Mother,Child)
```

call-formula. We can see that in a traditional Prolog system SLD resolution tree is evolved from left to right due the depth-first search and backtrack.

Table 1 The original Prolog program.

```
mother(alice, bob).
father(bob, cynthia).
father(bob, carmen).
mother(cynthia, daniel).
mother(cynthia, david).
father(edward,flora).
mother(flora, george).
```

In the previous sections if some agent knew some part of the question then he uttered it and forwarded the reduced question. At Horn clauses, instead of substitutions with $\top$ and $\bot$, we can use unification, but unfortunately this doesn't work. Let us distribute the previous Prolog program between five agents as in Table 2. Here $Listen_{ij}$ is defined as in Fig. 2.

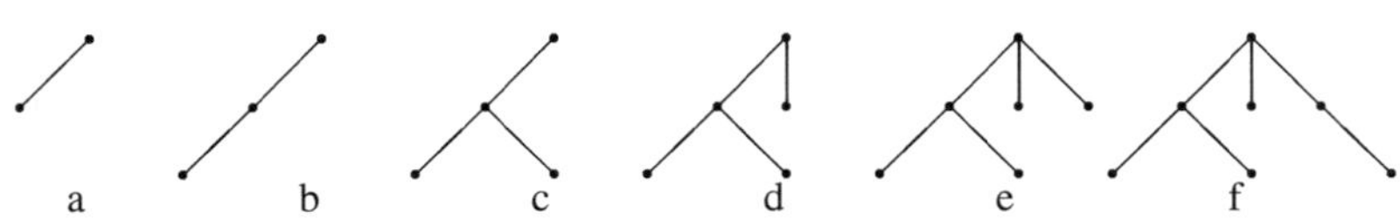

Fig. 3 Evolution of a SLD resolution tree

Table 2 The distributed Prolog program.

Agent 1	
	`mother(cynthia, daniel).`
Agent 2	
	`father(bob, cynthia).`
	`mother(flora, george).`
Agent 3	
	`father(edward,flora).`
Agent 4	
	`father(bob, carmen).`
	`mother(cynthia, david).`
Agent 5	
	`mother(alice, bob).`

Let us assume that agent 4 would like to find the answer for the previous call-formula. As he knows that Bob is father of Carmen, he uses this fact and the $(Grandfather, Mother \| bob, carmen)$ substitution, he asks

```
?- mother(carmen, Child)
```

but he cannot get an answer for this question. If agent 3 uses the same method for the same question, then he needs to ask `?- mother(flora, Child)` and he will get one solution but will lose two others. Therefore the agents need to transfer the original question. Agents can transfer their reduced questions as well, but in this case a huge network traffic might be generated.

So our idea is the following: the agents transfer only the original question and if some agent has a clause whose head unifies with some part of the call-formula (question) then he utters it. With this method the agent which asked the question can acquire all the information the other agents know (about this topic). If this agent knows the topology of the network and the speed of spreading of information (although these are not in our assumptions) then he can estimate the time point when he get all the answers for his question. But this does not mean that he can be sure to get all the information he needs. For example if agent 3 asks the question `grandpa(X,Y)` and only agent 5 knows that `grandpa(X,Y):- father(X,Z),mother(Z,Y)`, then having obtained this piece of information agent 3 needs to acquire information about predicates `father/2` and `mother/2`, too.

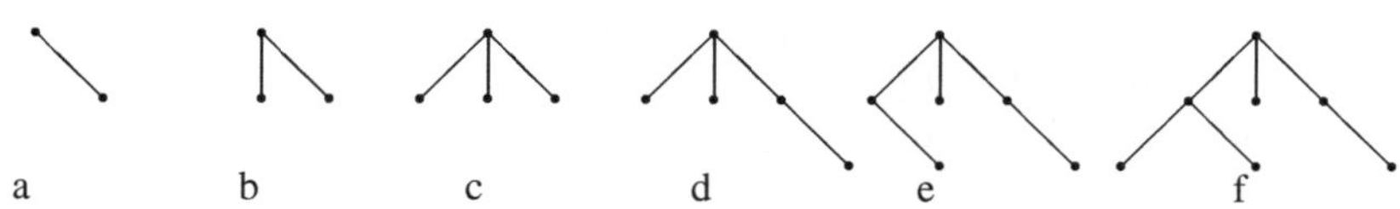

Fig. 4 Evolution of agent's SLD resolution tree

Hence there is no reason to wait for all the information, agents can start to build the SLD resolution tree (Fig. 4) In the following we show how such a tree is built. Let us assume at time point 0 agent 3 asks the question:

```
?-father(Grandfather,Mother), mother(Mother,Child).
```

At time point 1 agent 2 and agent 4 transfer this question to agent 1 and agent 5. Moreover at the same time they can answer this question, so they say all their knowledge about the `father/2` and `mother/2` predicates to agent 3. Meantime agent 3 realizes the he knows something suitable for his question so he can start to build the resolution tree (Fig. 4.a).

At time point 2 agent 3 gets agent 2's and agent 4's answer and he can start two other branches (Fig. 4.b-c). Moreover he can continue the old one, so he gets a solution (Fig. 4.d). Meantime agent 1 and agent 5 transmit their knowledge.

At time point 3 agent 2 and agent 4 transmit agent 1's and agent 5's knowledge about `father/2` and `mother/2` and time point 4 agent 3 can use this information to find the final solution (Fig. 4.e-f).

Now let us define a model for this example. This model is similar to the propositional case. Here instead of set of atoms we have a set of Horn rules *Rules* and set of call formulae *Call*. Now we do not have negated atoms, so we don't need to use ϑ_A, hence our model is $\vartheta = \langle \vartheta_Q, \vartheta_S, \vartheta_K \rangle$, where $\vartheta_S, \vartheta_K \subseteq AGT \times Rules \times \mathbb{N}$ associate mental attitudes with agents, Horn rules, and time points. At questions we have call formulae, so $\vartheta_Q \subseteq AGT \times Call \times \mathbb{N}$. As before we need to require the following conditions for all $i \in AGT$, $R \in Rules$, $C \in Call$ and $l \in \mathbb{N}$:

- $\vartheta_S \subseteq \vartheta_K$.
- If $\langle i,R,l \rangle \in \vartheta_K$ then $\langle i,R,n \rangle \in \vartheta_K$ for all $n > l$.
- If $Listen(i,j)$ and $\langle j,R,l \rangle \in \vartheta_S$ then $\langle i,R,l+1 \rangle \in \vartheta_K$. (These three points are the same as before)
- If $Listen(i,j)$, $\langle j,C,n \rangle \in \vartheta_Q$ for some $n < l$, $\langle i,R,l \rangle \in \vartheta_K$, $\langle i,R,m \rangle \notin \vartheta_S$ for all $m < l$, $R \in Rule$ is the Horn rule $\mathtt{r:-p_1,\ldots,p_s}$ (maybe $s = 0$), $C \in Call$ is the call formula $\mathtt{q_1,\ldots,q_t}$ and for some k $(1 \le k \le t)$ q_k can be unified with r then $\langle i,R,l \rangle \in \vartheta_S$. (If some neighbor has asked a call formula of which the agent knows some part, then he will utter the corresponding fact or rule, unless he had uttered it before.)
- If $Listen(i,j)$, $\langle j,C,n \rangle \in \vartheta_Q$ for some $n < l$, and $\langle i,C,m \rangle \notin \vartheta_Q$ for all $m < l$ then $\langle i,C,l \rangle \in \vartheta_Q$. (The agent needs to transfer the original question in any case. Of course only once.)

The truth of a formula in the model ϑ at time point l can defined as follows:

1. $\vartheta, l \models K_i R$ iff $\langle i,R,l \rangle \in \vartheta_K$
2. $\vartheta, l \models S_i R$ iff $\langle i,R,l \rangle \in \vartheta_S$
3. $\vartheta, l \models Q_i C$ iff $\langle i,C,l \rangle \in \vartheta_Q$
4. $\vartheta, l \models HA$ iff for all $n < l$, $\vartheta, n \models A$
5. $\vartheta, l \models GA$ iff for all $n > l$, $\vartheta, n \models A$
6. $\vartheta, l \models XA$ iff $\vartheta, l+1 \models A$

Note that we does not have negation, and that we treat rules as one unit, so the definition is simpler than before.

Based on this semantics, we can now formally analyze our example. Let H be the hypotheses about the agent databases as expressed in Table 2, such as $K_2(\texttt{father(bob,cynthia)})$. With the *Listen* relation as in Fig. 2 we can prove now that

$$H \wedge Q_3(\texttt{father(X,Y)}) \rightarrow XS_2(\texttt{father(bob,cynthia)})$$

and

$$H \wedge Q_3(\texttt{father(X,Y)}) \rightarrow XXK_3(\texttt{father(bob,cynthia)})$$

are valid.

In standard Prolog, if we cannot continue a branch then we need to apply backtracking, because there is no chance that we can continue this branch later. Here the situation is quite different. In a big network we might get useful information to continue the branch several time points after we got stuck.

If we don't know the topology of the network we cannot estimate the time point when we can decide that an open path in the resolution tree is hopeless. This means that it is a hard problem to find all the solutions in a MAS. Hopefully only some solutions are sufficient in most of the cases. In this situation it is very handy if we have a special message which commands the agents to abandon the searching of suitable clauses. This helps to avoid spreading a call formula over the whole network. Another useful feature is to use 'delayed questions', i.e. to let questions travel at lower speed through the network.

6 Discussion and Conclusion

In this article we have studied a peer-to-peer information sharing protocol allowing to answer queries in multi-agent logical databases. We have proposed a temporal multi-modal logic within which we can speak about what agents in multi-agent systems know, ask, and say. Our peer-to-peer protocols is 'hard-wired' in the semantics of the logic, in the sense that the agent's behavior is entirely determined. The logic allows to reason about knowledge and communication, and to prove properties of the protocol.

Our logic is related to Dynamic Epistemic Logics (DEL) that were popularized recently [4, 5, 10, 11]. These logics have event models which can account for varying perception of events by the agents. Our *Listen* relation could be encoded in such event models. However, it is not clear how to model questions in DEL.

Our logic is related to that of peer-to-peer inference systems (P2PIS) [1, 2]. These are systems made up of autonomous agents that can communicate in order to perform an inference task such as query answering and consequence finding. In these systems it is often supposed that some propositional variables are 'owned' by a given agent, while some others play the role of an 'interface' with the agents it is connected to.

Several distributed logic programming systems have been proposed in the literature. As far as we are aware all of them differ from our proposal in that they suppose that the different processes all have the entire program at their disposal. An example is the KL1 programming language, which was developed in the Japanese Fifth Generation Computer Systems project [7]. This system uses communication, but for scheduling and not for sharing the program.

References

1. Adjiman, P., Chatalic, P., Goasdoué, F., Rousset, M.-C., Simon, L.: Scalability Study of Peer-to-Peer Consequence Finding. In: Proceedings of the International Joint Conference on Artificial Intelligence, IJCAI (2005)
2. Adjiman, P., Chatalic, P., Goasdoué, F., Rousset, M.-C., Simon, L.: Distributed Reasoning in a Peer-to-Peer Setting: Application to the Semantic Web. Journal of Artificial Intelligence Research 25, 269–314 (2006)
3. Aszalós, L.: Said and Can Say in Puzzles of Knights and Knaves. In: Chaib-draa, B., Enjalbert, P. (eds.) Proc. 1ères Journées Francophones des Modèles formels pour l'interaction, Toulouse, May 2001, vol. 3, pp. 353–362 (2001)
4. Baltag, A., Moss, L.: Logics for Epistemic Programs. Synthese 139(2), 165–224 (2004)
5. Baltag, A., Moss, L., Solecki, S.: The Logic of Common Knowledge, Public Announcements, and Private Suspicions. In: Proceedings of the seventh Theoretical Aspects of Rationality and Knowledge conferene (TARK), pp. 43–46. Morgan Kaufmann Publishers Inc., San Francisco (1998)
6. Chellas, B.F.: Modal logic: an introduction. Cambridge University Press, Cambridge (1980)
7. Chikayama, T., Fujise, T., Sekita, D.: A Portable and Efficient Implementation of KL. In: Penjam, J. (ed.) PLILP 1994. LNCS, vol. 844, pp. 25–39. Springer, Heidelberg (1994)
8. Fagin, R., Halpern, J.Y., Moses, Y., Vardi, M.Y.: Reasonning about Knowledge. MIT Press, Cambridge (1995)
9. Kröger, F.: Temporal Logic of Programs. Springer, Berlin (1987)
10. van Benthem, J., van Eijck, J., Kooi, B.: Logics of Communication and Change. Information and Computation 204(11), 1620–1662 (2006)
11. van Ditmarsch, H., van der Hoek, W., Kooi, B.: Dynamic Epistemic Logic. Synthese Library, vol. 337. Springer, Heidelberg (2007)

A Framework for Agent-Based Evaluation of Genetic Algorithms

David F. Barrero, David Camacho, and María D. R-Moreno

Abstract. Genetic Algorithms (GA) are a set of algorithms that use biological evolution as inspiration to solve search problems. One of the difficulties found when working with GA are the several parameters that have to be set and the many details that can be tunned in the GA. Usually it leads to the execution of several experiments in order to study how the GA behaves under different circumstances. In general it requires several computational resources and time to code the same algorithm with slight differences several times. In this paper we propose a framework based on agent technology able to parallelize the experiment and to split it into several components. It is complemented with a description of how this framework can be used in the evolution of *regular expressions*.

1 Introduction

A common problem that system administrators, programmers or the data mining community must address is the retrieval of strings that satisfy a given pattern. Logs processing, spam detection and data mining are typical problems that need pattern matching tools. One of the best known string matching and extraction tool is regular expressions (*regex*). Several tools and languages provides support for regex, from UNIX commands such as *grep* or *sed*, to an extense list of programming languages such as PERL. Regex provide a powerful tool to match and extract strings. However, they have an obscure syntax and require a skilled person in order to use it.

David F. Barrero · María D. R-Moreno
Computer Engineering Department, Universidad de Alcalá, Madrid, Spain
e-mail: `{david,mdolores}@aut.uah.es`

David Camacho
Computer Science Department, Universidad Autónoma de Madrid, Madrid, Spain
e-mail: `david.camacho@uam.es`

G.A. Papadopoulos and C. Badica (Eds.): Intelligent Distributed Computing III, SCI 237, pp. 31–41.
springerlink.com © Springer-Verlag Berlin Heidelberg 2009

Automatic generation of languages is a well known problem in Machine Learning named *language induction* [9]. There are several approaches from the Formal Languages and Theoretical Computer Science perspectives [11], however it is still an open problem. A recent approach to language induction is provided by the Evolutionary Computation community, where regex are evolved by means of evolutionary algorithms like Genetic Programming (GP) [7]or Genetic Algorithms (GAs) [8].

In this paper we present a method based on GA able to generate automatically regex from a set of positive and negative samples and we propose a new chromosome codification based on messy Genetic Algorithms (mGA) [5] and crossover operators. In order to evaluate and compare the proposed codifications and operators some experiments were implemented in a framework based on a multiagent system (MAS) named Searchy [2]. The agent-based framework has a double goal. On one hand, to reduce the execution time of the experiments, on the other hand to improve the search capacity in the space problem considered, allowing agents to find better solutions.

The paper is structured as follows. First, a review of Variable Length Chromosomes and mGA is provided. Sections 3 and 4 present the codification and crossover operators respectively. The agent-based framework used in the experimentation is presented in section 5. Then, they are evaluated in section 6. Finally, some conclusions and future research lines are outlined.

2 Variable Length Chromosomes and Messy GA Review

GAs are part of the Evolutionary Computation, a computing paradigm inspired in the biological process of evolution. It can be considered as a stochastic search algorithm that represents the solutions of a problem as chromosomes using some codification. Chromosomes explore the search space through two genetic operators: mutation and sexual reproduction (*crossover*). The metric of how well a chromosome solves a problem is given by a fitness function. The genetic material contained in the chromosome is named genotype meanwhile the realization of such genetic material is referred as phenotype.

The number of variables involved in a GA problem is closely related to the chromosome length. Sometimes it is possible to determine the number of parameters that certain problem requires, and therefore to determine the chromosome length to introduce in the GA. However, there are many problems in which such approach is not an adequate solution because it is not possible to limit the size of the solution. The number of nodes in a neural network or the size of an evolved program in GP [7] are not easy to set prior to the execution of the algorithm. Of course, it is possible to set a maximum number of variables. But this approach sets an arbitrary limitation to the complexity of the solution. A more desirable solution is to use an algorithm able to adapt the size of the chromosome dynamically. This is the goal of the Variable-Length Genomes (VLGs) [6].

The main difference between fixed-length chromosomes and VLGs is the crossover operator. The simplest crossover used in VLGs is *cut and splice*. Given two chromosomes (likely with different lengths), this crossover operator selects a random point in each chromosome and use it to divide them in two parts, then the parts are interchanged.

An early work in VLGs is the one developed by Goldberg with the messy GA [5]. It is a variable-length, gene position independent representation. Basically, mGA decouples the position of the genes from the phenotype that they represent and thus any gene can be placed anywhere within the chromosome. It is done representing each bit with a tuple $(locus, value)$ where the position, or locus, of the bit is specified as well as its value. Common genetic operators such as bit inversion mutation and cut and splice crossover are then applied to the chromosome constructed in this way.

Codification in mGA may generate two special situations that must be handled. Since the locus is coded within the allele, it might happen that not all the genes are defined, generating a problem called *underspecification*. Original mGA handles this situation by means of templates. A second problem arises when there are several alleles coding the same gene, i.e., the *overspecification*. mGA solves overspecification by means of a first-come-first-served philosophy.

mGA defines three phases: initialization, primordial and juxtapositional phase. The initialization phase deals with the generation of the initial population that feed the GA. It is composed by the set of all chromosomes of size k, where k is the gene size. This algorithm generates an unnecessary number of individuals, so the second phase filters the individuals trying to select individuals with a high density of good building blocks. The selected individuals are used then as the initial population in the GA. The generational based evolution is done in the juxtapositional phase.

3 Chromosome Codification

There are a number of questions that must be answered in order to successfully implement a GA. One of these questions is how to represent in the chromosome the problem that is addressed. In this section three codifications able to represent a regex in a binary chromosome are presented: one based on a plain VLG and two codifications inspired in mGA.

The lexical approach that we have adopted requires an alphabet Σ of *atomic regex* x_i such as $\Sigma = \{x_0, x_1, ..., x_N\}$. Σ is constructed using the positive and negative samples. Atomic regex are identified applying Lempel-Ziv law [13]. This law states that texts are not composed by a uniform distribution of tokens, instead, a few tokens appear many times while many tokens have a reduced weight in the text. We build tokens using a set of symbols to divide the samples and then, those tokens that appear more times are selected to be part of the alphabet. The second subset of Σ is composed by a fixed set of symbols. This is an automatic and domain independent method that can be used with almost any codification schema.

3.1 Plain Variable-Length Genome

Despite the inherent difficulty to determine *a priori* the length of the regex, it is possible to imitate a VLG by means of an island model [1] with immigration of individuals. It can take benefits of a parallel algorithm implemented with agents, however the population must be increased and it imposes a maximum length for the chromosomes. In this context selecting a codification able to deal with VLG seems to be a natural solution.

A simple way to compose and evolve the set of atomic regex in Σ is the traditional VLG, a binary chromosome of arbitrary length that is recombined using cut and splice, as it was described in sec. 2. The correspondence between the genes and the atomic regex in Σ is done as follows. Each gene contains l_g bits that code an integer number $i < 2^{l_g}$, then the gene represents the element x_{imodN} of Σ. Initial population has chromosomes randomly created with lengths uniformly distributed between a minimum chromosome length l_{min} and a maximum length l_{max}. The rest of the paper will use the term plain VLG to mean the codification schema described in this section.

3.2 Modified Messy GA and Biased Messy GA

Plain VLG provides a simple variable-length genome coding, however variable-length genomes usually present some problems. One of them is the tendency to bloat the chromosome length, as Chu observed [3]. Another problem is the *genetic linkage*, i.e., the tendency of some alleles to remain joint due, for example, to crossover biases [10]. A solution to genetic linkage is mGA because it decouples the position of the gene in the chromosome from its semantics. It is still a simple codification and it seems to be a good choice to study how genetic linkage affects the regex evolution in VLGs. However, some modifications are needed in order to use mGA in the context of our work.

Like original mGA does, in our proposal genes are coded as $(locus, value)$, however *value* follows the same coding scheme as the one described in 3.1 instead of being a single bit. It represents a symbol $x_i \in \Sigma$. We have integrated the original mGA initialization and primordial phases into one phase that generates Σ from the data set using Lempel-Ziv law, following the same philosophy that the used with VLG codification. With this approach there is no need to implement the primordial phase because the building blocks are integrated in Σ.

Initial mGA and revised versions of the algorithm such as fast mGA [4], do not generate the initial population randomly. There is rather a slight control about how to initialize them, as it was described in sec. 2. We propose also a modification of the initialization. Given a random number l uniformly distributed between l_{min} and l_{max}, a chromosome with l/l_g genes is created. *value* is filled with a random value that codes an atomic regex following the same mechanism than plain VLG, while *locus* takes a value from 0 to $l/l_g - 1$.

A second modification to the mGA named biased messy GA (bmGA) is also proposed. The biased messy GA instead of initializing the loci field with positions from 0 to $l/l_g - 1$, they are initialized with a biased loci, and thus their values range from *bias* to $l/l_g + bias - 1$, where *bias* can take several values. However, we have used $bias = 2^{l^{locus}}/2$, i.e., alleles begin to be placed in the middle position.

4 Recombination Operators

The main role of the crossover is to recombine good chunks of chromosomes generating offspring [12] with better genetic information. Some authors have argued that the crossover performs better when it recombines two similar chromosomes [6], however this point is controversial and there are not a general consensus in the GA community. In the context of regex evolution, the disruptive properties is crossover is a main issue because of the rough nature of the fitness, a very small difference in the chromosome might lead to a dramatic change in the fitness. Following these ideas it seems natural to hypothesize that using a less destructive crossover operator will increase the performance of the GA in regex evolution.

The goal of the new crossover mechanism is to use the knowledge about the codification to recombine chromosomes in a less destructive way compared with the cut and splice crossover. Crossover is not directly performed with the chromosomes, instead an intermediate table is constructed. Our crossover proposal is divided in five phases as described.

1. *Integer chromosomes construction.* Alleles in the chromosomes (including their loci and values) are transformed into an integer representation. The order in which the alleles appear is respected.
2. *Intermediate table construction.* The intermediate table is a table composed by three columns and as many rows as the sum of not underspecified genes in the chromosomes. One column contains the sorted loci while the latter two columns contain each one the values (if any) defined for such locus.
3. *Crossover.* The intermediate table can be seen as two chromosomes, and thus any traditional crossover operators (one point, two points and uniform crossover) can be applied just interchanging the values of the chromosomes columns in the table.
4. *Recombined integer chromosomes construction.* Two integer chromosomes are constructed using the recombined intermediate table, it is the inverse operation of the phase two. It should be noticed that because of the lack of genetic linkage the position of the alleles is irrelevant for the crossover and thus their position can be changed without loss of relevant information.
5. *Recombined binary chromosomes construction.* The integer chromosomes are representated with a binary codification.

An example of modified one-point crossover is shown in Fig. 1. Two chromosomes are recombined in the example. Both use seven bits to code each gene, divided in three bits for the locus and four bits for the value. Chromosome A is

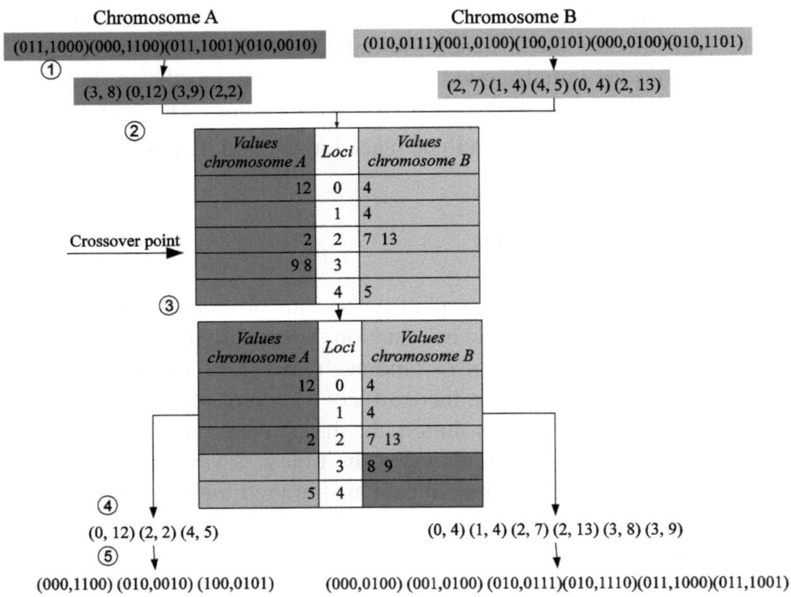

Fig. 1 Modified one point crossover example

composed by four alleles and chromosome B is composed by five alleles. Chromosome A and B present overspecification as well as underspecification. The intermediate table is constructed and, as it can be seen in the figure, underspecified genes correspond to empty cells. On the other hand, overspecified genes correspond to cells with several sorted values. A random point is used to interchange cells in the table, generating the recombined chromosomes. Any other traditional crossover mechanism may also be applied.

5 GA Evaluation Framework

Due to the high number of GA runs that must be performed and the parallel nature of the GA, the set of experiments were run in a MAS that decomposes the GA evolution in a sequence of operations performed by different agents. This MAS has been deployed using the Searchy platform [2]. In this way the experimentation can be divided into different simple operations that are composed and executed in parallel, increasing the performance and the search capability of the algorithm.

There are six roles defined in the MAS: control, population, crossover, fitness evaluation, codification and alphabet agent. Each role is implemented using a specialized agent and there are several agents to implement crossover and codification. A description of each role is briefly presented.

1. *Control Agent.* It is responsible for the execution of the experiment, and has to fulfill some tasks, such as the initialization of Population Agents and control the execution of the experiments. It also gathers measures from the populations and generate statistics, averaging the measures of all the GA executions.
2. *Population Agent.* This agent contains a population of individuals represented by a binary chromosome. It also performs the generational evolution of the population using the services provided by the crossover, fitness evaluation and coding agents services.
3. *Crossover Agent.* A crossover agent is an agent that performs a crossover between two chromosomes. Actually, there are four different crossover agents that implement the four crossover operators under study. Cut and splice crossover can be performed in any codification under study while modified one, two and any point crossover requires a mGA or bmGA.
4. *Fitness Evaluation Agent.* This is an agent that, given a string regex is able to evaluate its extraction capabilities using a training set. It should be noticed that since it takes a string as input, this agent in not affected by the chromosome codification.
5. *Codification Agent.* The codification generates the phenotype associated to a given chromosome, i.e., it transforms a chromosome into a string containg a regex. This regex is used by the Population Agent prior to evaluate any individual's fitness. There are two codification agents, the Plain VLG Coding Agent and the mGA Coding Agent. Since the only difference between mGA and bmGA is the initialization of the populations there is no need to use a bmGA Coding Agent.
6. *Alphabet Agent.* The alphabet agent takes as input the set of positive examples and using the Lempel-Ziv law identifies a set of tokens that are used to generate the atomic regex alphabet. The alphabet is used by the Codification Agents to generate the string regex.

Fig. 2 depicts the MAS architecture. First, the Control Agent initializes several Population Agents and associate each population with a Crossover Agent and a Codification Agent. In this way each Population Agent contains an experiment involving a certain crossover operator and codification. Once the Population Agents have been initialized they evolve their populations for a number of generations, then they return to the Control Agent several measures. The Control Agent repeats this process a given number of times and then averages the measures.

The Alphabet Agent reads the positive examples and generates the alphabet once, then it is provided to the Coding Agents which set a correspondence between each element in the alphabet and the codification used in the genome. It should be noticed that no agent with the exception of the Coding Agents need to know how the chromosome is coded, they manipulate the chromosome as a sequence of bits. The only agent that does not require a binary chromosome is the Fitness Agent because it receives the regex in form of string, instead of a binary chromosome.

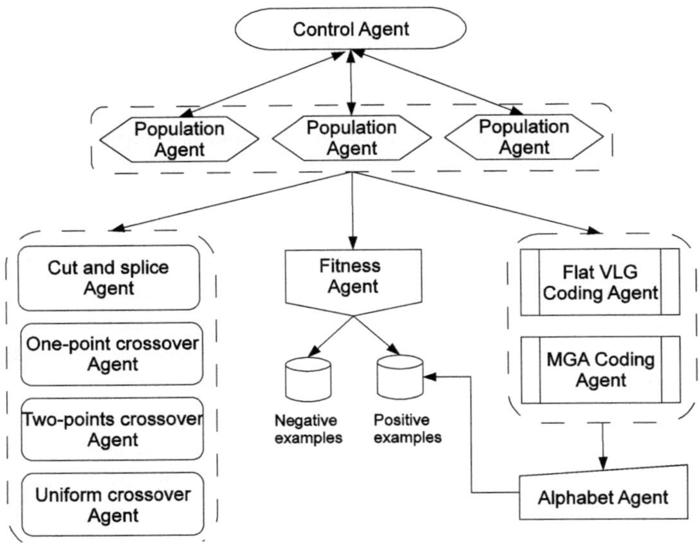

Fig. 2 MAS architecture used in the evaluation of the proposed crossover operators and codifications.

6 Experimental Study

This section describes the behavior and extraction capabilities of the coding and crossover mechanisms described in sections 3 and 4 using a MAS. A brief summary of the experiments results is shown.

Two case studies are used in the experiments where regex able to extract emails and phone numbers are evolved. These are two well known problems in data mining literature. Each study case uses a dataset with positive examples that have been divided into a training set and a testing set. Meanwhile the negative examples are shared among the study cases. Due to the stochastic nature of the GAs, each experiment has been run one hundred times, and the data has been averaged.

The fitness evaluation that has been used in all the experiments can take values from 0 to 1, where 1 is the maximum fitness that any chromosome can achieve. The calculus of the fitness is performed as follows. For each positive example the proportion of extracted characters is calculated. Then, the fitness is calculated subtracting the average proportion of false positives in the negative example set to the average of the characters correctly extracted. In this way, the maximum fitness that a chromosome can achieve is one, let us name it as ideal individual. Then, an ideal individual is able to extract correctly all the elements of positive examples while no element of the negative examples is matched.

Experimental study has been divided into three stages. The first one is an initial set of experiments whose aim is to select the GA parameters, the results are shown in Table 1. Parameters yield similar optimum values for all the investigated algorithms with one notable exception, the mutation probability. Algorithms that use

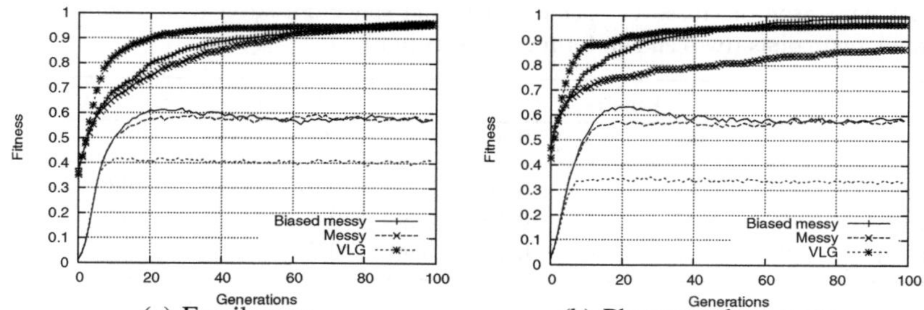

(a) Email regex (b) Phone numbers regex

Fig. 3 Comparison of codifications in regex evolution. Best individual and average fitness are shown.

cut and splice crossover operator (VLG and bmGA cs) have an optimum mutation probability arround one order of magnitude lower than the others algorithms (bmGA with any form of our proposed crossover). The higher disruptive capabilities of cut and splice operator compared to the proposed crossover operator may explain this difference.

A second set of experiments were executed to study the performance of the three described codifications. In order to obtain comparable results, a cut and splice recombination operator has been used in all the experiments belonging to this second experimental stage. The two case studies yield similar experimental results, as can be seen in Fig. 3. Results suggest that plain VLA archieve higher best fitness however Fig. 3(b) shows in generation 60 a slightly higher fitness for bmGA. In any case, plain VLG increase its best fitness faster than messy codifications due to its smaller chromosome: plain VLG does not need to codify the locus.

Comparing messy codifications, bmGA performs slightly better than mGA, specially in the phone numbers case study (see Fig. 3(b)). The better performance of bmGA compared to mGA in our experiments can be explained by the dynamics of the construction of the phenotype. Using a pure mGA, the first position of an atomic regex is 0, and thus the regex cannot be expanded to the left, because there is no

Table 1 Parameters for the experiments carried out using a basic VLG (VLG), bmGA with cut and splice crossover (mbGA cs), bmGA with modified one-point, two-points and uniform crossovers (bmGA one, bmGA two and bmGA uni)

Settings	VLG	bmGA cs	bmGA one	bmGA two	bmGA uni
Mutation probability (P_{mut})	0.005	0.002	0.02	0.01	0.015
Population size (n)	50	50	50	50	50
Min. chromosome length (l_{min})	4	9	9	9	9
Max. chromosome length (l_{max})	40	90	90	90	90
Gene length (l_g)	4	9	9	9	9
Crossover probability (P^c)	-	-	-	-	0.3

Table 2 Comparison of crossover operators: Cut and splice crossover (cs), modified one-point (one), two-points (two) and uniform crossovers (uni).

	Email				Phone			
	cs	one	two	uni	cs	one	two	uni
Best fitness	0.96	0.94	0.9	0.94	0.99	0.97	0.95	0.98
Avg. fitness	0.58	0.42	0.58	0.46	0.58	0.43	0.6	0.43
Prob. ideal	0.86	0.78	0.64	0.77	0.90	0.77	0.66	0.80

natural number lower than 0. BmGA places the first regex in *bias* and thus by means of mutation and crossover regex can grow to the left.

The third group of experiments aims to study the proposed crossover mechanisms for messy codifications. Table 2 shows the best fitness, mean fitness and probability of finding an ideal individual for both case studies being investigated. Results show that the crossover operator has a limited effect in the fitness. Cut and splice seems to outperform the other operators, however it would be desiderable to use hypothesis contrast to proof it.

7 Conclusions and Future Work

We have presented a method to generate regular expressions using supervised learning and an agent based testing framework. The distributed testing framework used has been a satisfactory approach due to the enhanced performance and easy composition of tasks involved in the GA. Additionally a brief empirical analysis of how different codifications and crossover mechanism influence the evolution of regex has been presented. The set of experiments carried out showed that the best performance is achieved with a direct codification of the alphabet using a plain VLG. These results leads us to conclude that there are some intrinsic limitations in the evolution of regex independently of the codification and crossover operator used. A natural step is the use of evolutionary techniques able to deal with non lineal codifications, such as Genetic Programming or Grammatical Evolution.

Acknowledgements. This work has been partially supported by the Spanish Ministry of Science and Innovation under the projects COMPUBIODIVE (TIN2007-65989), V-LeaF (TIN2008-02729-E/TIN), Castilla-La Mancha project PEII09-0266-6640 and HADA (TIN2007-64718).

References

1. Barrero, D.F., Camacho, D., R-Moreno, M.D.: Automatic Web Data Extraction Based on Genetic Algorithms and Regular Expressions. In: Data Mining and Multiagent Integration, August 2009. Springer, Heidelberg (2009)

2. Barrero, D.F., R-Moreno, M.D., López, D.R., García, Ó.: Searchy: A metasearch engine for heterogeneus sources in distributed environments. In: Proceedings of the International Conference on Dublin core and Metadata Applications, Madrid, Spain, September 2005, pp. 261–265 (2005)
3. Chu, D., Rowe, J.E.: Crossover operators to control size growth in linear GP and variable length GAs. In: Wang, J. (ed.) 2008 IEEE World Congress on Computational Intelligence, Hong Kong, June 1-6. IEEE Computational Intelligence Society. IEEE Press, Los Alamitos (2008)
4. Deb, K.: Binary and floating-point function optimization using messy genetic algorithms. PhD thesis, Tuscaloosa, AL, USA (1991)
5. Goldberg, D., Deb, K., Korb, B.: Messy genetic algorithms: motivation, analysis, and first results. Complex Systems 3(3), 493–530 (1989)
6. Harvey, I.: The saga cross: the mechanics of recombination for species with variable-length genotypes. In: Manner, R., Manderick, B. (eds.) Parallel Problem, pp. 269–278. North-Holland, Amsterdam (1992)
7. Koza, J.R.: Genetic Programming: On the Programming of Computers by Means of Natural Selection (Complex Adaptive Systems). The MIT Press, Cambridge (1992)
8. O'Neill, M., Ryan, C.: Grammatical evolution. IEEE Transactions on Evolutionary Computation 5(4), 349–358 (2001)
9. Parekh, R., Honavar, V.: Grammar inference, automata induction, and language acquisition. In: Handbook of Natural Language Processing, pp. 727–764. Marcel Dekker, New York (1998)
10. Rana, S.: The distributional biases of crossover operators. In: Proceedings of the Genetic and Evolutionary Computation Conference, pp. 549–556. Morgan Kaufmann Publishers, San Francisco (1999)
11. Sakakibara, Y.: Recent advances of grammatical inference. Theor. Comput. Sci. 185(1), 15–45 (1997)
12. Spears, W.M.: Crossover or mutation. In: Foundations of Genetic Algorithms 2, pp. 221–237. Morgan Kaufmann, San Francisco (1993)
13. Ziv, J., Lempel, A.: A universal algorithm for sequential data compression. IEEE Transactions on Information Theory 23(3), 337–343 (1977)

Efficient Broadcasting by Selective Forwarding

Doina Bein, Ajoy K. Datta, and Balaji ashok Sathyanarayanan

Abstract. A major challenge faced in MANETs is locating devices for communication, especially in the case of high node mobility and sparse node density. We propose to develop a strategy to reduce the redundant transmission of packets in normal flooding used in broadcasting and we describe strategies for choosing only a minimal set of nodes to re-broadcast in grid networks. Our strategies reduce the redundant transmission of packets, thus packets are forwarded with a minimal number of transmissions. To determine the minimal set of nodes we propose a new algorithm called Efficient Broadcasting by Selective Forwarding (EBSF) that uses a distance-based approach in selecting the nodes among all the nodes in a grid network.

1 Introduction

A mobile ad hoc network (MANET) is an autonomous system of mobile routers, connected by wireless links, the union of which forming an arbitrary graph (see Figure 1). The routers are free to move and organize themselves arbitrarily. Thus, the network's wireless topology may change rapidly and unpredictably. Such a network may operate in a stand-alone fashion.

Doina Bein
Applied Research Laboratory, The Pennsylvania State University,
University Park, PA 16802, USA
e-mail: siona@psu.edu

Ajoy K. Datta
School of Computer Science, University of Nevada, Las Vegas, NV 89154, USA
e-mail: datta@cs.unlv.edu

Balaji ashok Sathyanarayanan
School of Computer Science, University of Nevada, Las Vegas, NV 89154, USA
e-mail: ashoksathyan@yahoo.com

G.A. Papadopoulos and C. Badica (Eds.): Intelligent Distributed Computing III, SCI 237, pp. 43–52.
springerlink.com © Springer-Verlag Berlin Heidelberg 2009

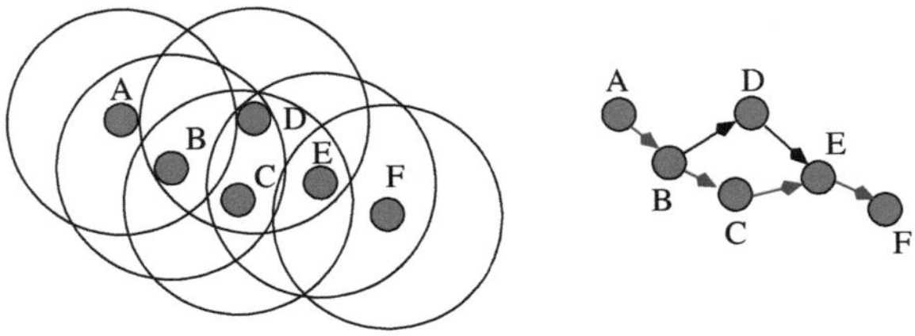

Fig. 1 Example of MANET

Broadcasting, the process by which one node needs to communicate a packet to all other nodes in the network, is a critical mechanism for information diffusion and maintaining consistent global network information. Additionally, it is an energy intensive function. Broadcasting or a derivation of it is used by routing protocols such as Dynamic Source Routing (DSR) [8], Ad Hoc On Demand Distance Vector (AODV) [15], Location Aided Routing (LAR) [9, 10], and Zone Routing Protocol (ZRP) [7], for establishing routing paths. Currently all these protocols rely on a simple form of broadcasting called *blind flooding*. In blind flooding, each node re-broadcast a packet whenever it receives it for the first time: Every incoming packet is sent out on every outgoing line except the one it arrived on. Blind flooding will generate many redundant transmissions, which may cause a more serious broadcast storm problem [12].

Given the expensive and limited nature of wireless resources such as bandwidth and battery power, minimizing the control message overhead for route discovery is a high priority in protocol design. Recently, a number of research groups have proposed more efficient broadcasting techniques. Centralized broadcasting schemes are presented in [1, 2, 6]. The algorithms in [11, 19, 13, 14, 16, 17] utilize neighborhood information to reduce the number of redundant messages.

In our protocol called Efficient Broadcasting by Selective Forwarding (EBSF), broadcasting is done by particular nodes which are allowed to broadcast based on so-called *threshold distance*, $threshold = n \times transmission\_radius$, where n is a real number. This threshold value can be tuned to show performance enhancement, thus minimizing the number of transmissions or retransmissions needed. If the threshold distance is set to 0, selective broadcasting will be exactly blind flooding. To check for message duplication, thus to reduce the number of redundant messages to be delivered, we implement a new data structure called *message cache*.

We have simulated the communication network using a network simulator called GLOMOSIM [5, 3, 4]. GLOMOSIM is a library-based sequential and parallel simulator for wireless networks. The library of GLOMOSIM is based on the parallel discrete-event simulation capability provided by Parsec, the compiler used in GLOMOSIM. Using GLOMOSIM we show the performance variation of our algorithm using different threshold factors.

In Section 2 we present an overview of various broadcasting schemes followed by the motivation of why blind flooding is not efficient for broadcasting in MANETs. In Section 3 we define efficient broadcasting, followed by a description of our protocol and its advantages. A discussion regarding the simulation results and the performance analysis is given in Section 4. We finish with concluding remarks in Section 5.

2 Preliminaries

In mobile ad hoc networks, it is often necessary to broadcast control information to all constituent nodes in the network. Blind flooding is often deployed to achieve the above objective. Its advantages, the simplicity and the guarantee that every destination in the network is reached, are downsized by the fact that it is expensive in terms of overhead and wastes valuable resources such as bandwidth and power:

- Some routers receive a packet multiple times.
- It leads to transmission of redundant packets.
- Packets can go in a loop forever.
- For dense networks, it causes significant contention and collisions – the so-called *broadcast storm problem*.

An improvement to blind flooding is to choose only a subset of nodes to re-broadcast and in this manner to reduce the number of data transmissions. Several alternatives are presented next. The probabilistic scheme [17, 18] is similar to blind flooding, except that nodes only re-broadcast with a predefined probability. Since some nodes do not re-broadcast, node and network resources are saved without having delivery effectiveness. In sparse networks, nodes will not receive all broadcast packets unless the probability parameter is high. When the probability is 100%, this scheme is identical to blind flooding.

There is an inverse relationship between the number of times a packet is received at a node and the probability of that node being able to reach some additional area on a broadcast. This result is the basis of their counter-based scheme [12, 17, 18]. Upon reception of a packet never received before, the node initiates a counter with value 1 and sets a RAD. During the RAD, the counter is incremented for each redundant packet received. If the counter is less than a threshold value when the RAD expires, then the packet is re-broadcast, otherwise the packet is simply dropped.

The probabilistic and counter-based schemes are simple and inherently adaptive to the local topologies. Their disadvantage is that the delivery is not guaranteed to all nodes even if the ideal MAC is provided. In other words, both schemes are unreliable.

Area-based methods [18] only consider the coverage area of transmission and do not consider whether there are nodes within that area. A node decides whether to re-broadcast purely based on its own information. There are two coverage area-based methods [17]: distance-based and location-based schemes. In this paper we have developed a new approach based on the distance [17] by which the data transmitted is reduced considerably to a greater extent, while ensuring that all data is received by all nodes in the network. In our approach, broadcasting is done by some selected nodes that are privileged to broadcast based on so-called *threshold distance*. If the threshold distance is set to 0 then we have blind flooding.

3 Efficient Broadcasting by Selective Forwarding

EBSF reduces the redundant transmission of packets used in broadcasting. An optimal selection is done to determine the minimal set of nodes to re-broadcast. To determine this minimal set of nodes we employ the distance-based approach [17]. We assume that the nodes are placed on a grid and all nodes have the same transmission range. We also assume that the transmission range is constant. The distance-based approach is implemented with the help of a threshold value that is taken to be $threshold = N \times transmission\_radius$, where N has values $0 < N < 1$.

In blind flooding, location discovery is achieved by allowing all nodes that receive some message to retransmit it. In our case we prevent any node within a distance factor which is the threshold to re-transmit. Thus a few selected nodes will be selected for retransmission, and subsequently we reduce the number of redundant transmissions (see Figure 2).

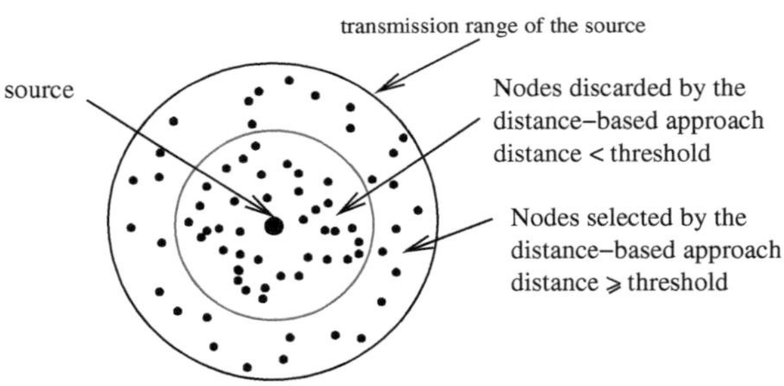

Fig. 2 Efficient Broadcasting by Selective Forwarding

This threshold value can be tuned as per requirements. This threshold value can be increased and decreased to show the performance variation. Higher the threshold, lesser is the number of packets transmitted by each node in the network and vice versa. If threshold is set to 0, then all packets are transmitted like in blind flooding; in this case there is no enhancement.

In the OSI model, upper layers deal with the application's issues and generally are implemented only in software. The highest layer, the application layer, is closest to the end user. The lower layers of the OSI model handle data transport issues, namely communications between network devices. When a packet is to be transmitted from source node to a destination node, the network layer (layer 3) is responsible for end-to-end (source to destination) packet delivery, whereas the data link layer (layer 2) is responsible for node-to-node (hop-to-hop) packet delivery. When a node receives a packet, the MAC layer of the data link layer sends the packet to the network layer for transmission. Efficient Broadcasting by Selective Forwarding (EBSF) is implemented in the network layer.

When a packet is forwarded to the network layer for transmission, the IP header in the Internet Protocol has the source node and the destination node coordinates. Based on these coordinates, the distance between the source and destination nodes cam be easily computed. In blind flooding, whenever a message is received and needs to be sent further, at the network layer the packet is broadcast to all the nodes within its transmission range once it is received from the MAC layer.

In EBSF protocol, when a packet reaches the network layer of a node for broadcasting some computation is performed as follows (see Figure 3). Based on the information about the source and destination taken from the IP header, EBSF computes the distance between the source and the destination. When this distance is less than the threshold distance, the node discards the message. If this distance is greater than the threshold distance, then EBSF broadcasts the message to the nodes at a distance equal to or greater than the threshold distance.

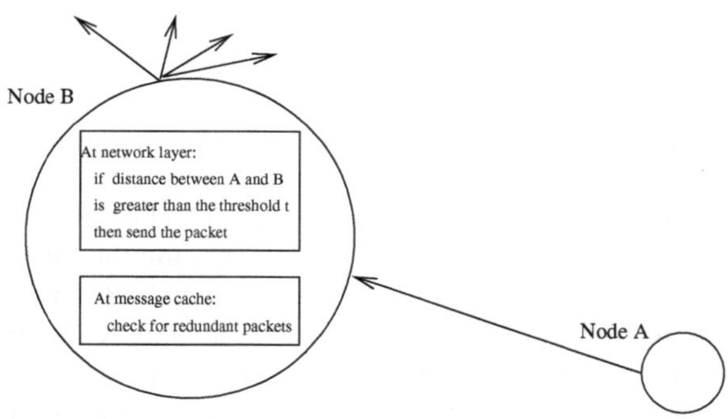

Fig. 3 EBSF protocol

When a node A broadcasts p packets to other nodes in the network, each node has a message cache, called MSGCACHE, which stores information about messages the node has received. When a node receives a packet, it checks with the MSGCACHE whether the packet has already been received. In affirmative, the packet is discarded. If the packet has not been received before, EBSF forwards it to the network layer for transmission. In this manner, the redundant packets are discarded. MSGCACHE is implemented as a linked-list data structure.

In EBSF, at any node every packet is checked for message redundancy using MSGCACHE and then send it to the network layer for broadcast. The network layer will decide, based on the threshold distance between the source node and destination nodes, whether to broadcast or not, as follows. Based on the information about the source and destination in the IP header, the distance between the source node and the destination node is calculated. When this distance is less than the threshold distance, the node discards the message. If this distance is greater than the threshold distance, the node will broadcast the message to the nodes at a distance equal or greater than the threshold distance. This process is repeated until all the nodes in the network receive all the packets originated from the source node.

For further enhancement, we partition the neighbors of a node into two types, *geographical neighbors* and *communicational neighbors*. Geographical neighbors of a node in a grid network are the nodes located at a distance of one grid unit: a corner node has two geographical neighbors and a border node has three geographical neighbors; all other nodes have four geographical neighbors. For example, in Figure 4, the geographical neighbors of node E are nodes B, D, F, H. The corner node G has two geographical neighbors, nodes D and H. The border node H has three neighboring nodes D, E, I.

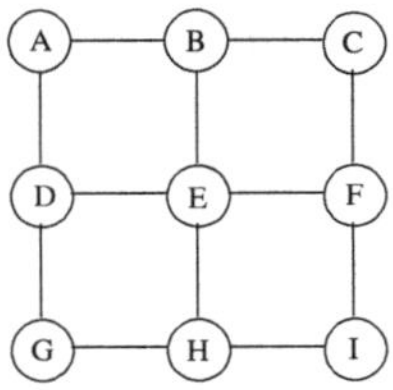

Fig. 4 Geographical Nodes

Communicational neighbors are the ones within the node's transmission range, excluding the geographical nodes.

We have enhanced EBSF to exclude to broadcast to the geographical neighbors of a node situated within the threshold distance, and thereby decreasing the number of transmissions. For example, given a sender node S, if some node R is at a distance greater than or equal to the threshold distance of node S, then more likely the geographical nodes of node R are also at threshold distance from node S. In this case in EBSF, node R and the geographical neighbors of R are allowed to broadcast the message received from node S. When node R and its geographical neighbors may have common nodes which are at a threshold distance from them, these nodes will receive the same messages from node R and its geographical neighbors, which would lead to redundant transmissions. Hence with the enhanced EBSF algorithm we restrict the geographical neighbors of a node to not broadcast and thereby reduce redundant transmission, which is evident from the simulation results.

4 Simulations

The tool we have used for the simulation is GLOMOSIM (Global mobile system simulator). GLOMOSIM is a scalable simulation environment for wireless and wired systems, designed using the parallel discrete-event simulation capability provided by Parsec. Parsec compiler is similar to the C compiler with some added functions. The simulation statistics will be stored in the "bin" directory of GLOMOSIM as "glomo.stat" file. The layers in GLOMOSIM are shown in Table 1.

All routing decisions are taken in the network layer, so we embed our protocol EBSF in the network layer of each node of the network. Whenever the packet enters

Table 1 Layers of GLOMOSIM

Layers	Protocols
Mobility	Random waypoint, Random drunken, Trace based
Radio Propagation	Two ray and Free space
Radio Model	Noise Accumulating
Packet Reception Models	SNR bounded, BER based with BPSK / QPSK modulation
Data Link (MAC)	CSMA, IEEE 802.11 and MACA
Networking (Routing)	IP with AODV, Bellman-Ford, DSR, Fisheye, LAR scheme 1, ODMRP
Transport	TCP and UDP
Application	CBR, FTP, HTTP and Telnet

the network layer, the packet will be handled by EBSF. For our protocol EBSF to work in the network layer of GLOMOSIM we have modified and included the files EBSF.H and EBSF.PC in the network directory of GLOMOSIM. The program written in the file EBSF.PC contains the code for threshold computation and the code to determine whether to broadcast or to discard a message.

Whenever the packet reaches the network layer, it will be handled by protocol EBSF as follows (see Figure 5). At the beginning, the initialization function *RoutingEBSFInit*() is called and defines a structure called *GlomoRoutingEBSF* and allocates memory to it. Also the stats (statistics to be printed to determine the enhancements) are initialized, the sequence table for the node, and message cache. The top value is initialized to 0.

A routing function *RoutingEBSFRouter*() determines the routing action to be taken and handles the packet accordingly (if it is from UDP or MAC). If the packet is from UDP, then the node is the source node and the data is sent, i.e., the function *RoutingEBSFSendData*() is called. If data comes from the MAC layer, then the decision is made as whether to send it to UDP or drop the packet. This decision is made in the function *RoutingEBSFHandleData*().

The function *Finalize*() initializes the statistics part of the protocol. When this function is called, it collects the statistics from the file "glomo.stat" in the "bin" directory and formats the statistics such that number of data transmitted, the number of data originated, and the number of data received for each node are printed.

The function *RoutingEBSFHandleData*() is called whenever the node receives the packet from the MAC layer. This function checks with the message cache and

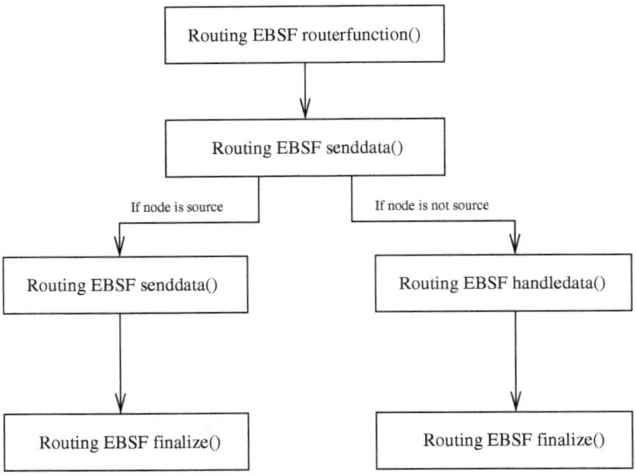

Fig. 5 Module Diagram of EBSF

it decides whether to transmit or discard the packet. The nodes that are with in the transmission range will receive the packet and only the nodes that are greater than threshold distance and lesser than transmission range will transmit the packet. If the node is at a distance lesser than a threshold value from a transmitted node, then the packet is discarded.

The message cache is implemented as a linked-list. The function LookupMessageCache searches whether the message already exists in the message cache using its sequence number. The function InsertMessageCache inserts a message into the cache if it is not already present there.

The threshold distance is defined as $N \times \mathit{transmission\_range}$ of a node, where N is a real number between 0 and 1.

We have implemented and tested our protocol for a network with $n = 49$ nodes (a perfect grid) and various values of the threshold (see Figure 6).

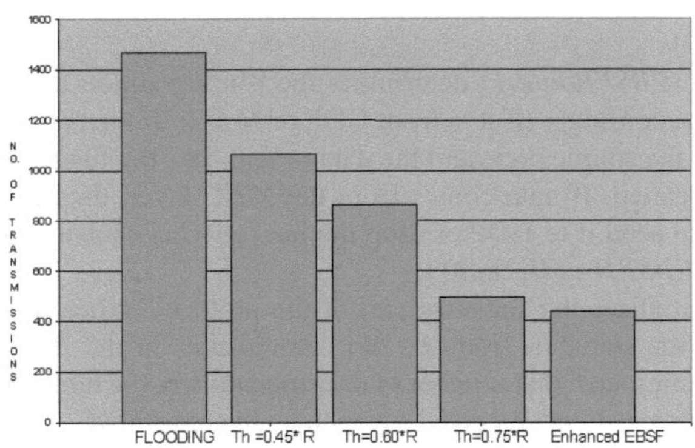

Fig. 6 Performance Analysis with Different Threshold Values

5 Conclusion

Building efficient protocols for ad hoc networks is a challenging problem due to the dynamic nature of the nodes. Efficient Broadcasting by Selective Forwarding has a number of advantages over other approaches considered in the literature. First of all, it selects a minimal number of nodes to retransmit, thus it utilizes the bandwidth efficiently. Secondly, it minimizes the number of unnecessary transmissions and therefore it reduces the number of redundant packets. Thirdly, it minimizes the overall power consumption.

The threshold value can be tuned to show the performance enhancements. Higher the threshold value, more optimized results are obtained. EBSF does not impose any bandwidth overhead and reduces the power consumption drastically. The efficiency of EBSF remains very high even in large networks. Overall, the proposed protocol shows that broadcasting can be enhanced greatly by choosing only an optimal set of nodes for transmission and thus avoiding redundant transmissions and at the same time ensuring data delivery to all the nodes in the network.

This protocol could be integrated with any routing protocol for finding a route in mobile ad-hoc networks with minimal power consumption and without imposing any bandwidth overhead.

Current research in wireless networks focuses on networks where nodes themselves are responsible for building and maintaining proper routing (self-configure, self-managing). Our algorithm does not adapt to topology changes; this is a topic of future research. If nodes are missing from the grid, the threshold value needs to be decreased; in case new nodes are added, the threshold value needs to increase to keep the performance of the protocol. This increasing or decreasing has to be done dynamically, and better in a non-centralized manner.

References

1. Alon, N., Bar-Noy, A., Linial, N., Peleg, D.: A lower bound for radio broadcast. J. Comput. Syst. Sci. 43, 290–298 (1991)
2. Gaber, I., Mansour, Y.: Broadcast in radio networks. In: Proceedings of SODA, January 1995, pp. 577–585 (1995)
3. GLOMOSIM, http://pcl.cs.ucla.edu/projects/GLOMOSIM/
4. GLOMOSIM, http://www.sm.luth.se/csee/courses/smd/161_wireless/glomoman.pdf
5. GLOMOSIM. Installation of glomosim 2.03 in windows xp, http://www.cs.ndsu.edu/~ahmed/GLOMOSIM.html
6. Guha, S., Khuller, S.: Approximation algorithms for connected dominating sets. In: Proceedings of ESA (1996)
7. Haas, J.: A new routing protocol for the reconfigurable wireless networks. In: Proceedings of 6th IEEE International Conference on Universal Personal Communications (ICUPC), October 1997, pp. 562–566 (1997)

8. Johnson, D.B.: Routing in ad hoc networks of mobile hosts. In: Proceedings of the Workshop on Mobile Computing Systems and Applications, December 1994, pp. 158–163 (1994)
9. Ko, Y.-B., Vaidya, N.H.: Location-aided routing (lar) in mobile ad hoc networks. In: Proceedings of ACM/IEEE Mobicom, October 1998, pp. 66–75 (1998)
10. Ko, Y.-B., Vaidya, N.H.: Location-aided routing (lar) in mobile ad hoc networks. Wireless Networks 6, 307–321 (2000)
11. Lim, H., Kim, C.: Multicast tree construction and flooding in wireless ad hoc networks. In: Proceedings of the ACM International Workshop on Modeling, Analysis and Simulation of Wireless and Mobile Systems, MSWIM (2000)
12. Ni, S.-Y., Tseng, Y.-C., Chen, Y.-S., Sheu, J.-P.: The broadcast storm problem in a mobile ad hoc network. In: Proceedings of ACM MOBICOM, August 1999, pp. 151–162 (1999)
13. Peng, W., Lu, X.: On the reduction of broadcast redundancy in mobile ad hoc networks. In: Proceedings of MOBIHOC (2000)
14. Peng, W., Lu, X.: Ahbp: an efficient broadcast protocol for mobile ad hoc networks. Journal of Science and Technology (2002)
15. Perkins, C., Das, S.: Ad hoc on-demand distance vector (aodv) routing (July 2003), `http://tools.ietf.org/html/rfc3561`
16. Qayyum, A., Viennot, L., Laouiti, A.: Multipoint relaying: an efficient technique for flooding in mobile wireless networks. Technical Report Technical Report 3898, INRIA - Rapport de recherche (2000)
17. Sucec, J., Marsic, I.: An efficient distributed networkwide broadcast algorithm for mobile ad hoc networks. Technical Report CAIP Technical Report 248, Rutgers University (September 2000)
18. Williams, B., Camp, T.: Comparison of broadcasting techniques for mobile ad hoc networks. In: Proceedings of the ACM MOBIHOC, pp. 194–205 (2002)
19. Wu, J., Dai, F.: Broadcasting in ad hoc networks based on self-pruning. In: Proceedings of IEEE INFOCOM (2003)

The Assessment of Expertise in Social Networks

Vincenza Carchiolo, Alessandro Longheu,
Michele Malgeri, and Giuseppe Mangioni

Abstract. Experts finding is a frequent issue in social (even web-based) networks. This work proposes a method to globally rank people according to their expertise in a set of topics . We also introduce contexts similarity to allow related contexts to be exploited during ranks assessment. These ideas are applied to the expertise network built from Epinions.com dataset, showing first significant results.

1 Introduction

Interpersonal information exchange underpins human society, not only in the past [1, 2], but also in tomorrow's world thanks to the spreading of web-based social networks. In this scenario, the question of identifying persons with expertise in a given context is essential for an effective and efficient information gathering [3, 4] and it is also useful in several real applications, e.g. trust and reputation management [5], the assignment of task in an enterprise, or paper reviewers in a conference [6].

In this work we present a method to rank people according to their expertise in a given set of topics. We perform this assessment in an *expertise network*, i.e. where the relationship among nodes is the expertise rank assigned in a given context. In particular, we aim at evaluating the *global* expertise a node v has in a network within a specific context based on *local* expertise ranks assigned by v's neighbor nodes.

The approach we endorse is someway similar to the one proposed by Eigen-Trust [7], where trust is replaced by expertise rank. Note that trust and expertise are generally considered as distinct concepts [8], however trust can be related to expertise, for instance [5] defines a context-dependent trust, and we also believe that the expertise rank a node v assigns to a node w in a given context significantly influences trust, and that the trustworthiness v assigns to w can actually be different depending

Vincenza Carchiolo · Alessandro Longheu · Michele Malgeri · Giuseppe Mangioni
Dip. Ingegneria Informatica e delle Telecomunicazioni, Facoltà di Ingegneria,
Università degli Studi di Catania, Catania, Italy
e-mail: car@diit.unict.it, alongheu@diit.unict.it

G.A. Papadopoulos and C. Badica (Eds.): Intelligent Distributed Computing III, SCI 237, pp. 53–62.
springerlink.com © Springer-Verlag Berlin Heidelberg 2009

on the context, it is indeed reasonable that in real life I could have much trust in my mechanic I used to go anytime my car is broken down, but this does not imply that I will trust him enough to put my children in his hands, thus I assign him a high trustworthiness in the "car repairing" context and a low one in the "baby sitting" context due to different expertise in such contexts.

The presence of context in real life however is not per se a limiting factor, for instance if I claim that a friend of mine is a good surgeon, probably I leverage his opinion about an ophthalmologist I need to contact, i.e. expertise ranks in related or similar context can influence themselves someway. We also aim at addressing this issue, introducing the contexts *similarity* to allow related contexts to be exploited during ranks assessment, hence for instance I could trust an ophthalmologist if some good surgeons and an expert optician I know plus my family doctor all provide a good expertise rank about him.

In the rest of paper, section 2 introduces the formalization of our approach, while in section 3 we apply the proposed model to the expertise network built from Epinions.com data set, defining and exploiting contexts similarity and also showing results. Section 4 presents our conclusion and further works.

2 Expertise Evaluation Model

The expertise network can be modeled as $G(V,L,lab)$, i.e. a labeled multi-digraph where the set of nodes V represent users[1], L is a set of oriented arcs, i.e. an arc (v, w) means that v assigned w at least an expertise rank in a context, and the labeling function $lab : L \rightarrow \{(C \times [0,1])\}$ associates to each arc (v,w) the set of pairs $\{(c_i, r_i)\}$ being $r_i \in [0,1]$ the expertise rank v assigned to w within context $c_i \in C$ (C is the set of all contexts). Note that for a given arc, the rank is at most one for each context. In the following we indicate $l^{\overline{c}}_{v,w}$ as the rank r associated to the arc (v, w) within context $\overline{c}$, assuming that $l^{\overline{c}}_{v,w} = 0$ for any context $\overline{c}$ when arc (v, w) does not exist, and $X^{\overline{c}} = [l^{\overline{c}}_{v,w}]$ as the *weighted adjacency* matrix. A transition matrix $P^{\overline{c}}$ is then defined starting from $X^{\overline{c}}$, as detailed later. Each element $p^{\overline{c}}_{vw}$ of $P^{\overline{c}}$ represents a normalized expertise rank v assigned to w in context $\overline{c}$.

Indeed, to assign a *global* expertise context-specific rank to users in the network we can initially focus on two generic users v and w evaluating how v can assign rank to w in a given context $\overline{c}$. In the real world, if w is one of v's neighbours, the *direct* rank ($p^{\overline{c}}_{vw}$) can be evaluated from direct experience v has about w, otherwise v can ask to his neighbours whether they have an opinion about w within $\overline{c}$. In this case, if each of v's neighbours (denoted as j in the following) directly knows w he can provide the direct rank $p^{\overline{c}}_{jw}$, and it is reasonable that v weights this values with ranks he assigned to his neighbours within the same context, thus having $r^{\overline{c}}_{vw} = \sum_j p^{\overline{c}}_{vj} \cdot p^{\overline{c}}_{jw}$. This one-step neighbours ranking can be written into matrix form as $(\mathbf{r}^{\overline{c}}_v)_{(1)} = (P^{\overline{c}})^T \cdot \mathbf{p}^{\overline{c}}_v$, where $\mathbf{r}^{\overline{c}}_v$ and $\mathbf{p}^{\overline{c}}_v$ are the vectors built from $r^{\overline{c}}_{vw}$ and $p^{\overline{c}}_{vj}$ respectively. User v can

[1] In the following, we will use the terms *user* and *node* interchangeably.

further extends its requests to two-steps neighbours getting to $(\mathbf{r}_v^{\overline{c}})_{(2)} = ((P^{\overline{c}})^T)^2 \cdot \mathbf{r}_v^{\overline{c}}$ and, therefore, at step $(k+1)$ can be expressed by eq. (1). If $P^{\overline{c}}$ is irreducible and aperiodic, $\mathbf{r}_v^{\overline{c}}$ will converge to the same vector for every v, specifically to $P^{\overline{c}}$ eigenvector associated with the principal eigenvalue $\lambda_1 = 1$, leading to a *global* expertise rank for w in $\overline{c}$.

$$(\mathbf{r}_v^{\overline{c}})_{(k+1)} = ((X^{\overline{c}})^T)^{k+1} \cdot (\mathbf{r}_v^{\overline{c}}) \tag{1}$$

This approach is frequently adopted in the context of Web or P2P networks, e.g. [7], [9], [10]. These works also offer a probabilistic interpretation of their method derived from the random walker graph model [11], a widely accepted mathematical formalization of a trajectory. We interpret the random walk as follow: *if an agent is searching for an expert within a given context, he can move along the network choosing the further node with probability* $p_{vw}^{\overline{c}} \in P^{\overline{c}}$; crawling with this method until a stable state is achieved, the agent is more likely to be at an expert node than at an unqualified node. From the random walker point of view, the first principal eigenvector of $X^{\overline{c}}$ correspond to the standing probability distribution of the Markov chain defined by $X^{\overline{c}}$, and network nodes are its states; thus, we define the expertise vector as the stationary point of the transformation given in (1) with non-negative components.

We also want to study contexts influence, i.e. even if the walker is biased by the context $\overline{c}$, it can walk towards an user in a context *similar* to $\overline{c}$, thus we have to choose how the walker moves along the network. To this purpose, we introduce two different walking models both taking into account that an user that has a large number of incoming links (within context $\overline{c}$) is considered an *expert* in $\overline{c}$, hence a walker that moves along a path connecting experts should enforce this quality.

Definition 1 (strong biased model). Given a topic $\overline{c} \in C$, a walker standing in a node v at step k moves to one of his outgoing neighbours w at step $k+1$ *if and only if* $l_{vw}^{\overline{c}} > 0$ (i.e. one of v's neighbours is expert within $\overline{c}$).

According to this definition, we define the transition matrix $P^{\overline{c}}$ according to eq. (2), where $outdeg(v)^{\overline{c}}$ is the number of arcs for which $l_{vw}^{\overline{c}} > 0$ (hence $\sum_w p_{vw}$ is always 0 or 1).

$$p_{vw}^{\overline{c}} = \begin{cases} 1/outdeg(v)^{\overline{c}} & \text{if} \quad l_{vw}^{\overline{c}} > 0 \\ 0 & \text{otherwise} \end{cases} \tag{2}$$

Definition 2 (smooth biased model). Given a topic $\overline{c} \in C$, a walker standing in a node v at step k moves to one of his outgoing neighbours w at step $k+1$ according to a probability distribution depending on similarity (relatedness) between $\overline{c}$ and all couples (c_i, r_i) labelling (v,w).

In smooth biased approach $p_{vw}^{\overline{c}}$ is defined accordingly to relatedness between topics pairs labelling the link $(v,w) \in L$ and $\overline{c} \in C$ (eq. (3)), in particular given a topic $\overline{c}$, we define a relatedness function $d : V \times V \times C \rightarrow [0,1]$ (denoted as $d(v,w,c)|(v,w) \in L, c \in C$) and

$$p^{\overline{c}}_{vw} = \begin{cases} d(v,w,\overline{c})/\sum_j d(v,j,\overline{c}) & \text{if} \quad l^{\overline{c}}_{vw} > 0 \\ 0 & \text{otherwise} \end{cases} \tag{3}$$

Note that both definitions may lead to a transition matrix $P^{\overline{c}}$ where all elements of some rows and/or some columns are 0 (therefore $P^{\overline{c}}$ is not irreducible), and sometimes the associated graph might be disconnected.

To find stationary vector the transition matrix is required to be strongly connected and aperiodic: the first condition implies that exists a directed path from each node to any other, whereas the second implies that for any users v and w, there are paths from v to w of any length except for a finite set of lengths. Both strong and smooth biased models do not work with dangling user and disconnected graphs; dangling users are those with no outgoing link that can be present in any real network. Moreover, in the strong biased case, users that have no outgoing links labelled by topic $\overline{c}$ also became dangling.

Several solutions for dangling users have been proposed [9, 12], we choose that a walker in a sink moves to any user according to a given probability distribution. We then define a new transition matrix $(P^{\overline{c}})'$ as $(P^{\overline{c}})' = P^{\overline{c}} + \delta \cdot \alpha^T$, where $\alpha = (1/n, \ldots, 1/n)$ and $\delta = [\delta_i]$ where $\delta_i = 1$ if i is a dangling user and 0 otherwise; this guarantee that $\sum_w p^{\overline{c}}_{vw} = 1, \forall v, w \in V$. The same trick is used to avoid users without ingoing links (that violates the aperiodic property), so achieving the following formula:

$$(P^{\overline{c}})'' = q \cdot (P^{\overline{c}})' + (1-q) \cdot A, \quad \text{where} \quad A = (1, \ldots, 1) \cdot \alpha^T, \quad q \in [0,1] \tag{4}$$

Thus from a non dangling user a walker follows one of the local outgoing links with probability q and jumps to some $w \in V$ with probability $(1-q)$; a common value for q is 0.05 [12].

3 Epinions.com: A Case Study

Epinions (http://www.epinions.com) is a recommendation system that *"helps people make informed buying decisions"* [13]. This goal is achieved through unbiased advice, personalized recommendations, and comparative shopping. Epinions allows to registered *users* to rate *products* writing a *review* in order to provide visitors with *opinions*. Registered users could also rate the reviews, thus providing a sort of expertise rank about other users. We use Epinions dataset to validate our approach because it is a *large* and *real* dataset and although it is mainly a recommendation network, the reviews voting actually implements an author's reviews reputation mechanism based on products categories (i.e. authors are assigned an expertise rank within contexts).

We however still need to investigate over the raw dataset the evaluation of (1) expertise and (2) contexts similarity (*semantic distance*). To test our approach over a significant network we need to define how nodes with reviews can be assigned an expertise rank: we consider an user w writing a review on a product (belonging to a given category, e.g. *electronics*), and another user v that can provide a rank to w's review, considering it *useful* or not; w can provide several reviews on products belonging to different categories, and v can rate all of them. Based on such information, we then build the arc (v,w) and label it with a set of pairs $\{(c_i, r_i)\}$, where we associate each context to exactly one products category, and the expertise rank with the rate v provided about w's review for the product belonging to that category; note that in the case w reviewed more products belonging to the same category, we evaluate the normalized average rate provided by v over all these products, so that r_i is within the $[0,1]$ range. Of course, we discard all users that did not provide any review.

Another issue is to define a metric to evaluate context similarity, (e.g. *TVs* category is intuitively much more related to *electronics* than *wellness and beauty*); this is needed by the random walker to exploit different yet related contexts. This semantic distance is a function we name $similarity(c_h, c_k) \in [0,1]$ where 0 means no similarity and 1 means that contexts c_h and c_k are identical terms. Measuring the semantic distance between terms (contexts) has been extensively considered in literature (Vector Space Model [14], Resnik [15], Lesk similarity [16]); since Epinions provides a hierarchical arrangement of contexts, e.g. *electronics* includes sub-contexts as *cameras & accessories, Home audio*, we can exploit this to simplify the semantic distance evaluation. We then adopt the following metric:

Definition 3. Find the concept c_3 which generalizes c_1 and c_2 with type T_3 such that T_3 is the most specific type which subsumes T_1 and T_2; the semantic distance between c_1 and c_2 is the sum of the distances from c_1 to c_3 and c_2 to c_3.

This metric is described in [17, 18] and it satisfies reflexivity, symmetry and triangle inequality properties. Moreover topics types are always the same, therefore our metric can be stated as the "sum of the distance between two concepts and first common ancestor" (along the hierarchical classification provided by Epinions). Finally, we normalize the semantic distance between contexts in order to have values into the $[0,1]$ range as shown in eq. (5), where *max_distance* is the length of the longest path between contexts.

$$similarity(c_i, c_j) = \frac{min(d(c_i, c_k) + d(c_j, c_k))}{max\_distance} \quad \forall c_k \prec c_i, c_j \tag{5}$$

Therefore, the similarity function defined in eq. (5) is used in smooth biased approach to calculate the relatedness between users (see eq. (6)).

$$d(v, w, \overline{c}) = \begin{cases} 1 & \text{if } l^{\overline{c}}_{vw} > 0 \\ \Sigma_j similarity(\overline{c}, c_j) * r_j \backslash \Sigma_j r_j & \text{otherwise} \end{cases} \tag{6}$$

Table 1 shows the characteristics of the dataset extracted from Epinions website we used in our experiments.

Table 1 Characteristics of the dataset extracted from Epinions website

Dataset extracted from www.epinions.com	
# nodes	37 321
# sink (out-degree = 0)	1 538
# source (in-degree = 0)	0
# link	460 504
average in-degree	12.13
average out-degree	8.81
average topics per node	17.71

3.1 Results

The expertise network built from Epinions dataset is used to validate the proposed expertise rank assessment model, in particular we evaluate the stationary point of transformation of the transition matrix in eq. (4) using $P^{\overline{c}}$ as defined for strong and smooth biased models, comparing them with an *unbiased* case (i.e. context independent) defined as follows:

$$p_{vw} = \begin{cases} 1/outdeg(v) & if\ outdeg(v) > 0 \\ 0 & otherwise \end{cases} \tag{7}$$

In the unbiased case (eq. (7)), the transition probability from a node v to a node w is independent from the context $\overline{c}$ hence the steady state probability vector depends only on the structure of the expertise network, i.e. the more links a node has, the more often it will be visited by a random walker. This also means that using the unbiased random walker model an user that has a low number of links will receive a low expertise rank value, even if he is the only one labelled as expert on a given topic $\overline{c}$.

In real life people expertise is always assigned within a given context $\overline{c}$ and our idea is to capture this behaviour using a random walker biased by context, as explained in the previous sections. In order to validate the strong and smooth biased random walker models presented in section 2, we will show that the probability of reaching nodes with expertise identical or similar to the target $\overline{c}$ grows with respect to the unbiased case.

In the following we report the results of a set of experiments performed using the network we extracted from Epinions. For each experiment we set a specific topic $\overline{c}$ and we evaluate the expertise vector for the unbiased random walker and for both the strong and the smooth biased random walker models. Therefore, for each topic c_i we sum all the expertise ranks (or steady state probability) of those users labelled with c_i obtaining the so-called *cumulative expertise* of topic c_i. It corresponds to the steady state probability that a random walker visits a node belonging to the topic c_i.

For the sake of simplicity, in the following all the Epinions' topics are indicated by a number instead of their names.

Figure 1 shows the comparison of percentage increment of biased models cumulative expertise with respect to unbiased (eq. (7)) considering a common and a rare topic with respect to the unbiased case. In particular, we focused on topic #30 which is very common (i.e. 14995 links associated to it over 460504 total, 14995 source nodes over 35783 and 5134 targets over 26000), and on topic #536 which is quite rare (5 links, 5 sources and 1 target, respectively). Results highlight that cumulative expertise on $\overline{c}$ always grows with respect to the unbiased case. Let us note that when expertise is biased by the common topic #30, the cumulative expertise related to some other topics (namely #322 and #426) also increase, whereas when the rare topic #536 is used only nodes labelled with such a topic are affected by biasing. The fact that biasing on topic #30 also affects the cumulative expertise of other topics is mainly due to the structure of Epinions network, indeed being topic #30 very common means that a large amount of nodes are somehow expert in that topic. Some of these nodes are also expert in topics #322 and #426 and a certain number of them have a high network degree, so there is a high probability that they are involved in most of paths followed by the random walker, hence the side effect of cumulative expertise increasing for topics #322 and #426 occurs.

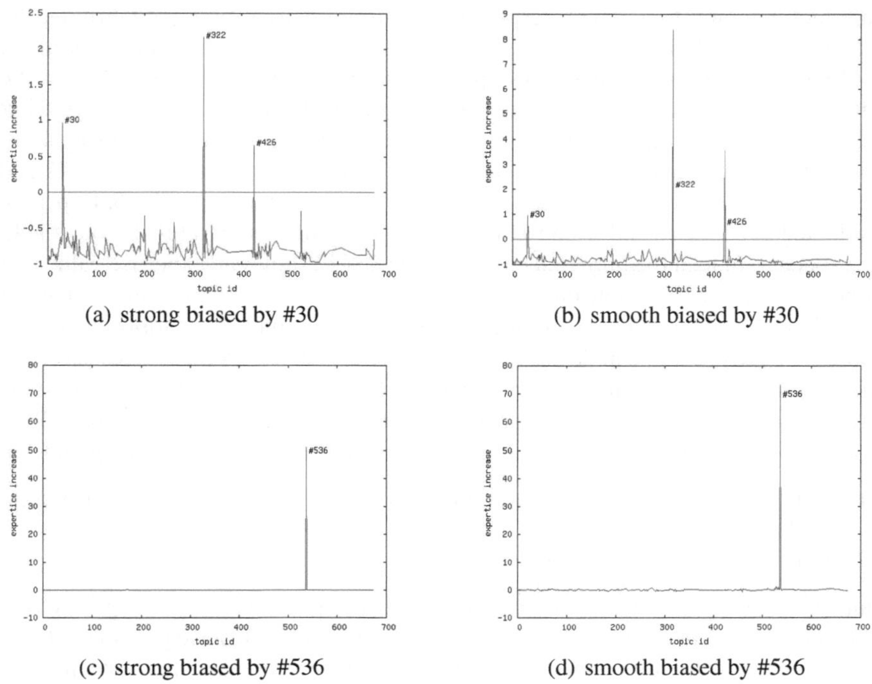

(a) strong biased by #30

(b) smooth biased by #30

(c) strong biased by #536

(d) smooth biased by #536

Fig. 1 Cumulative Expertise

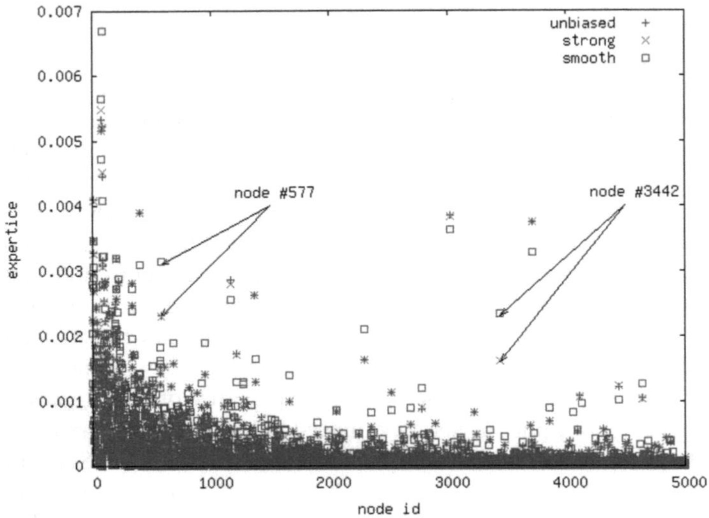

Fig. 2 User's expertise assessment

Also note that expertise in smooth biased model increases much more for both rare and common topics with respect to the strong biased model, confirming the advantage in exploiting similarity between topics during expertise rank assessment.

Another experiment focuses on Epinions' users, showing their expertise in the rare topic #536, where just one target node *w* is considered expert by just five other nodes. Figure 2 highlights each user's expertise on topic #536, evaluated using unbiased, strong and smooth biased models. In particular, we focus on users #3442 and #577, where the former is the only user labelled as expert in topic #536. Expertise evaluated in unbiased and strong-biased case slightly differs for all nodes but #3442 as expected. Indeed the unbiased case for node #3442 shows an expertise value that is nearly zero due to the low number of links such a node has. This confirms that our biased models are able to capture the expertise of a user on a given topic even if the topic is rare and also the node has few links with respect the average nodes degree in the network. The diagram also shows that user #577's expertise for unbiased case is the same as strong biased case since it has no in-links labelled with the topic #536.

The comparison of the smooth biased case with others is more interesting, indeed:

1. node #3442's expertise increases much more than the corresponding strong biased model
2. node #577's expertise increases also!

Item 1 is the expected behavior and confirms our hypothesis that the expertise of a node depends on the opinions of his/her neighbours. Item 2 instead puts in evidence the influence of highly connected nodes on the expertise evaluation. Specifically, node #577 is much more connected than the average node's connectivity having an out-degree of 1333 (versus a network average of 8.81), and in-degree of 241 (versus a network average of 12.13). This means that a random walker tends to visit such a node more frequently than the other nodes of the network, since it is included in many paths. In conclusion, the increasing of expertise not only trivially depends on the expertise in a given topic, but it is also affected by the structure of the network, i.e. the presence of hubs that can be somehow considered expert for (almost) any topic.

4 Conclusion

This paper introduces *expertise* as a property of the node of a social network and test the definition on a dataset extracted from *Epinions.com*. Expertise is defined using a biased random walker model and its corresponding probabilistic interpretation and has been applied to a dataset extracted from Epinions website, where a mechanism of expertise evaluation based on products review has been introduced, together with a *relatedness* function used to exploit contexts similarity for expertise rank assessment. Results confirmed that the expertise can be considered a network property that depend on network structure and direct (local) users experience. Further works are: to investigate different similarity function (e.g. when more, different ontologies are present), to integrate into a unique framework expertise, reputation and trust in order to provide effective and efficient search algorithms for distributed applications.

References

1. Katz, E., Lazarsfeld, P.: Personal Influence: The Part Played by People in the Flow of Mass Communications. Transaction Publishers (October 2005)
2. Brown, J.J., Reingen, P.H.: Social ties and word-of-mouth referral behavior. The Journal of Consumer Research 14(3), 350–362 (1987)
3. Zhang, J., Tang, J., Li, J.: Expert Finding in a Social Network. In: Kotagiri, R., Radha Krishna, P., Mohania, M., Nantajeewarawat, E. (eds.) DASFAA 2007. LNCS, vol. 4443, pp. 1066–1069. Springer, Heidelberg (2007)
4. Fang, H., Zhai, C.: Probabilistic models for expert finding. In: Amati, G., Carpineto, C., Romano, G. (eds.) ECiR 2007. LNCS, vol. 4425, pp. 418–430. Springer, Heidelberg (2008)
5. Grandison, T., Sloman, M.: A survey of trust in internet application (2000)
6. John, A., Seligmann, D.: Collaborative tagging and expertise in the enterprise. In: Proc. WWW 2006 (2006)
7. Sepandar, D., Kamvar, M.T., Schlosser, H.G.M.: The eigentrust algorithm for reputation management in P2P networks. In: Proceedings of the Twelfth International World Wide Web Conference (2003)
8. Artz, D., Gil, Y.: A survey of trust in computer science and the semantic web. Web Semantics: Science, Services and Agents on the World Wide Web 5(2), 58–71 (2007)
9. Page, L., Brin, S., Motwani, R., Winograd, T.: The pagerank citation ranking: Bringing order to the web. Technical report, Stanford InfoLab (1999)

10. Kleinberg, J.M.: Authoritative sources in a hyperlinked environment. J. ACM 46(5), 604–632 (1999)
11. Motwani, R., Raghavan, P.: Randomized algorithms. Cambridge University Press, New York (1995)
12. Berkhin, P.: A survey on pagerank computing. Internet Mathematics 2, 73–120 (2005)
13. Shopping.com Network: Epinions.com ©(1999-2000)
14. Salton, G., Wong, A., Yang, C.S.: A vector space model for automatic indexing. Commun. ACM 18(11), 613–620 (1975)
15. Resnik, P.: Using information content to evaluate semantic similarity in a taxonomy. In: IJCAI, pp. 448–453 (1995)
16. Banerjee, S., Pedersen, T.: An adapted lesk algorithm for word sense disambiguation using wordnet. Computational Linguistics and Intelligent Text Processing, 117–171 (2002)
17. Garner, B., Lukose, D., Tsui, E.: Parsing natural language through pattern correlation and modification. In: Proc. of the 7th International Workshop on Expert Systems & Their Applications, Avignon, France, pp. 1285–1299 (1987)
18. Foo, N., Garner, B.J., Rao, A., Tsui, E.: Semantic distance in conceptual graphs. Ellis Horwood, Upper Saddle River (1992)

Web Services for Search Integration

Mirel Cosulschi, Oana Nicolae, Mihai Gabroveanu, and Nicolae Constantinescu

Abstract. This paper is closely related to the problem of finding elements having certain positions resulted after merging between two or more result sets. The problem is encountered frequently nowadays, a typical situation being a business distributed environment i.e. SOA approach, where as a direct consequence of a Web search, each participating entity will provide a result set. The focus of paper is to describe a web service for optimally returning the *k*-th element obtained from joining web search services results into a composite result [3]. We introduce an optimized algorithm for solving the problem, based on a divide-and-conquer technique and we discuss, by means of examples, its performances as related to other approaches.

Keywords: Web services, Search services, integration algorithm, business approach, SOA/SOC paradigms, BPMN, graphical notation.

1 Introduction and Motivation

Web service technologies play a crucial role in the Web area being a result of the standardization efforts (OASIS[1], W3C[2], WS-I[3]) which is built upon more than a decade of work and experiences. This evolutionary process has had over time

Mirel Cosulschi · Mihai Gabroveanu · Nicolae Constantinescu
Department of Computer Science, Faculty of Mathematics and Computer Science,
University of Craiova, Romania
e-mail: {mirelc,mihaiug,nikyc}@central.ucv.ro

Oana Nicolae
Department of Internet Technology, Institute of Informatics,
Brandenburg Technical University of Cottbus, Germany
e-mail: nicolae@tu-cottbus.de

[1] OASIS - http://www.oasis-open.org/
[2] W3C - http://www.w3.org/
[3] WS-I - http://www.ws-i.org/

G.A. Papadopoulos and C. Badica (Eds.): Intelligent Distributed Computing III, SCI 237, pp. 63–73.
springerlink.com
© Springer-Verlag Berlin Heidelberg 2009

different, emerging aims i.e. obtaining a seamless and automated infrastructure for managing the complexity of distributed information systems, solving the interoperability problem, overcoming the issues from Enterprise Application Integration and workflow by means of service composition, making available/retrieving information via Web services.

The problem of finding information on the Web has become tedious due to the diversity and abundance of information. In order to find and use accurate results a user has to know what kind of search engine type is best suited for his current need (e.g. a large-scale search engine or a domain-specific search engine) and where it is located.

Moreover, the need of answering complex queries ranging over multiple domains, revealed the weakness of actual Search services. Either general-purposes or domain-specific, search engines fail when it comes to provide answers for cross-domain queries. The solution is to give clients the possibility of invoking multiple Web services, in a transparent manner, so that one query result will constitute the input for querying another search engine.

Trends in Search services technology challenge the assumption that research still has to be made i.e. customizing or improving algorithms in the area of retrieving Web information. We propose an enhanced algorithm for retrieving the k-th merged element from two result sequences, using a *divide-et-impera* technique. Another goal is to introduce and to consider our work also in the context of Service Oriented Architecture (SOA) and Service Oriented Computing (SOC) paradigms.

The rest of this paper is organized as follows: Section 2 provides a review on previous research works. Section 3 proposes a SOA-based showcase whose purpose is to encapsulate the `Select` algorithm inside the business logic of a business process. Section 4 discusses the terminology and defines the main concepts implied in the topic area of Search services. Section 5 describes the `Select` algorithm we propose and the last section discusses conclusions and future works.

2 Previous Work

In [16] the authors introduce a *Web Service Management System (WSMS)* as a general-purpose system that enables clients to query multiple web services simultaneously in a transparent and integrated manner. The architecture of WSMS has its roots in the earlier mediators from data integration systems [4], [7], [12].

In NGS [2] the authors developed an overall framework for multi-domain queries on the Web. The problems tackled here range from expressing multi-domain queries with an abstract formalism, separating the processing of "search" services within the model, till emphasizing their differences from "exact" web services.

Another approach, based on a Web 2.0 concept - "mash up", allowing to compose queries over distributed data sources, can be seen in Yahoo Pipes[4] and IBM Damia[5].

[4] http://pipes.yahoo.com/pipes/

[5] http://services.alphaworks.ibm.com/damia/

The problem of finding the kth smallest element among a set of n values is a classic problem, also known as the *selection* problem, which has been studied in a distributed ([11], [13]) and a non-distributed configuration ([1], [15]).

3 Web Search Services

From the business perspective, the literature provides many insights regarding the main language specifications used on the business market that claim to enable:

- Web services related methodologies that involve the control of the business process workflow and later its enactment i.e. *orchestration* technique - BPEL[6].
- Web services composition techniques that should provide long-lived, peer-to-peer conversations between business partners i.e. *choreography* technique - WS-CDL[7].

Choreography is a collaborative effort focused on exchange of messages. All participants in choreography need to be aware of the business process, operations to execute, messages to exchange, and the timing of message exchanges. Therefore, in the context of Search services, solving the problem of composing queries ranging over multiple domains resumes at invoking the appropriate *Search services* and furthermore, at managing the search results by the internal workflow of service orchestrator, in order to obtain the aimed results set.

In our case, BPEL language describes the complete, internal workflow of the business process and also expresses the conditional behavior i.e. a Search service invocation (i.e. query) depends on the value of a previous invocation. Also, the BPEL language provides constructs that can specify which Web services must be executed in a sequence and which can be executed in parallel, similar to the presence and absence of precedence constraints in the Web service joining problem. Remaining in the SOC context, another Web trend, that actually tends to generalize, is the movement towards the syntagma of providing the *software as a service* e.g. enabling the execution of the `Select` Algorithm proposed in this paper, as part of the business process workflow. More concrete, the `Select` algorithm is incorporated as a BPEL activity that will receive its input as two ordered sets of elements and will return as output the required element whose index we also specify as an input.

3.1 Quering Multi-domains Search Service

To exemplify our approach, we use the main graphical language on the business market for representing business processes i.e. Business Process Modelling Notation - BPMN [9].

[6] BPEL - http://www.oasis-open.org/committees/wsbpel/

[7] WS-CDL - http://www.w3.org/TR/ws-cdl-10/

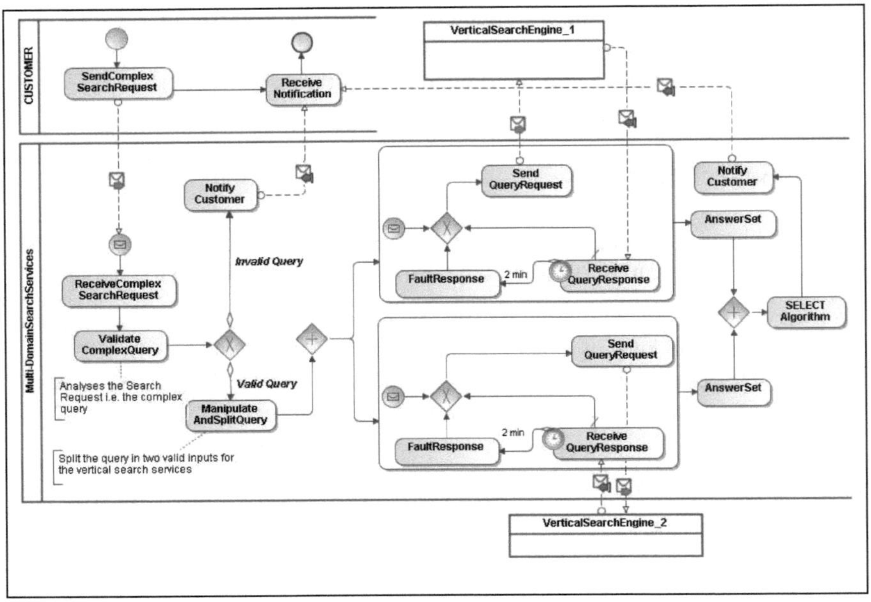

Fig. 1 Querying Search Services and applying SELECT algorithm.

Further on, we briefly describe a scenario that will enable clients to query the business process which, at its turn will perform appropriate queries on a set of Search services e.g. we focus on the simplest case: two distinct Search services. We describe this scenario using the top-level BPMN graphical construct, Business Process Diagram (BPD).

For the sake of simplicity, we omit to specify in the Figure 1 some of the model elements that would be needed to model the complete functionality of the described process (i.e. selecting the appropriate Search services, dealing with eventual, unexpected faults).

Our showcase includes a Customer role, which starts the process by requesting a complex query from the Multi-DomainSearchService. In Figure 1, the Multi-DomainSearchService receives the request from the Customer (can be human or automatic request - another Search service) and further communicates the query to appropriate Search services. The complex query is first *validated*, meaning that the `Validate` activity is performed, i.e. analyses if the complex query can be further split in two appropriate, valid queries for the Search services. This operation is performed at the middleware level and also involves some necessary query manipulation, such that they represent valid inputs for the vertical Search services. Each query will be then sent as a flow message on appropriate Search service. The process ends after the successful accomplishment of the service task that employs the `Select` algorithm with an appropriate reply task that will send to the customer the expected answer. The interactions suppose synchronous messages, meaning the requester blocks until the response's arrival. This situation implies some disadvantages, as the calling thread must wait some time to complete and waiting is not a good practise when invoking a Search service.

Following SOA-approach, the tasks are orchestrated in form of a BPEL process and will be further executed by a business process execution engine which will invoke the appropriate WSDL-based Search services.

4 Problem Statement

Through *selectivity* of a web service we understand the average number of tuples produced as output corresponding to the input tuple(s) [16]. A group of web services has *independent selectivities* if the selectivity of a web service does not depend on the previous invoked web services. A web service is called *selective* if its associated selectivity is less than or equal to 1 and it is called *proliferative* when its selectivity is greater than 1. If the independence condition is not preserved, then the web service has *correlated selectivities*. A *search* service returns a set of tuples organized in a ranked list. The order of elements is determined by a relevance measure.

Based on the modality of the results grouping, web services can be classified as *chunked* or *bulk*: a *chunked* web service returns the output results organized in chunks or pages, while a *bulk* web service will return all results at once.

Search services allow chunking of their results. A chunked web service has the advantage of a constant cost for randomly accessing selected element: the cost of bringing to the client the result chunk where the desired element resides. Accessing the needed information, e.g. the element situated on the k-th position, from a bulk web service incurrs waiting for arrivals of the first k elements. If the k-th element is the last one in ranking and the number of elements from output is huge, then this strategy is unfeasible: either the hardware capabilities from the client do not allow the storage of entire amount of fetched data or the time for transporting over the network all the result set is too costly. Due to these limitations, all proliferative search services have implemented at least the chunked operation.

We must consider that making a web service call will involve some amount of overhead including the network transmission costs or parsing the SOAP headers. Due to this, transportation of each tuple from the whole result set is out of question. Regarding the size of input data, a web service can receive and process an input set of tuples one-by-one or can receive it in batches while the processing can be done independently. Some web services allow to set the size of chunked data returned in terms of number of tuples or packet size. The client of those web services must adapt the size of theirs input data according to the different parameters values: network traffic, the overall loading of the web service, the overall loading of the client etc.

Having two or more proliferative search web services (i.e. *Google SOAP Search API*[8] or *Yahoo! Search BOSS*[9]), lets suppose that one wants to obtain the k-th element resulted after merging between the corresponding two or more result sets. The problem can be encountered frequently nowadays, a typical situation being a

[8] http://code.google.com/apis/soapsearch/
[9] http://developer.yahoo.com/search/boss/

distributed environment, where as a direct consequence of a search, each participating entity will provide a result set. We have investigated and proposed an optimized algorithm for the case of two proliferative web services in order to find the k-th merged element from the two result sets, based on a divide-and-conquer technique.

A prerequisite condition for applying the algorithm is the compatibility of the ranking functions used by the two separated search services. Following the process of collecting the search results from the local search engines, the metasearch engine must combine those results into a single ranking: this issue is known as *the collection fusion problem* and is thoroughly treated in the papers [8], [10], [6].

5 The Select Approach

The algorithm presented tackles a special case of the distributed k-selection, e.g. the selection among two sites when we deal with a small graph composed of two connected nodes. The algorithm is similar to *Halving* protocol [14].

5.1 The Description of the Proposed Algorithm

The procedure **BSearch**$(M, left, right, x)$ performs a binary search in the ordered sequence M for the value x, between the lower and upper bounding indexes, $left$ and $right$, returning the position where the element x should be inserted into M, in case of failure.

The procedure **Select** (algorithm 1) computes the k-th element of the result obtained from merging the sequence A having as the lower and upper limit, $left_a$ and $right_a$ respectively, with the sequence B having $left_b$ and $right_b$ as bounding limits.

The general outline of the algorithm is the following: it is taken one of the two sequences, then it is divided in two halves, and the element from the middle is placed/inserted in the other sequence. Depending of the position (index) obtained, it is selected which one of the two sub-sequences will be explored further. The computational complexity of this part is logarithmic ($O(\log(n))$) as the binary search algorithm has the main role and its complexity overcomes the other operations complexity. First, if one of the two sequences is empty (lines 4 and 7) then the procedure returns the element located at index k from the other sequence (lines 5 and 8).

If the first sequence, A, has the size equal with 1 (line 11) then it will be searched the position of the element $A[left_a]$ inside the sequence B as if this element should be inserted in B. If the position where the element $A[left_a]$ should be inserted is greater than k (line 13) the procedure will return the element from the second sequence B, located at the position $left_b + k - 1$ (line 14).

In case the sum of the two sequences' sizes, A and B, is equal with k (line 26), then the k-th element of the merged result will be the maximal element between the last elements of each of the two arrays: $max(A[right_a], B[right_b])$.

In case the sequence A has more elements than the sequence B (line 29), it will be chosen an element of the sequence A having a pivot role: $pivot \leftarrow (left_a + right_a)/2$. The position of this *pivot* element will be searched inside the sequence B (line 31). *length* is the size of the array obtained after merging the subsequence $left_a..pivot$ of A with the subsequence $left_b...poz$ of B.

The situation when *length* is less than k will draw to the conclusion that the k-th element is located in the other part of the sequence that could be obtained after a merge of the subsequence $pivot+1...right_a$ of A with the subsequence $poz+1...right_b$ of B (line 33): **return** $Select(A, pivot+1, right_a, B, poz, right_b, k-length)$.

5.2 Numerical Experiments

For better understanding of the proposed algorithm, we choose a simple numerical example, where details concerning web services were stripped off.

Example 1. Lets suppose that we have the following values for the input sequences:

$$A = \begin{bmatrix} 2\ 6\ 12\ 15\ 17\ 20\ 28\ 33\ 37\ 46 \end{bmatrix}\ B = \begin{bmatrix} 1\ 5\ 6\ 8\ 10\ 20\ 21\ 24\ 27\ 30 \end{bmatrix}$$

and we want to find out the element located at index 14 from the final result.

The lengths of the two arrays are equals (10), so the algorithm execution will reach the command **else** from the line 39: $pivot \leftarrow (left_b + right_b)/2$. The computed value of the *pivot* variable is 5. It is determined then the position at which the value of $b[pivot] = 10$ should be inserted inside the array A:

$$A = \begin{bmatrix} 2\ 6\ 12\ 15\ 17\ 20\ 28\ 33\ 37\ 46 \end{bmatrix}$$

The response obtained as a result of the subroutine call $poz \leftarrow BSearch(A, left_a, right_a, B[pivot])$ is 3. At this step, the two initial sequences are virtually separated, each one in two distinct partitions:

$$\begin{array}{llll} A_{left} = \begin{bmatrix} 2\ 6 \end{bmatrix} & A_{right} = \begin{bmatrix} 12\ 15\ 17\ 20\ 28\ 33\ 37\ 46 \end{bmatrix} \\ B_{left} = \begin{bmatrix} 1\ 5\ 6\ 8\ 10 \end{bmatrix} & B_{right} = \begin{bmatrix} 20\ 21\ 24\ 27\ 30 \end{bmatrix} \end{array}$$

After that, it is computed the number of elements located to the left of the position *poz* (representing the number of elements lesser than $b[pivot] = 10$) in the resulting array if the merge has been performed: $length = poz - left_a + pivot - left_b + 1 = 7$.

How $7 < 14$, the algorithm will continue searching the element from the position $k - length = 7$ resulted after merging the following input sequences:

$$A_{right} = \begin{bmatrix} 12\ 15\ 17\ 20\ 28\ 33\ 37\ 46 \end{bmatrix}\ B_{right} = \begin{bmatrix} 20\ 21\ 24\ 27\ 30 \end{bmatrix}$$

In this moment, the length of the sequence A is greater than the length of the sequence B: $pivot \leftarrow (left_a + right_a)/2$. *pivot* receives the value 6. The algorithm will search for the position where the element $a[pivot] = 20$ could be inserted in the sequence B_{right}:

$$B_{right} = [20\ 21\ 24\ 27\ 30]$$

The sequences A_{right} and B_{right} will be split again, each one, in two distinct partitions:

$$A_{right_{left}} = [12\ 15\ 17\ 20]\quad A_{right_{right}} = [28\ 33\ 37\ 46]$$
$$B_{right_{left}} = []\quad B_{right_{right}} = [20\ 21\ 24\ 27\ 30]$$

(with $B_{right_{left}}$ we denote the left part of B_{right} while with $B_{right_{right}}$ we denote the right part of B_{right} sequence). We shall determine now the element having the index $k - length = 3$ after the merging of the sequences

$$A_{right_{right}} = [28\ 33\ 37\ 46]\quad B_{right_{right}} = [20\ 21\ 24\ 27\ 30]$$

pivot is assigned with the value $(left_b + right_b)/2 = 8$, and the algorithm searches for the location where the element with value 24 should be inserted inside the sequence $A_{right_{right}} = [28\ 33\ 37\ 46]$. The sequences $[28\ 33\ 37\ 46]$ and $[20\ 21\ 24\ 27\ 30]$ are separated for the last time into:

$$A_{left} = []\quad A_{right} = [28\ 33\ 37\ 46]$$
$$B_{left} = [20\ 21\ 24]\quad B_{right} = [27\ 30].$$

The algorithm will reach the instruction 5: **return** $b[left_b + k - 1]$. In conclusion, after 4 calls of the function *Select* the final result (24) is obtained.

In table 1 and table 2 is presented a comparison of the running times obtained as a result of the tests performed with a classic algorithm [5] and our Select Algorithm implemented as Java applications. The computer used for testing was a Pentium IV Celeron at 1,7 Ghz, having 1 GB RAM, and a SATA WD HDD with 120 GB, 5400 rpm.

Table 1 75000 elements

K value	The classic algorithm(ms)	Algorithm Select(ms)
1000	0.010	0.004
5000	0.040	0.004
10000	0.080	0.004
20000	0.160	0.005
40000	0.330	0.005
50000	0.400	0.004
60000	0.480	0.005
75000	0.610	0.004

Table 2 200000 elements

K value	The classic algorithm(ms)	Algorithm Select(ms)
1000	0.010	0.004
5000	0.040	0.004
10000	0.090	0.004
20000	0.160	0.004
40000	0.330	0.006
50000	0.420	0.005
60000	0.490	0.005
75000	0.640	0.006
100000	0.840	0.004
120000	1.000	0.006
150000	1.270	0.004
180000	1.510	0.004
199000	1.660	0.004

Algorithm 1. Algorithm Select

INPUT:

A - the first ordered sequence.
$left_a$ - the index of the first element from the sequence A.
$right_a$ - the index of the last element from the sequence A.
B - the second ordered sequence.
$left_b$ - the index of the first element from the sequence B.
$right_b$ - the index of the last element from the sequence B.
k - the index of the element needed.

```
procedure SELECT(A, left_a, right_a, B, left_b, right_b, k)
    length_a ← right_a − left_a + 1                    ▷ the length of the first sequence
    length_b ← right_b − left_b + 1                    ▷ the length of the second sequence
    if (length_a < 1) then
        return b[left_b + k − 1]
    else
        if (length_b < 1) then
            return a[left_a + k − 1]
        end if
    end if
    if (length_a = 1) then
        poz ← BSearch(b, left_b, right_b, A[left_a])
        if (poz − left_b ≥ k) then
            return B[left_b + k − 1]
        else
            if (poz − left_b + 1 < k) then
                return B[left_b + k − 2]
            else
                return A[left_a]
            end if
        end if
    else
        if (length_b = 1) then
            ...                                        ▷ simmetric with the case of length_a = 1
        else
            if (length_a + length_b = k) then
                return max(A[right_a], B[right_b])
            else
                if (length_a > length_b) then
                    pivot ← (left_a + right_a)/2
                    poz ← BSearch(B, left_b, right_b, A[pivot])
                    length ← pivot − left_a + 1 + poz − left_b
                    if (length < k) then
                        return Select(A, pivot + 1, right_a, B, poz, right_b, k − length)
                    else
                        return Select(A, left_a, pivot, B, left_b, poz − 1, k)
                    end if
                else
                    pivot ← (left_b + right_b)/2
                    poz ← BSearch(A, left_a, right_a, B[pivot])
                    length = poz − left_a + pivot − left_b + 1
                    if (length < k) then
                        return Select(A, poz, right_a, B, pivot + 1, right_b, k − length)
                    else
                        return Select(A, left_a, poz − 1, B, left_b, pivot, k)
                    end if
                end if
            end if
        end if
    end if
end procedure
```

6 Conclusions

In this paper we have described a Web service for optimally returning the element located on k-th position after joining two web search services results into a composite one. The strength of the presented method relies on the fact that it is not necessary to bring the entire results sequences, thus reducing computational time and network load. Usage of a divide-and-conquer technique and experiments performed leads us to assume a logarithmic costs for complexity of the computations. The first experiments showed us that the approach is feasible and we will focus on the integration of this service in a framework that we are developing. There are major points that are not addressed i.e. how the appropriate domain web search services are selected, and these will constitute the subject of our future work.

Acknowledgements. The work reported was partly funded by the Romanian National Council of Academic Research (CNCSIS) through the grants CNCSIS 55/2008 and CNCSIS 79/2007.

References

1. Blum, M., Floyd, R.W., Pratt, V.R., Rivest, R.L., Tarjan, R.E.: Time Bounds for Selection. Journal of Computer and System Sciences 7(4), 448–461 (1973)
2. Braga, D., Calvanese, D., Campi, A., Ceri, S., Daniel, F., Martinenghi, D., Merialdo, P., Torlone, R.: R. Ngs: a framework for multi-domain query answering. In: Proc. of the Workshop on Information Integration Methods, Architectures, and Systems, IIMAS (2008)
3. Braga, D., Campi, A., Ceri, S., Raffio, A.: Joining the results of heterogeneous search engines. Information Systems 33(7-8), 658–680 (2008)
4. Casati, F., Dayal, U.: Special issue on web services. IEEE Data Eng. Bull. 25(4) (2002)
5. Cormen, T.H., Leiserson, C.E., Rivest, R.L., Stein, C.: Introduction to Algorithms, 2nd edn. MIT Press and McGraw-Hill (2001)
6. Fagin, R., Kumar, R., Mahdian, M., Sivakumar, D., Vee, E.: Comparing and aggregating rankings with ties. In: PODS, pp. 47–58 (2004)
7. Garcia-Molina, H., Papakonstantinou, Y., Quass, D., Sagiv, Y., Ullman, J., Vassalos, V., Widom, J.: The tsimmis approach to mediation: data models and languages. Journal of Intelligent Information Systems, 117–132 (1997)
8. Gravano, L., Garcia-Molina, H.: Merging ranks from heterogeneous internet sources. In: Twenty-Third International Conference on Very Large Databases, pp. 196–205 (1997)
9. OMG (BPMN), Business Process Modeling Notation version 1.2, Technical report (2009)
10. Renda, E.M., Straccia, U.: Web metasearch: rank vs. score based rank aggregation methods. In: Proc. of the ACM symposium on Applied computing, pp. 841–846 (2003)
11. Rodeh, M.: Finding the Median Distributively. J. Computer and System Science 24(2), 162–166 (1982)
12. Roth, M., Schwarz, P.: Don't scrap it, wrap it! a wrapper architecture for legacy data sources. In: Proc. of the International Conference on Very Large Data Bases, pp. 266–275 (1997)

13. Santoro, N., Sidney, J.B.: Order Statistics on Distributed Sets. In: Proc. 20th Allerton Conf. Comm., Control, and Computing, pp. 251–256 (1982)
14. Santoro, N.: Design and Analysis of Distributed Algorithms. Wiley Interscience, Hoboken (2006)
15. Schonhage, A., Paterson, M., Pippenger, N.: Finding the median. Journal of Computer and System Sciences 13, 184–199 (1976)
16. Srivastava, U., Munagala, K., Widom, J., Motwani, R.: Query optimization over web services. In: Proc. of the International Conference on Very Large Data Bases, pp. 355–366 (2006)

Fusing Approximate Knowledge from Distributed Sources*

Barbara Dunin-Kęplicz, Linh Anh Nguyen, and Andrzej Szałas

Abstract. In this paper we investigate a technique for fusing approximate knowledge obtained from distributed, heterogeneous information sources. We use a generalization of rough sets and relations [14], which depends on allowing arbitrary similarity relations. The starting point of this research is [2], where a framework for knowledge fusion in multi-agent systems is introduced. Agent's individual perceptual capabilities are represented by similarity relations, further aggregated to express joint capabilities of teams. This aggregation, allowing a shift from individual to social level, has been formalized by means of dynamic logic. The approach of [2] uses the full propositional dynamic logic, not guaranteeing the tractability of reasoning. Therefore the results of [11, 12, 13] are adapted to provide a technical engine for tractable approximate database querying restricted to a Horn fragment of serial PDL. We also show that the obtained formalism is quite powerful in applications.

1 Similarities and Approximate Reasoning

In this paper we investigate a technique for fusing approximate knowledge obtained from distributed information sources. This issue is substantial in modeling

Barbara Dunin-Kęplicz
Institute of Informatics, Warsaw University, Poland and ICS,
Polish Academy of Sciences, Poland
e-mail: `keplicz@mimuw.edu.pl`

Linh Anh Nguyen
Institute of Informatics, Warsaw University, Poland
e-mail: `nguyen@mimuw.edu.pl`

Andrzej Szałas
Institute of Informatics, Warsaw University, Poland and IDA, Linköping University, Sweden
e-mail: `andsz@mimuw.edu.pl`

* Supported by the MNiSW grants N N206 399134 and N N206 399334.

G.A. Papadopoulos and C. Badica (Eds.): Intelligent Distributed Computing III, SCI 237, pp. 75–86.
springerlink.com © Springer-Verlag Berlin Heidelberg 2009

multi-agent systems, where a group of loosely coupled heterogeneous agents cooperate in achieving a common goal. Information exchange, leading ultimately to knowledge fusion, is a natural and vital ingredient of this process. We use a generalization of rough sets and relations [14], which depends on allowing arbitrary similarity relations, while in [14] only equivalence relations are considered. In order to approximate relations one uses here a covering of the underlying domain by similarity-based neighborhoods. Such approximate relations have been shown to be useful in many application areas requiring the use of approximate knowledge structures [3].

There are many choices as to possible constraints to be placed on the similarity relation used to define approximations. For example, one might not want the relation be transitive since similar objects do not naturally chain in a transitive manner (see, e.g., [1, 6, 3, 10]).

The basic requirement as to approximations is that the lower approximation is included in the upper one of any set/relation. This is equivalent to the seriality of similarity relations (see [5]), which we set as the only requirement.

The focus of this paper is approximate knowledge fusion based on the idea of approximations. Our starting point is [2], where a framework for knowledge fusion in multi-agent systems is introduced. Agent's individual perceptual capabilities are represented by similarity relations, further aggregated to express joint capabilities of teams. The aggregation expressing a shift from individual to social level of agents' activity has been formalized by means of dynamic logic. The approach of [2], as using the full propositional dynamic logic, does not guarantee tractability of reasoning [9]. Our idea is to adapt the techniques of [11, 12, 13] to provide an engine for tractable approximate database querying restricted to a Horn fragment of serial PDL, denoted by SPDL.

To distinguish between extensional and intensional database, we use the terminology of description logics:

- ABox (*assertion box*) stands for the extensional database (containing facts)
- TBox (*terminological box*) stands for the intensional database (containing rules).

In this paper we present an algorithm that, given a Horn query $\mathscr{P}$ expressed in SPDL, and an ABox $\mathscr{A}$, constructs a least SPDL model $\mathscr{M}$ of $\mathscr{P}$ and $\mathscr{A}$. This model has the property that for every positive formula φ and for every individual a, $\varphi(a)$ is a logical consequence of $\mathscr{P}, \mathscr{A}$ in SPDL (denoted by $\mathscr{P}, \mathscr{A} \models_s \varphi(a)$) iff it is true in $\mathscr{M}$ (i.e. $a^{\mathscr{M}} \in \varphi^{\mathscr{M}}$). The least model $\mathscr{M}$ is constructed in polynomial time in the size of $\mathscr{A}$ (and has a polynomial size in the size of $\mathscr{A}$). As a consequence, the problem of checking whether $\mathscr{P}, \mathscr{A} \models_s \varphi(a)$ has PTIME data complexity (measured in the size of $\mathscr{A}$).

The paper is structured as follows. In Section 2 we recall Propositional Dynamic Logic, show its relationship to approximate reasoning and approximate databases, and justify the requirement as to seriality. Section 3 is devoted to showing the PTIME data complexity of the selected Horn fragment of SPDL. Section 4 shows an example illustrating the potential of the introduced machinery in real-world applications.

2 Serial Propositional Dynamic Logic

2.1 *Language and Semantics of* SPDL

Let us now define *serial propositional dynamic logic* (SPDL). The key idea is to provide calculus on similarity relations rather than on programs. This somehow unusual move allows us to reason about similarities using the whole apparatus of dynamic logic, where "programs" are replaced by similarity relations.

Let $\mathscr{MOD}$ denote the set of *similarity relation symbols*, and $\mathscr{PROP}$ denote the set of *propositions*. We use letters like σ to indicate elements of $\mathscr{MOD}$, and letters like p, q to indicate elements of $\mathscr{PROP}$.

Definition 1. *Formulas* and *similarity expressions* are respectively defined by the two following BNF grammar rules:

$$\varphi ::= \top \mid p \mid \neg\varphi \mid \varphi \wedge \varphi \mid \varphi \vee \varphi \mid \varphi \to \varphi \mid \langle\alpha\rangle\varphi \mid [\alpha]\varphi$$
$$\alpha ::= \sigma \mid \alpha;\alpha \mid \alpha \cup \alpha \mid \alpha^* \mid \varphi?$$
◁

We use letters like α, β to denote similarity expressions; φ, ψ to denote formulas; and a, b, c to denote *individuals*. Intuitively,

- $\alpha_1;\alpha_2$ stands for a sequential composition of relations α_1 and α_2
- $\alpha_1 \cup \alpha_2$ stands for set-theoretical union of relations α_1 and α_2
- α^* stands for the transitive closure of α
- $\varphi?$ stands for the test operator.

$\langle\alpha\rangle$ and $[\alpha]$ are modal operators of the dynamic logic with the meaning:

- $\langle\alpha\rangle\varphi$: there is an object similar w.r.t. α to a given object and satisfying formula φ
- $[\alpha]\varphi$: all objects similar w.r.t. α to a given object satisfy φ.

The following definitions naturally capture these intuitions. Observe, however, that rather than the set of worlds, the sets of objects are domains of Kripke structures.

Definition 2. A Kripke structure is a pair $\mathscr{M} = \langle\Delta^{\mathscr{M}}, \cdot^{\mathscr{M}}\rangle$, where $\Delta^{\mathscr{M}}$ is a set of *objects*, and $\cdot^{\mathscr{M}}$ is an interpretation function that maps each individual a to an element $a^{\mathscr{M}}$ of $\Delta^{\mathscr{M}}$, each proposition p to a subset $p^{\mathscr{M}}$ of $\Delta^{\mathscr{M}}$, and each similarity relation symbol σ to a binary relation $\sigma^{\mathscr{M}}$ on $\Delta^{\mathscr{M}}$. ◁

The interpretation function is extended for all formulas and similarity expressions:

$$\top^{\mathscr{M}} = \Delta^{\mathscr{M}},\ (\neg\varphi)^{\mathscr{M}} = \Delta^{\mathscr{M}} \setminus \varphi^{\mathscr{M}},\ (\varphi \wedge \psi)^{\mathscr{M}} = \varphi^{\mathscr{M}} \cap \psi^{\mathscr{M}}$$
$$(\varphi \vee \psi)^{\mathscr{M}} = \varphi^{\mathscr{M}} \cup \psi^{\mathscr{M}},\ (\varphi \to \psi)^{\mathscr{M}} = (\neg\varphi \vee \psi)^{\mathscr{M}}$$
$$(\langle\alpha\rangle\varphi)^{\mathscr{M}} = \{x \in \Delta^{\mathscr{M}} \mid \exists y\,[\alpha^{\mathscr{M}}(x,y) \wedge \varphi^{\mathscr{M}}(y)]\}$$
$$([\alpha]\varphi)^{\mathscr{M}} = \{x \in \Delta^{\mathscr{M}} \mid \forall y\,[\alpha^{\mathscr{M}}(x,y) \to \varphi^{\mathscr{M}}(y)]\}$$

$$(\alpha;\beta)^{\mathscr{M}} = \alpha^{\mathscr{M}} \circ \beta^{\mathscr{M}} = \{(x,y) \mid \exists z\,[\alpha^{\mathscr{M}}(x,z) \wedge \beta^{\mathscr{M}}(z,y)]\}$$
$$(\alpha \cup \beta)^{\mathscr{M}} = \alpha^{\mathscr{M}} \cup \beta^{\mathscr{M}},\ (\alpha^*)^{\mathscr{M}} = (\alpha^{\mathscr{M}})^*,\ (\varphi?)^{\mathscr{M}} = \{(x,x) \mid \varphi^{\mathscr{M}}(x)\}.$$

We sometimes write $\mathcal{M},x \models \varphi$ to denote $x \in \varphi^{\mathcal{M}}$. For a set Γ of formulas, we write $\mathcal{M},x \models \Gamma$ to denote that $\mathcal{M},x \models \varphi$ for all $\varphi \in \Gamma$. If $\mathcal{M},x \models \Gamma$ for all $x \in \Delta^{\mathcal{M}}$ then we call $\mathcal{M}$ a *model of* Γ.

When dealing with the data complexity of the instance checking problem, without loss of generality we can assume that both the sets $\mathcal{MOD}$ and $\mathcal{PROP}$ are finite and fixed. Under this assumption, the *size of a Kripke structure* $\mathcal{M}$ is defined to be the number of elements of the set $\Delta^{\mathcal{M}}$.

For every $\sigma \in \mathcal{MOD}$, we adopt the *seriality axioms* $[\sigma]\varphi \rightarrow \langle\sigma\rangle\varphi$. It is well known (see, e.g., [5]) that seriality axioms correspond to the *seriality property* $\forall x \exists y [\sigma^{\mathcal{M}}(x,y)]$. By an *admissible interpretation for* SPDL we understand any Kripke structure $\mathcal{M}$ with all $\sigma \in \mathcal{MOD}$ satisfying the seriality property. We call such Kripke structures *serial*.

2.2 SPDL as a Query Language in Approximate Databases

Let us now explain how SPDL is used as a query language. First observe that interpretations assign sets of objects to formulas. Thus it is natural to consider each formula as the query selecting all objects satisfying the formula.

Example 1. Let, in a given interpretation $\mathcal{M}$, $\Delta^{\mathcal{M}} = \{o_1,o_2,o_3,o_4,o_5\}$, $red^{\mathcal{M}} = \{o_1,o_3,o_4\}$ and $small^{\mathcal{M}} = \{o_1,o_2,o_4,o_5\}$. Then the $(red \wedge small)^{\mathcal{M}} = \{o_1,o_4\}$, thus the query $(red \wedge small)$ returns the set $\{o_1,o_4\}$. Similarly, the query $(red \rightarrow small)$ returns $\{o_1,o_2,o_4,o_5\}$. ◁

Let us now recall the notion of approximations.

Definition 3. Let Δ be a set of objects and α be a similarity expression representing a serial binary relation on Δ. For $a \in \Delta$, by the *neighborhood of a w.r.t.* α, we understand the set of elements similar to a, $n^{\alpha}(a) \stackrel{\text{def}}{=} \{b \in \Delta \mid \alpha(a,b)\}$. For $A \subseteq \Delta$, the *lower and upper approximations of* A *w.r.t.* α, denoted respectively by A^{+}_{α} and $A^{\oplus}_{\alpha}$, are defined by $A^{+}_{\alpha} = \{a \in \Delta \mid n^{\alpha}(a) \subseteq A\}$, $A^{\oplus}_{\alpha} = \{a \in \Delta \mid n^{\alpha}(a) \cap A \neq \emptyset\}$. ◁

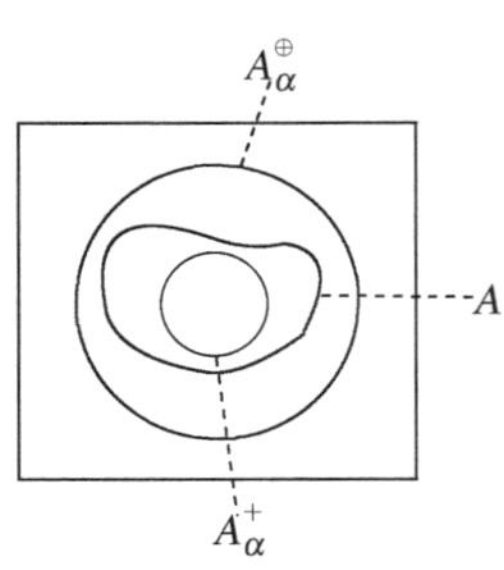

Fig. 1 Lower approximation A^{+}_{α} and upper approximation $A^{\oplus}_{\alpha}$ of a set A.

The meaning of those approximations is illustrated in Figure 1. Intuitively, assuming that the perception of an agent is modeled by similarity expression α,

- $a \in A^{+}_{\alpha}$ means that, from the point of view of the agent, a surely is in A, since all objects indiscernible from a are in A
- $a \in A^{\oplus}_{\alpha}$ means that, from the point of view of the agent, a possibly is in A, since there are objects indiscernible from a which are in A.

Now observe that:

$$[\alpha]A \text{ expresses the lower approximation of } A \text{ w.r.t. } \alpha, \text{ i.e., } A^{+}_{\alpha}, \tag{1}$$

$$\langle\alpha\rangle A \text{ expresses the upper approximation of } A \text{ w.r.t. } \alpha, \text{ i.e., } A^{\oplus}_{\alpha}. \tag{2}$$

Remark 1. In the view of (1) and (2), the seriality axiom expresses the property that the lower approximation of a set w.r.t. any similarity expression is included in its upper approximation. This justifies seriality as the basic requirement on approximations. ◁

Example 2. Let $\mathcal{M}$ be the interpretation considered in Example 1. Let σ be the reflexive closure of relation $\{\langle o_1, o_2\rangle, \langle o_2, o_1\rangle, \langle o_3, o_4\rangle\}$. Then, for example,

$$red^{+}_{\sigma} = \{o_3, o_4\},\ red^{\oplus}_{\sigma} = \{o_1, o_2, o_3, o_4\}.$$ ◁

2.3 The Horn Fragment of SPDL

In order to express tractable queries we restrict the query language to the Horn fragment.

Definition 4. *Positive formulas* φ_{pos} are defined by the following BNF grammar:

$$\begin{aligned}
\varphi_{pos} &::= \top \mid p \mid \varphi_{pos} \wedge \varphi_{pos} \mid \varphi_{pos} \vee \varphi_{pos} \mid \langle \alpha_{pos\Diamond} \rangle \varphi_{pos} \mid [\alpha_{pos\Box}]\varphi_{pos}\\
\alpha_{pos\Diamond} &::= \sigma \mid \alpha_{pos\Diamond} ; \alpha_{pos\Diamond} \mid \alpha_{pos\Diamond} \cup \alpha_{pos\Diamond} \mid \alpha^{*}_{pos\Diamond} \mid \varphi_{pos}?\\
\alpha_{pos\Box} &::= \sigma \mid \alpha_{pos\Box} ; \alpha_{pos\Box} \mid \alpha_{pos\Box} \cup \alpha_{pos\Box} \mid \alpha^{*}_{pos\Box} \mid (\neg \varphi_{pos})?
\end{aligned}$$

Program clauses φ_{prog} are defined by the following BNF grammar:[1]

$$\begin{aligned}
\varphi_{prog} &::= \top \mid p \mid \varphi_{pos} \to \varphi_{prog} \mid \varphi_{prog} \wedge \varphi_{prog} \mid \langle \alpha_{prog\Diamond} \rangle \varphi_{prog} \mid [\alpha_{prog\Box}]\varphi_{prog}\\
\alpha_{prog\Diamond} &::= \sigma \mid \alpha_{prog\Diamond} ; \alpha_{prog\Diamond} \mid \varphi_{prog}?\\
\alpha_{prog\Box} &::= \sigma \mid \alpha_{prog\Box} ; \alpha_{prog\Box} \mid \alpha_{prog\Box} \cup \alpha_{prog\Box} \mid \alpha^{*}_{prog\Box} \mid \varphi_{pos}?
\end{aligned}$$

A *positive logic program* is a finite set of program clauses. ◁

Example 3. Observe that the Horn fragment of SPDL is quite expressive. For example, it allows one to express a variant of default rules (discussed, e.g., in [3]). Namely, a typical default rule can be expressed as $A^{+}_{\sigma}, B^{\oplus}_{\sigma} \vdash C^{+}_{\sigma}$, with intuitive meaning "if A is surely true and B might be true then accept C as surely true". ◁

Let us now formally link SPDL with databases.

[1] Notice the two occurrences of φ_{pos} in the grammar. We do not allow formulas of the form $\langle\alpha \cup \beta\rangle\varphi$ or $\langle\alpha^{*}\rangle\varphi$ to be program clauses because they cause non-determinism.

Definition 5. An *individual assertion* is an expression of the form $p(a)$, where p is a proposition and a is an individual. A *similarity assertion* is an expression of the form $\sigma(a,b)$, where σ is a similarity relation symbol and a, b are individuals. An *ABox* is a finite set of individual assertions and similarity assertions. ◁

Comparing to description logics, individual assertions play a role of concept assertions, while similarity assertions play a role of assertions.

Definition 6

- Given a Kripke structure $\mathscr{M}$ and an ABox $\mathscr{A}$, we say that $\mathscr{M}$ is a *model of* $\mathscr{A}$, denoted by $\mathscr{M} \models \mathscr{A}$, if $a^{\mathscr{M}} \in p^{\mathscr{M}}$ for every individual assertion $p(a) \in \mathscr{A}$ and $(a^{\mathscr{M}}, b^{\mathscr{M}}) \in \sigma^{\mathscr{M}}$ for every similarity assertion $\sigma(a,b) \in \mathscr{A}$.
- Given a positive logic program $\mathscr{P}$, an ABox $\mathscr{A}$, a positive formula φ and an individual a, we say that *a has the property φ w.r.t. $\mathscr{P}$ and $\mathscr{A}$ in* SPDL (or $\varphi(a)$ is a logical consequence of $\mathscr{P}, \mathscr{A}$ in SPDL), denoted by $\mathscr{P}, \mathscr{A} \models_s \varphi(a)$, if for every serial Kripke structure $\mathscr{M}$, if $\mathscr{M}$ is a model of $\mathscr{P}$ and $\mathscr{A}$ then $a^{\mathscr{M}} \in \varphi^{\mathscr{M}}$.
- By the *instance checking* problem of the Horn fragment of SPDL we mean the problem of checking $\mathscr{P}, \mathscr{A} \models_s \varphi(a)$. The *data complexity* of this problem is measured when $\mathscr{P}$, φ and a are fixed (and compose a query), while $\mathscr{A}$ varies as input data. ◁

3 Computational Aspects

A Kripke structure $\mathscr{M}$ is *less than or equal to* $\mathscr{M}'$, $\mathscr{M} \leq \mathscr{M}'$, if for every positive formula φ and every individual a, $a^{\mathscr{M}} \in \varphi^{\mathscr{M}}$ implies $a^{\mathscr{M}'} \in \varphi^{\mathscr{M}'}$.

Let $\mathscr{P}$ be a positive logic program and $\mathscr{A}$ be an ABox. We say that a Kripke structure $\mathscr{M}$ is a *least* SPDL *model of $\mathscr{P}$ and $\mathscr{A}$* if $\mathscr{M}$ is an SPDL model of $\mathscr{P}$ and $\mathscr{A}$ and is less than or equal to any other SPDL model of $\mathscr{P}$ and $\mathscr{A}$.

Let us now present an algorithm that, given a positive logic program $\mathscr{P}$ and an ABox $\mathscr{A}$, constructs a finite least SPDL model of $\mathscr{P}$ and $\mathscr{A}$. The algorithm constructs the following data structures:

- Δ is a set of objects. We distinguish the subset Δ_0 of Δ that consists of all the individuals occurring in the ABox $\mathscr{A}$. In the case $\mathscr{A}$ is empty, let $\Delta_0 = \{\tau\}$ for some element τ.
- H is a mapping that maps every $x \in \Delta$ to a set of formulas, which are the properties that should hold for x. When the elements of Δ are treated as states, $H(x)$ denotes the content of the state x.
- *Next* is a mapping such that, for $x \in \Delta$ and $\langle\sigma\rangle\varphi \in H(x)$, we have $Next(x, \langle\sigma\rangle\varphi) \in \Delta$. The meaning of $Next(x, \langle\sigma\rangle\varphi) = y$ is that:
  - $\langle\sigma\rangle\varphi \in H(x)$ and $\varphi \in H(y)$,
  - the "requirement" $\langle\sigma\rangle\varphi$ is realized for x by going to y via a σ-transition.

Using the above data structures, we define a Kripke structure $\mathcal{M}$ such that:

- $\Delta^{\mathcal{M}} = \Delta$,
- $a^{\mathcal{M}} = a$ for every individual a occurring in $\mathcal{A}$,
- $p^{\mathcal{M}} = \{x \in \Delta \mid p \in H(x)\}$ for every $p \in \mathcal{PROP}$,
- $\sigma^{\mathcal{M}} = \{(a,b) \mid \sigma(a,b) \in \mathcal{A}\} \cup \{(x,y) \mid Next(x, \langle\sigma\rangle\varphi) = y$ for some $\varphi\}$ for every $\sigma \in \mathcal{MOD}$.

The *saturation* of a set Γ of formulas, denoted by $\mathsf{Sat}(\Gamma)$, is defined to be the smallest superset of Γ such that:

- $\top \in \mathsf{Sat}(\Gamma)$ and $\langle\sigma\rangle\top \in \mathsf{Sat}(\Gamma)$ for all $\sigma \in \mathcal{MOD}$,
- if $\varphi \wedge \psi \in \mathsf{Sat}(\Gamma)$ or $\langle\varphi?\rangle\psi \in \mathsf{Sat}(\Gamma)$ then $\varphi \in \mathsf{Sat}(\Gamma)$ and $\psi \in \mathsf{Sat}(\Gamma)$,
- if $\langle\alpha;\beta\rangle\varphi \in \mathsf{Sat}(\Gamma)$ then $\langle\alpha\rangle\langle\beta\rangle\varphi \in \mathsf{Sat}(\Gamma)$,
- if $[\alpha;\beta]\varphi \in \mathsf{Sat}(\Gamma)$ then $[\alpha][\beta]\varphi \in \mathsf{Sat}(\Gamma)$,
- if $[\alpha \cup \beta]\varphi \in \mathsf{Sat}(\Gamma)$ then $[\alpha]\varphi \in \mathsf{Sat}(\Gamma)$ and $[\beta]\varphi \in \mathsf{Sat}(\Gamma)$,
- if $[\alpha^*]\varphi \in \mathsf{Sat}(\Gamma)$ then $\varphi \in \mathsf{Sat}(\Gamma)$ and $[\alpha][\alpha^*]\varphi \in \mathsf{Sat}(\Gamma)$,
- if $[\varphi?]\psi \in \mathsf{Sat}(\Gamma)$ then $(\varphi \to \psi) \in \mathsf{Sat}(\Gamma)$.

The *transfer of Γ through σ* is defined by $\mathsf{Trans}(\Gamma,\sigma) \stackrel{\text{def}}{=} \mathsf{Sat}(\{\varphi \mid [\sigma]\varphi \in \Gamma\})$. We use procedure $\mathsf{Find}(\Gamma)$ defined as:

> if there exists $x \in \Delta \setminus \Delta_0$ with $H(x) = \Gamma$ then return x,
> else add a new object x to Δ with $H(x) = \Gamma$ and return x.

The algorithm shown in Figure 2 constructs a least SPDL model for a positive logic program $\mathcal{P}$ and an ABox $\mathcal{A}$.

Theorem 1

1. *Let $\mathcal{P}$ be a positive logic program and $\mathcal{A}$ be an ABox. The Kripke structure $\mathcal{M}$ constructed by the algorithm shown in Figure 2 for $\mathcal{P}$ and $\mathcal{A}$ is a least* SPDL *model of $\mathcal{P}$ and $\mathcal{A}$.*
2. *Let $\mathcal{P}$ be a positive logic program, $\mathcal{A}$ an ABox, φ a positive formula, and a an individual. Then checking $(\mathcal{P},\mathcal{A}) \models_s \varphi(a)$ can be done in polynomial time in the size of $\mathcal{A}$ (by constructing a least* SPDL *model $\mathcal{M}$ of $\mathcal{P}$ and $\mathcal{A}$ using our algorithm, and then checking whether $a^{\mathcal{M}} \in \varphi^{\mathcal{M}}$). That is, the data complexity of the Horn fragment of* SPDL *is in* PTIME. ◁

4 An Exemplary Scenario

Consider the following scenario:

> Two robots, R_1 and R_2, have the goal to move objects from one place to another. Each robot is able to move objects of a specific signature,[2] and together they might be able to move objects of a combined signature. Some objects,

[2] For example, dependent on weight, size and type of surface.

Input: A positive logic program $\mathscr{P}$ and an ABox $\mathscr{A}$.
Output: A least SPDL model $\mathscr{M}$ of $\mathscr{P}$ and $\mathscr{A}$.

1. let Δ_0 be the set of all individuals occurring in $\mathscr{A}$;
   if $\Delta_0 = \emptyset$ then $\Delta_0 := \{\tau\}$;
   set $\Delta := \Delta_0$, $\mathscr{P}' := \mathsf{Sat}(\mathscr{P})$;
   for $x \in \Delta$, set $H(x) := \mathscr{P}' \cup \{p \mid p(x) \in \mathscr{A}\}$;
2. for every $x \in \Delta$ reachable from Δ_0 and for every formula $\varphi \in H(x)$
   a. case $\varphi = \langle\sigma\rangle\psi$: if $Next(x, \langle\sigma\rangle\psi)$ is not defined then
      $Next(x, \langle\sigma\rangle\psi) := \mathsf{Find}(\mathsf{Sat}(\{\psi\}) \cup \mathsf{Trans}(H(x), \sigma) \cup \mathscr{P}')$;
   b. case $\varphi = [\sigma]\psi$:
      i. for every $y \in \Delta_0$ such that $\sigma^{\mathscr{M}}(x,y)$ holds and $\psi \notin H(y)$
         $H(y) := H(y) \cup \mathsf{Sat}(\{\psi\})$;
      ii. for every $y \in \Delta \setminus \Delta_0$ such that $\sigma^{\mathscr{M}}(x,y)$ holds and $\psi \notin H(y)$
         A. $y_* := \mathsf{Find}(H(y) \cup \mathsf{Sat}(\{\psi\}))$;
         B. for every ξ such that $Next(x, \langle\sigma\rangle\xi) = y$
            $Next(x, \langle\sigma\rangle\xi) := y_*$;
   c. case $\varphi = (\psi \to \xi)$: if $x \in \psi^{\mathscr{M}}$ and $Next(y, \langle\sigma\rangle\top)$ is defined for every y reachable from x and every $\sigma \in \mathscr{MOD}$ then
      i. if $x \in \Delta_0$ then $H(x) := H(x) \cup \mathsf{Sat}(\{\xi\})$
      ii. else
         A. $x_* := \mathsf{Find}(H(x) \cup \mathsf{Sat}(\{\xi\}))$;
         B. for every y, σ, ζ such that $Next(y, \langle\sigma\rangle\zeta) = x$
            $Next(y, \langle\sigma\rangle\zeta) := x_*$;
3. if some change occurred, go to Step 2;
4. delete from Δ every x unreachable from Δ_0 and delete from H and $Next$ all elements related with such an x.

Fig. 2 Algorithm constructing the least SPDL model for a positive logic program and an ABox.

when attempted to be moved, may cause some damages for robots. Robots are working independently, but sometimes have to cooperate to achieve their goals.

To design such robots one has to make a number of decisions as described below.

4.1 Formalizing Movability of Objects

We assume that the signature of movable objects for each robot is given by its specification together with a similarity relation defining the range of these objects as follows:

$$spec_1 \stackrel{\text{def}}{\equiv} (light \land smooth) \lor (heavy \land rough) - \text{ for robot } R_1 \tag{3}$$

$$spec_2 \stackrel{\text{def}}{\equiv} small \vee medium - \text{ for robot } R_2. \tag{4}$$

Movable objects are then specified by

$$spec_i \rightarrow movable_i, \tag{5}$$

where $i \in \{1,2\}$ and $movable_i$ is true for objects that can be moved by R_i.

The idea is that all objects similar to movable ones are movable too.[3] Let σ_1 and σ_2 be similarity relations reflecting perceptual capabilities of R_1 and R_2, respectively (for a discussion of such similarity relations based on various sensor models see [4]). Now, in addition to (5), movable objects are characterized by

$$\langle \sigma_i \rangle spec_i \rightarrow movable_i. \tag{6}$$

Remark 2 Note that rather than (6) one could assume $[\sigma_i]spec_i \rightarrow movable_i$. This would mean that, in addition to (5), we would consider objects which are similar only to movable objects. In some applications this choice would indeed be reasonable and perhaps less risky. ◁

Observe that in general it is impossible to automatically derive combined signatures that specify what robots can move together. Therefore, we introduce $spec_3$ and α_3 as a specification of such joint capabilities. An example of $spec_3$ can be given by

$$spec_3 \stackrel{\text{def}}{\equiv} large \wedge rough. \tag{7}$$

Objects movable by robots working together are then defined by

$$(spec_3 \vee \langle \alpha_3 \rangle spec_3) \rightarrow movable\_by\_two. \tag{8}$$

Observe that α_3 is usually computed on the basis of σ_1 and σ_2, since we do not assume any observer other than R_1, R_2, and σ_1, σ_2 reflects their perceptual capabilities. We shall assume that

$$\alpha_3 \stackrel{\text{def}}{=} \sigma_1 \cup \sigma_2 \cup (\sigma_1 ; \sigma_2) \cup (\sigma_2 ; \sigma_1). \tag{9}$$

The meaning of this expression is that an object o is similar w.r.t. α_3 to o' whenever

- o is perceived similar to o' by R_1 or by R_2, or
- there is an object o'' such that
  - o is perceived similar to o'' by R_1 and o'' is perceived similar to o' by R_2, or
  - o is perceived similar to o'' by R_2 and o'' is perceived similar to o' by R_1.

Let us emphasize that one can use much more complex expressions, reflecting particular algorithms for computing α_3, since our operators are those accepted in PDL. Of course, there are some restrictions, if one wants to stay in a tractable framework. Recall that we have accepted Definition 4 to guarantee tractability.

[3] This is a very natural and quite powerful technique, allowing one to express the inheritance of particular properties of objects by similar objects.

4.2 The Database

Let $\mathscr{A}$ be the ABox consisting of the assertions about objects $o_1, \ldots, o_5$ that are illustrated in Figure 3. It contains, for example, $light(o_1)$, $smooth(o_1)$, $\sigma_1(o_1,o_2)$, $\sigma_1(o_2,o_1)$, etc. Let $\mathscr{P}$ be the positive logic program consisting of the clauses (5) and (6) for $i \in \{1,2\}$, and (8). We consider here the database consisting of $\mathscr{P}$ and $\mathscr{A}$ in SPDL (which adopts the axioms $\langle\sigma_1\rangle\top$ and $\langle\sigma_2\rangle\top$ for all the objects of the domain).

Fig. 3 An ABox for the example considered in Section 4.2.

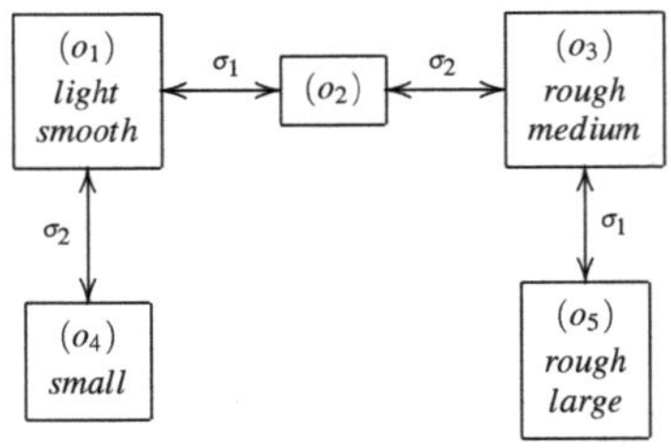

In Figure 4 we present the least SPDL model $\mathscr{M}$ of $\mathscr{P}$ and $\mathscr{A}$ constructed by the algorithm. The object o_6 is the only additional object, not satisfying any proposition.

Fig. 4 The least SPDL model of the database considered in Section 4.2.

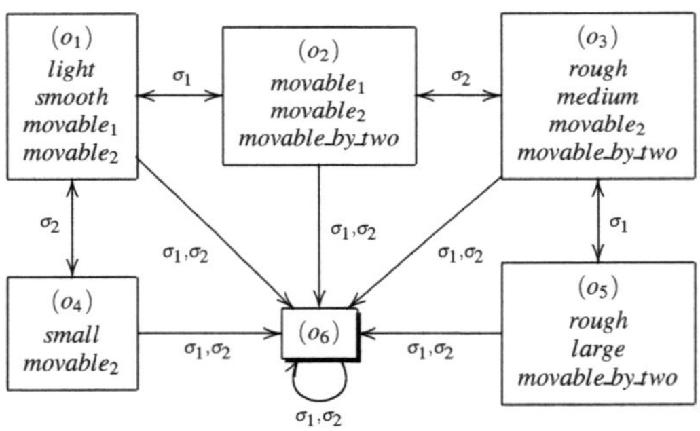

4.3 Some Queries

Recall that, having a least SPDL model $\mathscr{M}$ of $\mathscr{P}$ and $\mathscr{A}$, to check whether an individual a has a positive property φ w.r.t. $\mathscr{P}$ and $\mathscr{A}$ in SPDL, it suffices to check whether $a \in \varphi^{\mathscr{M}}$. We have that $movable_1^{\mathscr{M}} = \{o_1,o_2\}$, $movable_2^{\mathscr{M}} = \{o_1,o_2,o_3,o_4\}$, $movable\_by\_two^{\mathscr{M}} = \{o_2,o_3,o_5\}$, $(movable\_by\_two \wedge \langle\sigma_1\rangle movable_1)^{\mathscr{M}} = \{o_1\}$, $(movable\_by\_two \wedge [\sigma_1]movable_1)^{\mathscr{M}} = \emptyset$.

5 Conclusions

In this paper we have presented a powerful formalism for approximate knowledge fusion, based on adaptation of Propositional Dynamic Logic. We have shown that restricting this logic to its suitably chosen Horn fragment results in tractable querying mechanism which can be applied in application, where approximate knowledge from various sources is to be fused, e.g., in robotics and multi-agent systems.

Importantly, serial PDL, denoted by SPDL, is also useful as a description logic for domains where seriality condition appears naturally.[4] For example, in reasoning about properties of web pages one can assume that every considered web page has a link to another page (or to itself).

We plan to extend the framework to deal with other operations on similarity relations, which would allow expressing even more subtle approximations and fused knowledge structures applicable in different stages of teamwork in multi-agent systems as discussed in [7, 8].

References

1. Demri, S.P., Orłowska, E.S.: Incomplete Information: Structure, Inference, Complexity. In: Monographs in Theoretical Computer Science. An EATCS Series. Springer, Heidelberg (2002)
2. Doherty, P., Dunin-Kęplicz, B., Szałas, A.: Dynamics of approximate information fusion. In: Kryszkiewicz, M., Peters, J.F., Rybiński, H., Skowron, A. (eds.) RSEISP 2007. LNCS (LNAI), vol. 4585, pp. 668–677. Springer, Heidelberg (2007)
3. Doherty, P., Łukaszewicz, W., Skowron, A., Szałas, A.: Knowledge Representation Techniques. A Rough Set Approach. Studies in Fuziness and Soft Computing, vol. 202. Springer, Heidelberg (2006)
4. Doherty, P., Łukaszewicz, W., Szałas, A.: Communication between agents with heterogeneous perceptual capabilities. Journal of Information Fusion 8(1), 56–69 (2007)
5. Doherty, P., Szałas, A.: On the correspondence between approximations and similarity. In: Tsumoto, S., Słowiński, R., Komorowski, J., Grzymała-Busse, J.W. (eds.) RSCTC 2004. LNCS (LNAI), vol. 3066, pp. 143–152. Springer, Heidelberg (2004)
6. Dunin-Kęplicz, B., Szałas, A.: Towards approximate BGI systems. In: Burkhard, H.-D., Lindemann, G., Verbrugge, R., Varga, L.Z. (eds.) CEEMAS 2007. LNCS (LNAI), vol. 4696, pp. 277–287. Springer, Heidelberg (2007)
7. Dunin-Kęplicz, B., Verbrugge, R.: Collective intentions. Fundamenta Informaticae 51(3), 271–295 (2002)
8. Dunin-Kęplicz, B., Verbrugge, R.: A tuning machine for cooperative problem solving. Fundamenta Informaticae 63, 283–307 (2004)
9. Harel, D., Kozen, D., Tiuryn, J.: Dynamic Logic. MIT Press, Cambridge (2000)
10. Maluszyński, J., Szałas, A., Vitória, A.: Paraconsistent logic programs with four-valued rough sets. In: Chan, C.-C., Grzymala-Busse, J.W., Ziarko, W.P. (eds.) RSCTC 2008. LNCS (LNAI), vol. 5306, pp. 41–51. Springer, Heidelberg (2008)

[4] Related work on Horn fragments of description logics can be found in [12]. The papers by other authors do not consider PDL but only versions of the description logic $\mathcal{SHI}$.

11. Nguyen, L.A.: On the deterministic Horn fragment of test-free PDL. In: Hodkinson, I., Venema, Y. (eds.) Advances in Modal Logic, vol. 6, pp. 373–392. King's College Publications (2006)
12. Nguyen, L.A.: Weakening Horn knowledge bases in regular description logics to have PTIME data complexity. In: Ghilardi, S., Sattler, U., Sofronie-Stokkermans, V., Tiwari, A. (eds.) Proceedings of ADDCT 2007, pp. 32–47 (2007)
13. Nguyen, L.A.: Constructing finite least Kripke models for positive logic programs in serial regular grammar logics. Logic Journal of the IGPL 16(2), 175–193 (2008)
14. Pawlak, Z.: Rough Sets. Theoretical Aspects of Reasoning about Data. Kluwer Academic Publishers, Dordrecht (1991)

Case-Study for TeamLog, a Theory of Teamwork

Barbara Dunin-Kęplicz, Rineke Verbrugge, and Michał Ślizak

Abstract. This article presents a case-study of a theoretical multi-agent system designed to clean up ecological disasters. It focuses on the interactions within a team of agents, outlines their goals, and establishes the necessary distribution of knowledge and commitment throughout the team. In addition, cooperation between teams is treated. These aspects of teamwork are presented in the TEAMLOG formalism. We will show how to make a bridge between theoretical foundations of a BDI system and an application. Complex team attitudes are justified to be necessary in the course of teamwork. At the same time we show how to tune TEAMLOG to the case-study by establishing sufficient, but still minimal levels for the team attitudes.

1 Defining Teamwork

When constructing BDI systems, firstly a model of an agent as an *individual, autonomous* entity [21] has to be constructed. Nowadays a key point is to organize agents' cooperation in a way allowing the achievement of their common goal, while preserving their individual autonomy (see [18, 23, 15, 24, 1, 6, 7, 14] for some logical approaches to teamwork). The BDI model comprises agents' individual beliefs, goals, and intentions. However in teamwork, when a team of agents needs to work

Barbara Dunin-Kęplicz
Institute of Informatics, Warsaw University, Banacha 2, 02-097 Warsaw, Poland
Institute of Computer Science, Polish Academy of Sciences,
Ordona 21, 1-237 Warsaw, Poland
e-mail: keplicz@mimuw.edu.pl

Michał Ślizak
Institute of Informatics, Warsaw University, Banacha 2, 02-097 Warsaw, Poland
e-mail: Michal.Slizak@mimuw.edu.pl

Rineke Verbrugge
Department of Artificial Intelligence, University of Groningen, P.O. Box 407,
9700 AK Groningen, The Netherlands
e-mail: rineke@ai.rug.nl

G.A. Papadopoulos and C. Badica (Eds.): Intelligent Distributed Computing III, SCI 237, pp. 87–100.
springerlink.com
© Springer-Verlag Berlin Heidelberg 2009

together in a planned and coherent way, agents' individual attitudes are not enough: the group needs to present a common attitude over and above individual ones. This group attitude is a necessary condition for a loosely-coupled group of agents to become a strictly cooperative team. In this case-study we focus on full cooperation, where agents' attitudes are considered on the individual, social (i.e. bilateral) and collective level.

A theory of individual and group beliefs has been formalized in terms of epistemic logic [12, 19, 20]. General, common, and distributed knowledge and belief were defined in terms of agents' individual knowledge and belief. Different axiom systems express various properties of knowledge and belief, while the corresponding semantics naturally reflect these properties.

As regards motivational attitudes, the situation is much more complex as the bilateral and collective notions cannot be viewed as a sort of sum of individual ones. Additional subtle and diverse aspects of teamwork need to be isolated and appropriately defined. The *static, descriptive* theory of collective motivational attitudes TEAMLOG [6, 7, 11] has been formed on the basis of individual goals, beliefs and intentions of cooperating agents. It addresses the question what it means for a group of agents to have a *collective intention*, and then a *collective commitment* to achieve a common goal. While collective intention consolidates a group as a strictly cooperating team, collective commitment leads to team action, i.e., to coordinated realization of individual actions by committed agents according to a plan. The social plan can be constructed from first principles, or may be chosen from a repository of pre-constructed plans. Both collective intentions and collective commitments allow to fully express the potential of strictly cooperative teams [6, 7].

When modelling group attitudes, agents' awareness about the overall situation needs to be taken into account. Awareness is understood here as a limited form of consciousness: it refers to the state of an agent's beliefs about *itself* (*intra-personal*), about *others* (*inter-personal*) and about *the environment* (*group awareness*). Thus, various epistemic logics and different gradations of group information (from distributed belief to common knowledge) are adequate to formalize agents' awareness [12, 7, 20].

In TEAMLOG, group awareness is usually expressed in terms of *common belief*, (C-BEL_G) fully reflecting collective aspects of agents' behavior. Due to its infinitary flavor, this concept has a high complexity: its satisfiability problem is EXPTIME-complete [11]. There are general ways to reduce the complexity by restricting the language, by allowing only a small set of atomic propositions or restricting the modal context in formulas, as proved in [11, 10]. However, when building MAS applications, it may be more profitable to use domain-specific means to tailor TEAMLOG to the circumstances in question, calling for weaker forms of awareness [9]. In this case-study of prevention of ecological disasters, we will illustrate how to adjust TEAMLOG to a specific environment. Our aim is to show how the infinitary definitions of collective attitudes can be reduced in a real-world situation.

This paper is structured into several sections. This one introduces the problem in general terms. Next, some definitions and assumptions regarding the environment are presented, including an outline of the interactions within and between teams.

This is followed in section 3 by definitions of plans to be executed. In section 4, a short reminder of part of TEAMLOG is given. In section 5 we explore the minimal requirements for successful teamwork in the case-study from the theoretical perspective. A short discussion sums up this case-study.

2 The Case-Study: Ecological Disasters

The case-study deals with prevention and repair of ecological disasters caused by specific poisons, by means of heterogeneous multi-agent teams. They work in situations where time is critical and resources are bounded [17, 22]. The maintenance goal *safe* is to keep a given region *REG* safe or to return it to safety if it is in danger.

Possible hazards are two kinds of poison, X_1 and X_2, which are dangerous in high concentrations. They may be explosive if they react with one another to form compound $X_1 \oplus X_2$, which happens at high concentrations. Three functions f_1, f_2 and f_3 reflect the influence of temperature $t(A)$, pressure $p(A)$ and concentrations $c_1(A)$ and $c_2(A)$ of poisons X_1 and X_2 at location A on the possible danger level at that location. The function ranges are divided into three intervals, as follows:

The first poison X_1:

- safe_1 iff $f_1(p(A),t(A),c_1(A)) \in [0,v_1]$;
- risky_1 iff $f_1(p(A),t(A),c_1(A)) \in (v_1,n_1]$;
- $\mathit{dangerous}_1$ iff $f_1(p(A),t(A),c_1(A)) \in (n_1,\infty)$;

The second poison X_2:

- safe_2 iff $f_2(p(A),t(A),c_2(A)) \in [0,v_2]$;
- risky_2 iff $f_2(p(A),t(A),c_2(A)) \in (v_2,n_2]$;
- $\mathit{dangerous}_2$ iff $f_2(p(A),t(A),c_2(A)) \in (n_2,\infty)$;

The compound poison $X_1 \oplus X_2$:

- safe_3 iff $f_3(p(A),t(A),c_1(A),c_2(A)) \in [0,v_3]$;
- risky_3 iff $f_3(p(A),t(A),c_1(A),c_2(A)) \in (v_3,n_3]$;
- $\mathit{explosive}$ iff $f_3(p(A),t(A),c_1(A),c_2(A)) \in (n_3,\infty)$;

We define $\mathit{safe} := \mathit{safe}_1 \wedge \mathit{safe}_2 \wedge \mathit{safe}_3$ and refer to it as a goal and as a predicate. There are also thresholds ε_1 and ε_2: when the concentration of a poison X_i exceeds ε_i, the respective function f_i is computed.

2.1 Starting Point: The Agents

This model reflects cooperation between humans, software agents, robots, unmanned aerial vehicles (*UAVs*) [4, 3], and a helicopter steered by a pilot. The whole process is coordinated by one *coordinator*, who initiates cooperation, coordinates teamwork between different teams, is responsible for dividing the disaster zone into sectors and assigning a team to each sector to perform clean-up. Several teams of similar make-up work in parallel, aiming to prevent or neutralize a contamination. Each of these teams consist of:

- one *UAV* - responsible to the coordinator for keeping assigned sectors in a safe state. Cannot carry heavy load, but has considerable computational capabilities for planning and is capable of mapping terrain and observation;
- one regular helicopter steered by the human *pilot*, can independently choose the order in which it will clean up assigned areas;
- n identical neutralizing robots $rob_1, \ldots, rob_n$ - responsible to their *UAV* for cleaning up a zone.

2.2 Cooperation between Teams

The entire disaster zone is divided into sectors by the coordinator, based on terrain type, team size and known hot spots. Teams are responsible for (possibly many) sectors. Each team's *UAV* prepares a plan to keep its sectors safe. Each plan is judged based on a fitting function *fit*, which must take into account:

1. available robots, current task, load, capacity and position of each one,
2. whether the plan relies on robots from other teams to help the current one,
3. task priorities,
4. the minimum amount of time it takes to implement,
5. the minimum amount of robots it requires.

The *UAVs* communicate and cooperate. If performing tasks requires more robots than are currently available, an *UAV* can call for reinforcements from another *UAV*. Of course fulfilling one's team's objectives has a priority over helping others.

2.3 A Bird'S-Eye View on Cases

To maintain the goal *safe*, the situation is monitored on a regular basis with frequency *freq*. During *situation recognition*, in the risky cases monitoring is performed twice as frequently. Depending on the mixture and density of poisons in a location, some general cases followed by the relevant procedures are established. All remedial actions are to be performed relative to the contaminated area:

Case *safe***:**
true $\longrightarrow$ *situation recognition*
Case $dangerous_1$**:**
rain $\longrightarrow$ *liquid* L_1 *to be poured on the soil*
normal or dry $\longrightarrow$ *liquid* L_2 *to be sprayed from the air*
Case $dangerous_2$**:**
rain $\longrightarrow$ *solid* S_1 *to be spread*, followed by *liquid catalyst* K_1 *to be poured*
normal or dry $\longrightarrow$ *solid* S_1 *to be spread*
Case *explosive***:**

before explosion $\longrightarrow$ *evacuation*
after explosion $\longrightarrow$ *rescue action*

Due to the space limit, we cannot present too many details of plans. Failure handling will not be discussed for the same reason.

3 Global Plans

In order to control the amount of interactions and decrease the time needed to establish beliefs, the accepted team model is hierarchical. A coordinator views a team as a single cleaning robot, even though the *UAVs* use many autonomous neutralizing robots to perform their work.

3.1 The Social Plan $\langle Cleanup \rangle$

The social plan for which the coordinator and *UAVs* are responsible, is designed with respect to location A. It is a while-loop, in which observation is interleaved with treatment of current dangers by level of priority, from most to least dangerous. The goal (denoted as $Clean$) is to keep locations in a *safe* state.

begin
  freq := a; $\{freq$ - interval between two checks of the environment$\}$
  while true do
    $\langle$ Plan *SR* $\rangle$ {Compute the situation at A, with frequency *freq*}
    if *explosive* then do $\langle$ Plan E $\rangle$ end;
    elif $dangerous_1$ and rain then do $\langle$ Plan D_1R $\rangle$ end;
    elif $dangerous_1$ then do $\langle$ Plan D_1N $\rangle$ end;
    elif $dangerous_2$ and rain then do $\langle$ Plan D_2R $\rangle$ end;
    elif $dangerous_2$ then do $\langle$ Plan D_2N $\rangle$ end;
    elif $risky_1 \vee risky_2 \vee risky_3$ then *freq* := $\frac{a}{2}$ end
    else {safe situation} *freq* := a end;
  end
end.

3.2 The Social Plan $\langle SR \rangle$

This plan performs situation recognition at location A.

begin
  $C_1 := c_1(A)$ $\{C_1$ is the measured concentration of poison X_1 at $A\}$
  $C_2 := c_2(A)$ $\{C_2$ is the measured concentration of poison X_2 at $A\}$
  $T := t(A)$ $\{T$ is the measured temperature at $A\}$
  $P := p(A)$ $\{P$ is the measured air pressure at $A\}$
  {Computation of the situation at A}

if $C_1 > \varepsilon_1$ then compute $f_1(C_1, T, P)$ end;
if $C_2 > \varepsilon_2$ then compute $f_2(C_2, T, P)$ end;
if $C_1 > \varepsilon_1$ and $C_2 > \varepsilon_2$ then compute $f_3(C_1, C_2, T, P)$ end;
end.

After an explosion, evacuation and rescue of people should take place. This subject is discussed in many studies [22, 17] and will not be elaborated here. See the Appendix for the other plans included in $\langle Cleanup \rangle$.

4 TeamLog: A Logical Theory for Teamwork

It does not suffice for agents to only have an individual intention (INT) towards their projection of the social plan. They would still act as individuals, so if something new appears in their region or the circumstances change calling for re-planning, the group would be helpless to adapt, as not being formed properly. Thus, group attitudes such as *collective intentions* and *collective commitments* become relevant. Due to the space limit, only a short reminder of TEAMLOG can be given here. For extensive explanations and discussion, see [6, 7].

4.1 Definitions of Beliefs in TeamLog: General and Common Belief

For the individual part, we adopt a standard $KD45_n$ system for n agents governing individual belief operator BEL, as explained in [12]. Additionally, for group beliefs, with $G \subseteq \{1, \ldots, n\}$, as in [12]:

C1 $\text{E-BEL}_G(\varphi) \leftrightarrow \bigwedge_{i \in G} BEL(i, \varphi)$ *(general belief: "everyone believes")*

And in more general iterated form:

$$\text{E-BEL}^k_G(\varphi) \leftrightarrow \text{E-BEL}^{k-1}_G(\text{E-BEL}_G(\varphi)), \text{ where } \text{E-BEL}^1_G(\varphi) \equiv \text{E-BEL}_G(\varphi)$$

$\text{C-BEL}_G(\varphi)$ - "it is common belief in the group G that φ is true":

C2 $\text{C-BEL}_G(\varphi) \leftrightarrow \text{E-BEL}_G(\varphi \wedge \text{C-BEL}_G(\varphi))$
RC1 From $\varphi \rightarrow \text{E-BEL}_G(\psi \wedge \varphi)$ infer $\varphi \rightarrow \text{C-BEL}_G(\psi)$ *(induction)*

4.2 Definitions of Intentions in TeamLog: Collective Intention

For operator INT, the TEAMLOG axioms comprise the system KD_n, including the intention consistency axiom D. In addition, the system developer may choose whether or not to add positive and negative introspection of intentions (see [6]).

The focus in TEAMLOG on strictly cooperative teams makes the definition of collective intention rather strong. It is certainly not sufficient that all members of the team G have the associated individual intention $\text{INT}(i, \varphi)$ to achieve φ, i.e. a *general intention*. To exclude competition, all agents should *intend* all members to have the associated individual intention, as well as the intention that all members have the individual intention, and so on; we call this a *mutual intention* ($\text{M-INT}_G(\varphi)$). Furthermore, all team members are aware of this mutual intention by a common belief: $\text{C-BEL}_G(\text{M-INT}_G(\varphi))$. Of course, team members remain autonomous in maintaining their other motivational attitudes, and may compete about other issues.

M1 $\text{E-INT}_G(\varphi) \leftrightarrow \bigwedge_{i\in G} \text{INT}(i,\varphi)$ *(general intention: "everyone intends")*

We also iteratively define E-INT_G^k, similar to higher-order general beliefs:

M1′ $\text{E-INT}_G^k(\varphi) \leftrightarrow \text{E-INT}_G^{k-1}(\text{E-INT}_G(\varphi))$, where $\text{E-INT}_G^1(\varphi) \equiv \text{E-INT}_G(\varphi)$

Mutual and collective intentions are governed by:

M2 $\text{M-INT}_G(\varphi) \leftrightarrow \text{E-INT}_G(\varphi \wedge \text{M-INT}_G(\varphi))$ *(mutual intention)*
RM1 From $\varphi \to \text{E-INT}_G(\psi \wedge \varphi)$ infer $\varphi \to \text{M-INT}_G(\psi)$ *(induction)*
M3 $\text{C-INT}_G(\varphi) \leftrightarrow \text{M-INT}_G(\varphi) \wedge \text{C-BEL}_G(\text{M-INT}_G(\varphi))$ *(collective intention)*

4.3 Awareness and Tuning Collective Attitudes

The definitions of collective attitudes from TEAMLOG allow to calibrate the strength of agents' *awareness*. In this case-study we adapt the strength of collective attitudes to the specific domain, and to set relevant group attitudes at the minimal level ensuring the effective team operation. Here the notion of awareness [9] comes in very useful. In general, the question regarding the level of awareness about each specific aspect of teamwork needs to be addressed.

The level of awareness is defined in [7, 9] by the strength of agents' beliefs. Instances of $awareness_G$ in definitions of motivational attitudes can be anything from $\emptyset$, through individual beliefs, different levels of E-BEL_G^k, to common belief C-BEL_G. Stronger levels of belief may increase communication, since beliefs have to be propagated. It has been argued that in ideal teamwork, $awareness_G$ is taken to be C-BEL_G [6]. Supposing that the communication medium is perfect, it is possible to attain this. In practical implementations (in an asynchronous, uncertain medium) common knowledge (C-KNOW_G) has been proven to be impossible to achieve [16], and common belief (C-BEL_G) under extremely restricted constraints [8]; usually only a finite approximation E-BEL_G^k can be achieved. Here follows the flexible scheme for collective intentions:

M3′ $\text{C-INT}_G(\varphi) \leftrightarrow \text{M-INT}_G(\varphi) \wedge awareness_G(\text{M-INT}_G(\varphi))$

4.4 Collective Commitment in TeamLog

After a group is constituted on the basis of collective intention, a *collective commitment* between the team members needs to be established. While a collective intention is an inspiration for team activity, the collective commitment reflects the concrete manner of achieving the goal. This concrete manner is provided by planning, and hinges on the allocation of actions according to a social plan. This allocation is concluded in bilateral (i.e. social) commitments to realize their individual actions. This way, our approach to collective commitments is plan-based. A bilateral commitment from agent i towards agent j to perform action α is represented as $\text{COMM}(i, j, \alpha)$; in this bilateral commitment is viewed as a primitive notion, but see [7] for its characterization and governing axiom.

Collective commitment schema
A flexible tuning schema for *collective commitments* is presented in [7]. In words, group G has a *collective commitment* to achieve goal φ based on social plan P ($\text{C-COMM}_{G,P}(\varphi)$) iff all of the following hold (in the corresponding definition below, parts between curly brackets may or may not be present): The group mutually intends φ (with or without being aware); moreover, successful execution of social plan P leads to φ ($cons(\varphi, P)$) (with or without the group being aware of this); and finally, for every one of the actions α from a plan P, there should be one agent in the group who is bilaterally committed to another agent in the group to fulfil the action ($\text{COMM}(i, j, \alpha)$) (with or without the group being aware of this):

$$\begin{aligned}&\text{C-COMM}_{G,P}(\varphi) \leftrightarrow \text{M-INT}_G(\varphi) \wedge \{awareness_G(\text{M-INT}_G(\varphi))\} \wedge \\ &cons(\varphi, P) \wedge \{awareness_G(cons(\varphi, P))\} \wedge \\ &\bigwedge_{\alpha \in P} \bigvee_{i,j \in G} \text{COMM}(i, j, \alpha) \wedge \{awareness_G(\bigwedge_{\alpha \in P} \bigvee_{i,j \in G} \text{COMM}(i, j, \alpha))\}\end{aligned}$$

Weak collective commitment
Different types of collective commitments related to different organization structures and environments have been introduced in terms of a 'tuning machine' [7]. One instantiation of the above scheme by tuning the 'awareness dial' is the *weak collective commitment* [7]. In this case, the team knows the overall goal, but does not know details of the plan: there is no collective awareness of the plan's correctness, so no $\text{C-BEL}_G(cons(\varphi, P)))$, even though there is a global awareness that things are under control. Weak collective commitment may be applicable in teams with a dedicated planner, who takes care of the plan's correctness $cons(\varphi, P)$:

$$\begin{aligned}&\text{W-COMM}_{G,P}(\varphi) \leftrightarrow \text{C-INT}_G(\varphi) \wedge cons(\varphi, P) \wedge \bigwedge_{\alpha \in P} \bigvee_{i,j \in G} \text{COMM}(i, j, \alpha) \\ &\wedge \text{C-BEL}_G(\bigwedge_{\alpha \in P} \bigvee_{i,j \in G} \text{COMM}(i, j, \alpha))\end{aligned}$$

5 Adjusting the TeamLog Definitions to the Case-Study

Why is a collective intention and a social plan still not enough to start team action in the case-study? Because agents may not feel *responsible* for their share. Thus, they need to commit to performing their part from the plan. Here, agents naturally commit to the coordinator, either directly or via an *UAV* acting as a 'middle manager'.

Now, what is the type of collective commitment fitting to the scenario? On the inner-team level, the *UAV* has the highest level of awareness within its team as it knows the entire social plan. There is no need for the others to know all the details. This corresponds to *weak collective commitment*, as defined above.

On the inter-team level, the controller has the highest awareness. The *UAVs* he oversees only need to know their part of the overall plan, and believe that it has been shared among all *UAVs*. The controller knows both the plan and task distribution.

5.1 *Organization Structure: Who Is Socially Committed to Whom?*

The coordinator is socially committed to achieving the main goal, with respect to the social plan as a whole. Other agents are committed to their share in the plan. The coordinator is committed towards itself and towards the relevant control authority, e.g. the national environmental agency for which he works.

The *UAV* is committed towards the coordinator with respect to achieving his part of the plan - keeping specified regions in a safe state.

The robots commit to perform their share to their leading *UAV*, which has the power to uncommit them. There is a clear hierarchy where the coordinator is the leader of all the groups, while the *UAVs* are 'middle-managers'.

5.2 *Minimal Levels of Group Intention and Awareness*

What are the minimal levels of awareness and group intention needed for the agents on both inner- and outer-team levels?

5.2.1 The Robots - Two Cases Are Applicable

1. They act only individually; this is the most limited (and economical) case;
2. They perform a limited form of cooperation, for example, they work together to clean up areas faster, or pitch in for other robots when these turn out to be unable to perform their part of the social plan.

The level of belief

1. In case 1, the robots need general belief about every group intention ($\text{E-BEL}_G(\text{E-INT}_G(\varphi))$) and about the distribution of the plan in bilateral commitments ($\text{E-BEL}_G(\bigwedge_{\alpha \in P} \bigvee_{i,j \in G} \text{COMM}(i, j, \alpha))$). This allows deliberation on actions of other robots and prevent them from doing all the work by themselves.
2. In case 2, E-BEL^2_G will be enough to allow deliberation about other robots' intentions and beliefs (especially $\text{E-BEL}^2_G(\text{E-INT}^2_G(\varphi))$. To see this, one may consider a pair of robots. With E-BEL^2_G, both robots have the same intention (this is the fact $\text{E-INT}_G(\varphi)$), believe they have the same intentions (the first-order belief $\text{E-BEL}_G(\text{E-INT}_G(\varphi))$), and believe that the other believes this (the second-order belief $\text{E-BEL}_G(\text{E-BEL}_G(\text{E-INT}_G(\varphi)))$). Therefore, the robots can reason about the beliefs and intentions of their partner.

The level of intention

1. In case 1, the robots need a general intention E-INT_G about the goals.
2. In case 2, E-INT^2_G will be enough to allow forming two-robot teams that are not competitive internally. (But see [6] for a counter-example showing that a two-

level intention is not sufficient to preclude competition among two-agent coalitions). If agents are supposed to be strictly cooperative, a two-level definition is also sufficient for larger teams: all agents intend to achieve the goal with the others included in their team.

Although robots sometimes individually compete for resources, in our application where fast real-time team reaction to dangers is substantial, we opt for strictly cooperative robots that use fixed protocols to load up on resources. The cleanup robots do not communicate with robots from other teams, and therefore do not need to have any beliefs, intentions and commitments about them.

5.2.2 The *UAVs*

The *UAVs* must sometimes work with each other. This requires at least E-BEL^2_G of other *UAVs'* intentions.

The level of belief - Within each team of *UAV* and robots, the *UAV* has the highest level of awareness, and acts as a coordinator. In order to facilitate this (make plans and reason correctly), it will require one level of belief more than its agents:

- in case 1 we require $\text{BEL}(UAV, \text{E-BEL}_G(\text{E-INT}_G(\varphi)))$ with regard to the inner-team group intention $\text{E-INT}_G(\varphi)$ as well as: $\text{BEL}(UAV, \text{E-BEL}_G(\bigwedge_{\alpha \in Cleanup} \bigvee_{i,j \in G} \text{COMM}(i,j,\alpha)))$,
- in case 2 we require $\text{BEL}(UAV, \text{E-BEL}^2_G(\text{E-INT}^2_G(\varphi)))$ with respect to the level of inner-team group intention $\text{E-INT}^2_G(\varphi)$ as well as: $\text{BEL}(UAV, \text{E-BEL}^2_G(\bigwedge_{\alpha \in Cleanup} \bigvee_{i,j \in G} \text{COMM}(i,j,\alpha)))$.

The level of intention - Within the team, the *UAV* must make sure that all agents are motivated to do their tasks. Therefore:

- in case 1 we require $\text{INT}(UAV, \text{E-INT}_G(\varphi))$ with regard to the inner-team group intention $\text{E-INT}_G(\varphi)$,
- in case 2 we require $\text{INT}(UAV, \text{E-INT}^2_G(\varphi))$ with regard to the level of inner-team group intention $\text{E-INT}^2_G(\varphi)$.

5.2.3 The Coordinator

The level of belief - One extra level of belief allows the coordinator introspection and reasoning about the joint effort of all *UAVs*. Therefore, since teams are cooperative in a limited way, we have $\text{BEL}(coordinator, \text{E-BEL}^2_G(\text{E-INT}^2_G(\varphi)))$ with respect to every group intention $\text{E-INT}^2_G(\varphi)$ as well as: $\text{BEL}(coordinator, \text{E-BEL}^2_G(\bigwedge_{\alpha \in Cleanup} \bigvee_{i,j \in G} \text{COMM}(i,j,\alpha)))$.

The level of intention - Similarly, the coordinator has one level of intention more than the *UAVs* it manages, therefore we have $\text{INT}(coordinator, \text{INT}^2_G(\varphi)))$.

Commands from the coordinator overrule temporary contracts between teams. He does not only know the plan, but also keeps track of all relevant environmental conditions. We assume that even in the safe situation, the robots, *UAVs* and the pilot are prepared to take action at any moment.

5.3 Complexity of the Language without Collective Attitudes

It seems that in the environmental case-study, the language used is richer than propositional modal logic. Fortunately, we can reduce most of the relevant part to a fixed finite number of propositional atoms (that may be combined and be the subject of attitudes), based on finitely many predicates and constants, as follows:

- a fixed number of relevant environmental states;
- a fixed number of pre-named locations;
- a fixed finite number of agents and teams;
- a fixed finite number of other objects (liquids, solids, catalyst, helicopter);
- a fixed number of relevant thresholds $n_1, n_2, n_3, \varepsilon_1, \varepsilon_2$.

The only possible source of unbounded complexity is the use of continuous intervals and real-valued functions $f_1, f_2, f_3, \mathit{fit}$. This can probably be simplified by using discretization.

6 Conclusion

In the case-study we have shown how to implement teamwork within a strictly cooperative, but still heterogenous group of agents in the TEAMLOG formalism. The heterogeneity is taken seriously here, as advocated in [13]. Natural differences in agents' shares, opportunities and capabilities when acting together, have been additionally reflected in different levels of agents' awareness about various aspects of their behaviour. The study dealt especially with cooperation and coordination. Having very generic definitions of common motivational and informational attitudes in TEAMLOG, it is challenging to choose a proper level of their complexity. We have shown that this is possible, by illustrating how to tailor complex definitions of intentions and commitments to a specific environment. For lack of space, not all the essential aspects of teamwork have been shown. Our focus was on building beliefs, intentions and, finally, commitments of all agents involved in teamwork on an adequate, but still minimal level. This way a bridge between theory and practice of teamwork has been effectively constructed for a specific application.

Future work will be to embed TEAMLOG into a form of approximate reasoning suitable for modeling perception, namely *similarity-based approximate reasoning*, which has intuitive semantics compatible with that of TEAMLOG [2, 5].

Acknowledgements. This research is supported by the Polish MNisW grant N N206 399334.

References

1. Aldewereld, H., van der Hoek, W., Meyer, J.-J.C.: Rational teams: Logical aspects of multi-agent systems. Fundamenta Informaticae 63 (2004)
2. Doherty, P., Dunin-Kęplicz, B., Szałas, A.: Dynamics of approximate information fusion. In: Kryszkiewicz, M., Peters, J.F., Rybiński, H., Skowron, A. (eds.) RSEISP 2007. LNCS (LNAI), vol. 4585, pp. 668–677. Springer, Heidelberg (2007)
3. Doherty, P., Granlund, G., Kuchcinski, K., Nordberg, K., Sandewall, E., Skarman, E., Wiklund, J.: The WITAS unmanned aerial vehicle project. In: Proc. of the 14th European Conference on Artificial Intelligence, pp. 747–755 (2000)
4. Doherty, P., Łukaszewicz, W., Skowron, A., Szałas, A.: Knowledge Representation Techniques. A Rough Set Approach. Studies in Fuzziness and Soft Computing, vol. 202. Springer, Heidelberg (2006)
5. Dunin-Kęplicz, B., Szałas, A.: Towards approximate BGI systems. In: Burkhard, H.-D., Lindemann, G., Verbrugge, R., Varga, L.Z. (eds.) CEEMAS 2007. LNCS (LNAI), vol. 4696, pp. 277–287. Springer, Heidelberg (2007)
6. Dunin–Kęplicz, B., Verbrugge, R.: Collective intentions. Fundamenta Informaticae 51(3), 271–295 (2002)
7. Dunin–Kęplicz, B., Verbrugge, R.: A tuning machine for cooperative problem solving. Fundamenta Informaticae 63, 283–307 (2004)
8. Dunin-Kęplicz, B., Verbrugge, R.: Creating common beliefs in rescue situations. In: Dunin-Keplicz, B., Jankowski, A., Skowron, A., Szczuka, M. (eds.) Proc. of Monitoring, Security and Rescue Techniques in Multiagent Systems (MSRAS), Berlin. Advances in Soft Computing, pp. 69–84. Springer, Heidelberg (2005)
9. Dunin-Kęplicz, B., Verbrugge, R.: Awareness as a vital ingredient of teamwork. In: Stone, P., Weiss, G. (eds.) Proc. of the Fifth Int. Joint Conference on Autonomous Agents and Multiagent Systems (AAMAS 2006), pp. 1017–1024. ACM Press, New York (2006)
10. Dziubiński, M.: Complexity of the logic for multiagent systems with restricted modal context. In: Dunin-Kęplicz, B., Verbrugge, R. (eds.) Proc. of the Third Int. Workshop on Formal Approaches to Multi-Agent Systems, FAMAS 2007, pp. 1–18. Durham University (2007)
11. Dziubiński, M., Verbrugge, R., Dunin–Kęplicz, B.: Complexity issues in multiagent logics. Fundamenta Informaticae 75(1-4), 239–262 (2007)
12. Fagin, R., Halpern, J., Moses, Y., Vardi, M.: Reasoning about Knowledge. MIT Press, Cambridge (1995)
13. Gold, N. (ed.): Teamwork. Palgrave McMillan, Basingstoke (2005)
14. Grant, J., Kraus, S., Perlis, D.: Formal approaches to teamwork. In: Artemov, S., et al. (eds.) We Will Show Them: Essays in Honour of Dov Gabbay, vol. 1, pp. 39–68. College Publications, London (2005)
15. Grosz, B.J., Kraus, S.: Collaborative plans for complex group action. Artificial Intelligence 86(2), 269–357 (1996)
16. Halpern, J., Moses, Y.: Knowledge and common knowledge in a distributed environment. Journal of the ACM 37, 549–587 (1990)
17. Kleiner, A., Prediger, J., Nebel, B.: RFID technology-based exploration and SLAM for search and rescue. In: Proc. of the IEEE/RSJ Int. Conference on Intelligent Robots and Systems (IROS 2006), Bejing, pp. 4054–4059 (2006)
18. Levesque, H., Cohen, P., Nunes, J.: On acting together. In: Proc. Eighth National Conference on AI (AAAI 1990), pp. 94–99. MIT Press, Cambridge (1990)
19. Meyer, J.-J.C., van der Hoek, W.: Epistemic Logic for AI and Theoretical Computer Science. Cambridge University Press, Cambridge (1995)

20. Parikh, R., Krasucki, P.: Levels of knowledge in distributed computing. Sadhana: Proc. of the Indian Academy of Sciences 17, 167–191 (1992)
21. Rao, A., Georgeff, M.: Modeling rational agents within a BDI-architecture. In: Fikes, R., Sandewall, E. (eds.) Proc. of the Second Conference on Knowledge Representation and Reasoning, pp. 473–484. Morgan Kaufmann, San Francisco (1991)
22. Sycara, K., Lewis, M.: Integrating intelligent agents into human teams. In: Salas, E., Fiore, S. (eds.) Team Cognition: Understanding the Factors that Drive Process and Performance, Washington (DC), pp. 203–232. American Psychological Association (2004)
23. Tambe, M.: Teamwork in real-world, dynamic environments. In: Tokoro, M. (ed.) Proc. Second Int. Conference on Multi-Agent Systems, pp. 361–368. AAAI Press, Menlo Park (1996)
24. Wooldridge, M., Jennings, N.: Cooperative problem solving. Journal of Logic and Computation 9, 563–592 (1999)

Appendix

In all plans we assume we start from the base B where neutralizers are stored.

The social plan $\langle D_1 R\rangle$

Goal $\psi_1(L_1)$: to apply liquid L_1 on all areas contaminated with poison X_1.

{Assumption: One portion of L_1 neutralizes poison X_1 at a single location.}
while *contaminated-area* $\neq$ *emptyset* do
begin
  $A := calculate(UAV, \{rob_i\})$; {*UAV* finds region A for rob_i to clean up}
  $get(rob_i, L_1, B)$; {rob_i retrieves a tank with liquid L_1 from location B}
  $path := get\_path(UAV, rob_i, B, A)$; {$rob_i$ requests a path to follow}
  $move(rob_i, path)$; {rob_i moves from location B to location A}
  $pour(rob_i, L_1, A)$;
  contaminated-area := *contaminated-area* $\setminus A$;
  $return\_path := get\_path(UAV, rob_i, A, B)$;
  $move(rob_i, return\_path)$;
end.

The social plan $\langle D_1 N\rangle$

Goal $\psi_2(L_2)$: to spray liquid L_2 on areas contaminated with poison X_1.

{Assumption: One portion of L_2 neutralizes poison X_1 at a single location.}
{Assumption: The helicopter can transport k portions of liquid L_2.}
while *contaminated-area* $\neq$ *emptyset* do
begin
  $request(UAV, coordinator, pilot, \psi(L_2)$;
  $confirm(pilot, UAV, coordinator, \psi(L_2)$;

$request(pilot, UAV, list_1, k)$;
$send(UAV, pilot, list_1)$; {$list_1$ has at most k contaminated areas}
$upload(helicopter, L_2)$; {*pilot* retrieves liquid L_2}
$take\text{-}off(helicopter, B)$; {*pilot* takes off from location B}
do $\langle plan\text{-}for\text{-}spraying(helicopter, L_2, l)\rangle$; {*pilot* sprays L_2 using his own invented plan}
$confirm(pilot, UAV, done(plan\text{-}for\text{-}spraying(helicopter, L_2, l))$;
$contaminated\text{-}area := contaminated\text{-}area \setminus list_1$;
$landing(helicopter, B)$;
$free(pilot, coordinator)$;
end.

The social plan $\langle D_2R\rangle$

Goal $\psi_3(S_1, K_1)$: to spread solid S_1 on all areas contaminated with poison X_2, followed by applying catalyst K_1 to all areas where S_1 is present.

{Assumption: One portion of S_1 and K_1 neutralize poison X_2 at a single location.}
while *contaminated-area* $\neq$ *emptyset* do
begin
$A := calculate(UAV, \{rob_i, rob_j\})$;
begin_parallel {operations are done in parallel}
a plan similar to $\langle D_1R\rangle$, but using S_1;
||
$wait\_for(transporting(rob_i, S_1, A))$; {$rob_j$ waits until rob_i is on the way to A}
$get(rob_j, K_1, B)$;
$path := get\_path(UAV, rob_j, B, A)$;
$move(rob_j, path)$;
$wait\_for(spread(S_1, A))$; {rob_j waits for *someone* to spread S_1 in A}
$pour(rob_j, K_1, A)$;
$return\_path := get\_path(UAV, rob_j, A, B)$;
$move(rob_j, return\_path)$;
end_parallel
$contaminated\text{-}area := contaminated\text{-}area \setminus A$;
end.

Plan $\langle D_2N\rangle$ is similar to $\langle D_1R\rangle$

Hot Topic Detection Based on Opinion Analysis for Web Forums in Distributed Environment

Changjun Hu, Yu Weng, Xuechun Zhang, and Chengxiu Xue

Abstract. For improving the cooperative detection ability of hot topics, this paper analyzes web topic spreading features and proposes a hot topic detection model based on opinion analysis for web forums in distributed environment (named TOAM). The model not only evaluates the topic opinion impacts of the single web forum, but also cooperatively schedules the opinion information of hot topics among different network domains. TOAM monitors the topics opinion of each web forum and generates the hot topics of local network domain periodically. Through scheduling the local hot topics between different network domains, the model effectively improves the ability of topic spreading analysis and optimizes the local topics information database. To validate the performance, the experiments on the data corpus about "Campus Alert Network Culture" demonstrate that TOAM has higher application validity and practicality of hot topic detection for web forums in distributed environment.

1 Introduction

As a kind of web text mining technology for public opinion monitoring, web topic opinion analysis has attracted more and more attention, and provides a powerful data basis for the decision analysis of related supervision department.

Web forum is a popular information communication platform of Internet. Each user could publish personal topics or comment the others. Besides the topic body text, every topic also consists of a series of comments and is ordered by the published time. Through analyzing the amount of comments and reviews periodically, the hot degree of web topic could be evaluated simply. However, due to neglecting the impacts of topic opinion, the method could not effectively analyze the supportive or negative status of the comments [1, 2]. It will influence the authenticity of web

Changjun Hu
University of Science and Technology Beijing, China
e-mail: huchangjun@ies.ustb.edu.cn

G.A. Papadopoulos and C. Badica (Eds.): Intelligent Distributed Computing III, SCI 237, pp. 101–110.
springerlink.com © Springer-Verlag Berlin Heidelberg 2009

topic evaluation. On the other hand, some traditional topic opinion analysis methods [3, 4] merely focus on the topic opinion monitoring of inner network domain (a network domain represents a web forum), and ignore the cooperative schedule of topic opinion information among inter-network domains. It could not meet the new demand for the generation and fast spreading of web topics. Oriented to the deficiencies above, a hot topic detection model based on opinion analysis for web forums in distribution environment (named TOAM) is proposed. The model not only considers the topic opinion influence of inner network domain, but also takes the spreading impacts of inter network domains into calculation. Through monitoring the amount of comments, reviews and the comment opinion of each topic periodically, TOAM calculates the topic spreading degree and evaluates the local hot topics of each network domain. By means of cooperatively scheduling the opinion information of local hot topics among different network domains, TOAM improves the analytical ability of topic spreading and provides the data supports for the fast detection of burst topics in distributed environment.

This paper is organized as follows. Section 2 outlines and analyzes the previous approaches of topic detection. In section 3, some problems and the general process are described. In section 4, each part of TOAM is presented in details. In section 5, experimental results on the corpus of "Campus Alert Network Culture" are given. Finally section 6 concludes the work with some possible extensions.

2 Related Work

Recent years, lots of efforts have been made to the research of text opinion analysis. Pang [3] (2002) classified the documents by sentiment analysis and showed that machine learning approaches on sentiment classification do not perform as well as that on traditional topic-based categorization at document level. Soo-Min Kim and Eduard Hovy [4] (2006) describe a sentence-level opinion analysis system. The experiment based on MPQA (Wiebe et al. [5], 2005) and TREC (Soboroff and Harman [6], 2003) showed that automatic method for obtaining opinion-bearing words can be used effectively to identify opinion-bearing sentences. Lun-Wei Ku, Hsiu-Wei Ho and Hsin-Hsi Chen [7] (2006) selected TREC, NTCIR [8], and some web blogs as the opinion information sources and proposes an algorithm for opinion extraction at word, sentence and document level. Ruifeng Wong and et al. [9](2008) Proposed an opinion analysis system based on linguistic knowledge which is acquired from small-scale annotated text and raw topic-relevant webpage. The system used a classifier based on support vector machine to classify the opinion features, identify opinionated sentences and determine their polarities. Veselin and Claire [10] (2008) presented a novel method for general-purpose opinion topic identification and evaluate the validity of this approach by the MPQA corpus.

These technologies above could be applied in the comments opinion analysis in single web forum successively. However, since neglecting the mutual influence of topic spreading among different network domains, they could not realize the fast detection of burst topics in distributed environment effectively. TOAM comprehensively considers the impacts of topic opinion in multi-web forums. Through cooperatively scheduling the topic opinion information of local network domain, TOAM greatly strengthens the comprehensive analysis ability of topic spreading.

3 Problem Description and General Process

Different from the topic detection models based on the topic content analysis (e.g. some high frequency or sensitive words), TOAM tracks the comment opinion trend of each topic, calculates the topic spreading degree in local network domain, and improves analysis ability of inter network domains by scheduling the opinion information of local hot topics. In fig.1, the general process of TOAM is given.

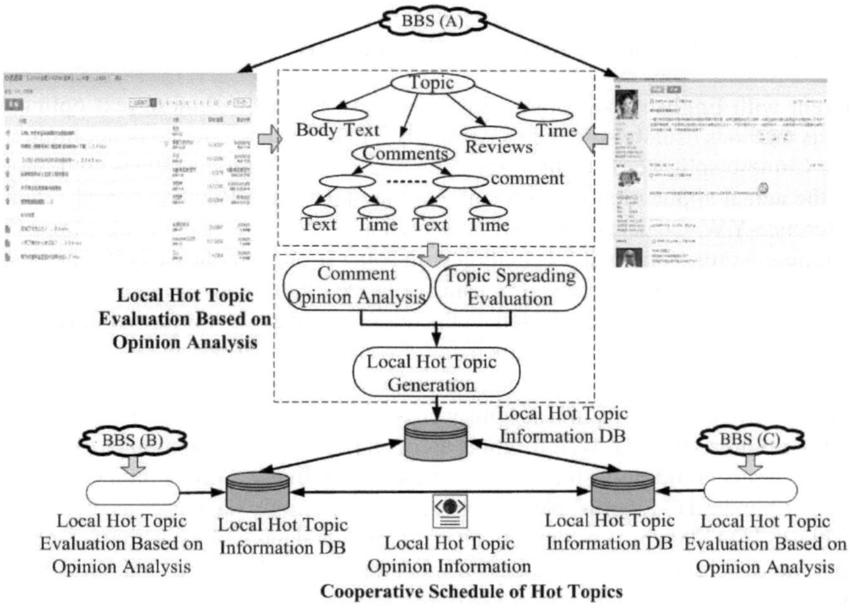

Fig. 1 The general process of TOAM

Comment Opinion Analysis. The aim of comment opinion monitoring is to analyze the attitudes of reviewers to topics (e.g. positive, negative or neutral), and evaluate the hot degree of topics more precisely. In calculation process, TOAM considers the comprehensive influence of the length of text, SO (Semantic Orientation) of characters, and distribution status of opinion characters and identifies the opinion of web forum comments.

Topic Spreading Evaluation. As the basis of hot topic detection, TOAM evaluates the popularity of topics and calculates the topic spreading degree by monitoring the amount of reviews, comments and the comment opinion of each topic periodically.

Hot Topic Cooperative Schedule. The hot topic generation usually does not merely rely on the occurring status in local network domain, but also is affected by the spreading impacts of multi-network domains. TOAM simulates the hot topic spreading process, and improves the topic detection ability by scheduling the opinion information of other network domains.

4 Hot Topic Detection Based on Comment Opinion Analysis for Web Forums

4.1 Topic Spreading Evaluation Based on Comment Opinion Analysis

4.1.1 Comment Opinion Analysis for Web Forums

Different with English text opinion analysis, the traditional Chinese text opinion analysis methods usually split the words firstly [11]. However, due to being limited by the Chinese split words technology, the precision of opinion analysis could not meet the actual application requirement. On the other hand, Raymond W.M. Yuen and Terence Y.W. Chan [12] (2004) presented a general strategy for inferring SO for Chinese words from their association with some strongly-polarized morphemes. The experimental results proved that using polarized morphemes is more effective than strongly-polarized words. Based on the reasons above, TOAM further improves the calculation model [8] (Liu-Wei Ku and Yu-Ting Liang 2006) and evaluates the text opinion by analyzing the semantic orientation of Chinese characters. The model avoids the dependency of Chinese split words technology and strengthens the precision of opinion analysis. Web forum is an open information communication environment. Users could publish or comment the topics freely. Some comments may consist of hundreds of words. Nevertheless, some ones only have dozens of words. In Table 1, TOAM calculates the semantic orientation degree of each character and adopts the different opinion calculation methods for the long text and short text, respectively.

Given a paragraph of comment text T,C_i represents the $i-th$ character of T, N_{count} is the amount of words of T. fp_{ci} and fn_{ci} stands for the occurring frequency of C_i in positive and negative corpus. Sci denotes the opinion degree of C_i. $OpDensity(S_{ci})$ is the distribution density of positive characters. $N(S_{c+})$is the amount of positive characters in T. m and n denote the total number of unique characters in positive and negative words corpus. $Th_{LongText}$ is the boundary threshold of long text and short text.

In Step2, to balance the quantitative difference of positive and negative words corpus, TOAM normalizes the weights of C_i as positive and negative characters and determines the semantic orientation of C_i by formula 3. If it is a positive value,

Table 1 Comment Opinion Analysis Algorithm for Web Forum

Title	Operation	Description
Input	Comment Text T,C_i, N_{count}, $Th_{LongText}$, m, n	
Output	Semantic Orientation of S(T)	
Step1	S(T)=0;	Initialize Inputs
Step2	Calculate the SO of each character	transverse all the characters of T
	For each character Ci	
	$$P_{ci}=\frac{fp_{ci}/\sum_{j=1}^{n}fp_{cj}}{(fp_{ci}/\sum_{j=1}^{n}fp_{cj})+(fn_{ci}/\sum_{j=1}^{m}fn_{cj})} \quad (1)$$	where Pci and Nci denote the weights of ci as positive and negative characters.
	$$N_{ci}=\frac{fn_{ci}/\sum_{j=1}^{m}fn_{cj}}{(fp_{ci}/\sum_{j=1}^{n}fp_{cj})+(fn_{ci}/\sum_{j=1}^{m}fn_{cj})} \quad (2)$$	
	$$S_{ci}=P_{ci}-N_{ci} \quad (3)$$	SO of character
Step3	Evaluate the Comment Opinion of T	
	if($N_{count} \leq Th_{LongText}$)	judge the length
	then	T is short text
	$$S(T)=\sum_{i=1}^{Ncount} S_{ci} \quad (4)$$	
	else	T is long text
	$$S(T)=\sum_{j=1}^{N(Sc+)} S_{cj}^{+}\times OpDensity(Sc^{+})-\sum_{j=1}^{N(Sc-)} S_{cj}^{-}\times OpDensity(Sc^{-}) \quad (5)$$	
Step4	Return S(T);	Output SO of T

then this character appears more times in positive words; and vice versa. A value close to 0 means that it is not a sentiment character or it is a neutral sentiment character. In Step3, TOAM analyzes the data feature, and calculates the SO of T by two different methods (long text and short text). If the length of T is less than the threshold $Th_{LongText}$, the opinion of target text is determined by the sum of semantic orientation degree of Chinese characters. Otherwise, the length of T is greater than the threshold, as shown in formula 5, the SO of T is evaluated by comprehensively considering the mutual influence of the semantic orientation of characters and the opinion distribution density. TOAM firstly sums all the semantic orientation degree of subjective characters. Then, the model evaluates the opinion distribution density of T, including the positive density and the negative one.

4.1.2 Topic Spreading Evaluation

Given a topic of web forum Tpt and $t+1$ represents two times of testing. As shown in formula 6, TOAM evaluates the topic spreading degree by analyzing the number variation of two times.

$$HotTopicDegree(Tp,\triangle t) = \|TpSpread(Tp,t+1) - TpSpread(Tp,t)\| \quad (6)$$

Here, $TpSpread(Tp,t+1)$ and $TpSpread(Tp,t)$ denote the topic spreading degree of two times, respectively. Formula 7 and 8 further expand the calculation process. In formula 7 the topic spreading degree at $t+1$ comprehensively considers the impacts of comment opinion and the diffusion effect of reviewing.

$$TpSpread(Tp,t+1) = \frac{Comment_{t+1}^{+} - Comment_{t+1}^{-}}{Comment_{t+1}} \times V_{t+1} \quad (7)$$

$$TpSpread(Tp,t) = \frac{Comment_{t}^{+} - Comment_{t}^{-}}{Comment_{t}} \times V_{t} \quad (8)$$

Where, $Comment_{t+1}^{+}$ and $\mathrm{Comment}_{t+1}^{-}$ represent the positive and negative comments at $t+1$. $Comment_{t+1}$ and V_{t+1} denote the amount of comments and reviews at the same time.Then,through monitoring the variation of positive comment rate, three cases are discussed as follows:

1. With the increment of positive comment rate, the topic spreading degree is increased. To be specific, it could consist of three possibilities.
   a. $TpSpread(Tp,t+1) > 0, TpSpread(Tp,t) > 0$ and $TpSpread(Tp,t+1) > TpSpread(Tp,t)$
   b. $TopicSpreading(Tp,t+1) > 0, TopicSpreading(Tp,t) < 0$
   c. $TpSpread(Tp,t+1) < 0, TpSpread(Tp,t) < 0$ and $\|TpSpread(Tp,t+1)\| < \|TpSpread(Tp,t)\|$
2. With the reduction of positive comment rate, the topic spreading degree is decreased.
   a. $TpSpread(Tp,t+1) > 0, TpSpread(Tp,t) > 0$ and $TpSpread(Tp,t+1) < TpSpread(Tp,t)$
   b. $TopicSpreading(Tp,t+1) < 0, TopicSpreading(Tp,t) > 0$
   c. $TpSpread(Tp,t+1) < 0, TpSpread(Tp,t) < 0$ and $\|TpSpread(Tp,t+1)\| > \|TpSpread(Tp,t)\|$
3. If the positive comment rates of two times are equal, the topic spreading degree keeps invariant.
   a. $TopicSpreading(Tp,t+1) = TopicSpreading(Tp,t)$

4.2 Hot Topics Cooperative Schedule

The local hot topic information generation is usually limited by the difference of cultures or regions. TOAM considers the topic spreading feature of Internet and improves the fast detection ability of burst topics in distributed environment by cooperatively scheduling the local hot information of multi-network domains.

Given a local hot topic α. $Domain_A$ and $Domain_B$ represent the topics sender and receiver network domain. $TIDB_{DomainA}$ and $TIDB_{DomainB}$ denote the local topics information database of $Domain_A$ and $Domain_B$, respectively. $\triangle t$ is the time interval of cooperative scheduling. β is any hot topic of $TIDB_{DomainB}$, and Th is the threshold of local hot topic. The cooperative schedule algorithm of hot topics is as follows:

Table 2 Hot Topic Cooperative Scheduling Algorithm

Title	Operation	Description
Input	α and β, $TIDB_{DomainA}$, $TIDB_{DomainB}$, $\triangle t$,Th	
Output	Target topics information database $TIDB_{DomainB}$	
Step1	$SendList_{DomainA}, Receive_{ListDomainB} \leftarrow \emptyset$;	Initialize Inputs
Step2	Local Hot Topics Distribution	
	For each topic α of $TIDB_{DomainA}$	Transverse all topics
	if($TpSpread(\alpha, \triangle t) \geq Th$)	Local hot topics evaluation
	then insert α into $Sendlist_{DomainA}$;	prepare to be scheduled by other network domains
Step3	Hot Topic Scheduling Periodically	Retrieving the hot topic α
	Insert α into $ReceivedList_{DomainB}$	
	For each topic β of $TIDB_{DomainB}$	Transverse all topics
	if($\alpha.title$ is similar $\beta.title$)	Comparing the similarity by analyzing the topic titles
	then Upgrade B.matching of TIDBDomainB	If they are similar topics, Update the topic frequency
	else Insert α into $TIDB_{DomainB}$	If they are not similar topics, insert new topic into $TIDB_{DomainB}$
Step4	Return $TIDB_{DomainB}$;	Output

In step2, TOAM calculates the spreading degree of each topic in local network domain, and generates the local hot topics. In step3, the model retrieves the local hot topics and queries the similar topic information of target domain network by the Top-k Structural Similarity Search over XML Documents [13]. If there exists a similar topic in target network domain, TOAM updates the topic occurring frequency. Otherwise, inserting the new one into target hot topics database.

5 Experiment

5.1 Experimental Corpus

We prepared an experimental corpus for the campus alert network culture (named CANC) that spans from March 1 to May 30, 2008 consisting of over 70,000 Web pages and constructed two web forums to evaluate the performance of TOAM. Four kinds of information sources are included in CANC: two Black Board Systems, one Blog Website and one Website for Campus Psychology Web Site. Table 3 presents the information of CANC corpus in details.

Table 3 Information on CANC corpus

Source	Type	Pages	Av(characters)/page
www.smth.edu.cn	BBS	19532	83.526
bbs.ccnu.edu.cn	BBS	20765	72.118
www.xiaonei.com	Blog	20451	108.651
student.ustb.edu.cn/page/xlkc/xlkc.asp	BBS	11419	108.332

5.2 Experimental Results

Testing 1: The Validity Testing of Comment Opinion Analysis To compare the validity of opinion analysis, we took a comparison testing among OSNB [7], Morpheme [12] and TOAM. The testing set (40,000 pages from CANC) is divided into long text corpus and short text corpus, respectively. Through calculating the Precision (P), Recall(R) F-measure (F) and average execution time(T)the performance of three methods will be evaluated.

Table 4 Comparison Analysis Results

		Long Text			Short Text	
	OSNB	Morpheme	TOAM	OSNB	Morpheme	TOAM
Precision	59.87%	73.85%	79.49%	54.22%	70.23%	72.55%
Recall	78.48%	74.11%	82.24%	65.79%	75.03%	74.28%
F-measure	67.92%	73.98%	80.84%	54.55%	72.55%	73.40%
T	3.5s	2.1s	1.6s	3.1s	1.5s	1.4s

From the results of the comparison testing, we noticed that, TOAM could effectively adapt the different features of long text and short text, and improve the validity and practicability of opinion analysis.For the short text corpus, TOAM avoided the limit of Chinese split words technology and had better performance than OSNB (P 54.22% , R 65.79%). TOAM comprehensively considers the mutual influence

Table 5 Alert Topics Statistics

	T1	T2	T3	T1	T2	T3
UP	15(+[a]),8(-[b]),26(all[c])	35(+),12(-),57(all)	64(+),21(-),96(all)	53(r[d])	84(r)	119(r)
BH	14(+),10(-),33(all)	31(+),16(-),72(all)	43(+),22(-),85(all)	42(r)	93(r)	133(r)
WS	15(+),12(-),31(all)	24(+),19(-),46(all)	36(+),28(-),73(all)	56(r)	121(r)	88(r)
CL	8(+),5(-),33(all)	19(+),14(-),54(all)	22(+),16(-),68(all)	72(r)	163(r)	135(r)
CV	12(+),11(-),33(all)	31(+),17(-),58(all)	35(+),37(-),79(all)	89(r)	86(r)	94(r)
CE	6(+),8(-),21(all)	14(+),11(-),32(all)	23(+),26(-),57(all)	74(r)	78(r)	82(r)

[a] Av Comments(+)/Page [b] Av Comments(-)/Page [c] AvComments(all) [d] Av Reviews/Page

(the semantic orientation of Chinese characters, the distribution density of positive and negative characters, and the opinion probability of target text). For the long text corpus, the precision of TOAM (P 79.49%, R 82.24%, 1.6 s) is much than OSNB (P 59.87% , R 78.48% , 3.5 s) and Morpheme (P 73.85% , R 74.11% , 2.1s).

Testing 2: The Hot Topics Cooperative Scheduling Ability Testing. Two web forums were constructed to evaluate the topics cooperative scheduling ability of TOAM. We divided the process of testing into three phases (t1, t2 and t3) and input different kinds of alert topics into two forums. Web Forum A: Unhealthy Psychology(UP, 311 topics), Bad Habits(BH, 165 topics) and Warning Speeches(WS, 264 topics); Web Forum B:Corruptible Learning(CL, 242 topics), Campus Violence(CV, 202 topics) and Campus Eroticism(CE, 153 topics). Then, through monitoring the number variation of alert topics in each local information database, the hot topics cooperative scheduling algorithm could be validated. In table.5, the cooperative scheduling results of six alert topics are illustrated.

The testing results showed that, with the local topics cooperative scheduling of inter-network domains, TOAM avoids the limits of local topics and strengthens topic spreading detection ability in distributed environment. In fig.2, TOAM schedules the

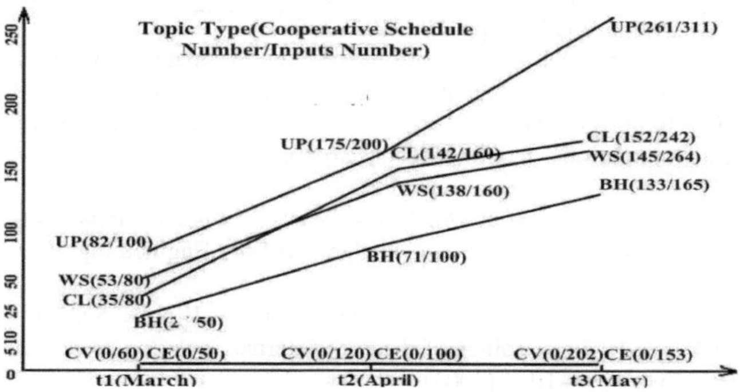

Fig. 2 Cooperative Scheduling Testing Results

local hot topics among multi-network domains by the actual spreading status of topics. Along with the increasing of the positive ratios, the cooperative scheduling number of UP and BH were rising gradually (Cooperative Scheduling Number/Inputs Number t1:80/100, 24/50; t2:175/200, 142/160 and t3:261/311, 151/242). Different with UP and BH, since the topic spreading degrees were not obvious, there was not any local hot topic of CV and CE to be transferred. Within the time interval of t1 and t2, the variation trends of WS and CL were same to UP and BH. However, following with the positive ratio is declining at t3, the rising slope of cooperative scheduling decreases.

References

1. Manquan, Y., Weihua, L., et al.: Research on Hierarchical Topic Detection in Topic Detection and Tracking. Journal of Computer Research And Development 43(3), 489–495 (2006)
2. Li, Y., Meng, X., Li, Q., Wang, L.: Hybrid Method for Automated News Content Extraction from the Web. In: proceeding of 7th International Conference on Web Information Systems Engineering, pp. 327–338 (2006)
3. Pang, B., Lee, L.: Vaithyanathan, Sentiment classification using machine learning techniques. In: Proceedings of the 2002 Conference on EMNLP, pp. 79–86 (2002)
4. Soo-Min, K., Eduard, H.: Extracting Opinions, Opinion Holders, and Topics Expressed in Online News Media Text. In: Proceedings of the ACL Workshop on Sentiment and Subjectivity in Text, pp. 1–8 (2006)
5. Wiebe, J., Wilson, T., Cardie, C.: Annotating expressions of opinions and emotions in language. Language Resources and Evaluation (2005)
6. Soboroff, I., Harman, D.: Overview of the TREC 2003 novelty track. In: The Twelfth Text REtrieval Conference, National Institute of Standards and Technology, pp. 38–53 (2003)
7. NTCIR Project. (Available via DIALOG, 2009), `http://research.nii.ac.jp/ntcir/index-en.html` (Cited 10 June 2009)
8. Ku, L.-W., Liang, Y.-T., Chen, H.-H.: Opinion extraction, summarization and tracking in news and blog Corpora. In: Proceedings of AAAI-2006 Spring Symposium on Computational Approaches to Analyzing Weblogs, pp. 100–107 (2006)
9. Xu, R., Wong, K.-F., et al.: Learning Knowledge from Relevant Webpage for Opinion Analysis. Web Intelligence and Intelligent Agent Technology, pp. 307–313 (2008)
10. Stoyanov, V., Cardie, C.: Topic Identification for Fine-Grained Opinion Analysis. In: Proceedings of the 22nd International Conference on Computational Linguistics, pp. 817–824 (2008)
11. Abbasi, A., Chen, H., Thoms, S., Fu, T.: Affect Analysis of Web Forums and Blogs Using Correlation Ensembles. IEEE Transactions on Knowledge and Data Engineering 20(9) (September 2008)
12. Yuen, R.W.M., Chan, T.Y.W., et al.: Morpheme-based Derivation of Bipolar Semantic Orientation of Chinese Words. In: Proceedings of the 20th International Conference on Computational Linguistics, pp. 1008–1014 (2004)
13. Xie, T., Sha, C., et al.: Approximate Top-k Structural Similarity Search over XML Documents. Frontiers of WWW Research and Development, 319–330 (2006)

A Case Study on Availability of Sensor Data in Agent Cooperation

Christian Johansson, Fredrik Wernstedt, and Paul Davidsson

Abstract. Multi-agent cooperation can in several cases be used in order to mitigate problems relating to task sharing within physical processes. In this paper we apply agent based solutions to a class of problems defined by their property of being predictable from a macroscopic perspective while being highly stochastic when viewed at a microscopic level. These characteristic properties can be found in several industrial processes and applications, e.g. within the energy market where the production and distribution of electricity follow this pattern. We evaluate and compare the performance of the agent system in three different scenarios, and for each such scenario it is shown to what degree the optimization system is dependent on the level of availability of sensor data.

1 Introduction

Schemes for sustaining cooperative behavior among agents are often dependent on a certain level of communication in order to establish and maintain a reciprocal sense of trust. However, in real-life applications it is not always possible to uphold the desired level of availability and quality of data being communicated among the agents, thus causing suboptimal cooperative behavior.

Information sharing, i.e. communication and its effect on overall performance is a well established area and has been studied by several researchers [5, 6, 13]. Also,

Christian Johansson
Blekinge Institute of Technology, 372 25, Ronneby, Sweden
e-mail: christian.johansson@bth.se

Fredrik Wernstedt
Blekinge Institute of Technology, 372 25, Ronneby, Sweden
e-mail: fredrik.wernstedt@bth.se

Paul Davidsson
Blekinge Institute of Technology, 372 25, Ronneby, Sweden
e-mail: paul.davidsson@bth.se

G.A. Papadopoulos and C. Badica (Eds.): Intelligent Distributed Computing III, SCI 237, pp. 111–120.
springerlink.com © Springer-Verlag Berlin Heidelberg 2009

the area of multi-sensor networks and sensor data quality and fusion has received a fair amount of interest [3, 9, 7]. However, the quality of information in combination with information sharing has so far, to our knowledge, only received little attention.

The problem domain is characterised by being predictable from a macroscopic perspective while being stochastic when viewed at a microscopic level. As the macroscopic behaviour is a reflection of a collection of highly stochastic microscopic events which in themselves cannot be predicted, it follows that although a process control system is able to foresee general trends and tendencies within the process, it must be able to handle the stochastic behaviour in order to actually manipulate the process.

When optimizing the operational production one tries to determine the financially and operationally most efficient way to combine the production resources, while satisfying consumer needs. This problem is often formalized by using the Economic Dispatch Problem (EDP) and the Unit Commitment Problem (UCP) [4]. By solving the EDP we find out how much load to generate by each of the different available production units at a given time, while the solving the UCP shows when to start and how long each unit should be committed to being in use.

The consumption, and thus the production, follow certain patterns which are predictable to some extent from a system wide perspective. These patterns are generated by a composition of highly stochastic microscopic behaviour among consumer entities, which, as long as their demand is fulfilled, are oblivious to their surroundings or any other part of the larger system. By reacting on these individual microscopic events and controlling and limiting the effect of them, the overall system can achieve several benefits for both the consumers and the suppliers of the utility. Trying to control the consumption in such a way is generally called Demand Side Management (DSM), and can in many cases be achieved by using agent technology or other distributed control schemes [14, 11, 15].

The problem is that the agent based solutions proposed for solving DSM in such environments are dependent on the availability of high-quality sensor data, which in practice can be hard to achieve due to limitations in underlying hardware and communication solutions. By using different levels of availability and quality of communicated sensor data among the agent system we try to quantify the impact on overall system performance.

2 The Agent System

The agent system we study in this paper is used to implement DSM strategies within district heating systems and its function has been described in previous work [14]. The agent system is based on distributed cooperative entities with an overall goal of combining the production and consumption in an optimal manner.

Every producer and consumer entity in the system is represented by an agent. A producer agent will try to minimize its own supply cost function while supplying enough utility to satisfy consumer demand. When a producer agent deems it necessary to implement an DSM action it will try to do so by sharing the task among the consumer agents in order to minimize the side effects of DSM on any individual consumer agent. A consumer agent will seek to implement these requests as long as its internal comfort constraints allow for this. The producer agent is responsible for supervising the continuous utility consumption and also for instigating and distributing DSM tasks when the measured consumption deviates from the desired DSM level. The task sharing is done by first decomposing the initial task into smaller tasks. This is done since the optimization action as a whole is usually too large for one single consumer agent to handle. The tasks are then allocated through a series of auctions. The DSM level is found beforehand by solving the optimization problem relating to the production units, and this is then used as input to the production agent. The producer agent needs to know the wanted consumption level in order to implement DSM. This is found by solving the EDP and the UCP. These solutions are then used as decision basis for the DSM strategy for the following time frame, normally the next twenty-four hour period. In order to solve the EDP the agent uses an objective function which is found in the smooth function described in Equation 1 and 2.

$$\text{Minimize} \sum_{i \in I} F_i(P_i) \tag{1}$$

$$F_i(P_i) = \alpha_i + \beta_i P_i + \gamma P_i \tag{2}$$

This is simply a summation of the utility cost in all supply units [1]. The value of α describes a fixed cost for starting and running the production unit, while the values of β and γ describe costs dependant on the level of production. The accompanying equality constraint is the utility balance which should be satisfied accordingly:

$$\sum_{i \in I} P_i = D + P_{loss} \tag{3}$$

where D represent the utility demand and P_{loss} indicates any production and distribution losses. The inequality constraints describes the production units working within their respective limits:

$$P_i, min \leq P_i \leq, max \qquad \forall i \in I \tag{4}$$

In practical settings these functions are normally not sufficient to describe many situations in utility production. Normally the production entity will have to treat the cost function as a set of piecewise quadratic functions which are defined as [8, 10]:

$$F_i(P_i) = \begin{cases} \alpha_{i1} + \beta_{i1} P_i + \gamma_{i1} P_i & if P_i^{min} \leq P_i \leq P_{i1} \\ \alpha_{i2} + \beta_{i2} P_i + \gamma_{i2} P_i & if P_{i2} \leq P_i \leq P_{i2} \\ \cdot & \cdot \\ \cdot & \cdot \\ \cdot & \cdot \\ \alpha_{im} + \beta_{im} P_i + \gamma_{im} P_i & if P_{im} - 1 \leq P_i \leq P_i^{max} \end{cases} \tag{5}$$

This behaviour is due to the fact that a utility provider usually has a range of different production units, using differently priced fuels. From a economical point of view there is no smooth transition when switching between the different fuels, which makes the resulting function non-differentiable. As demand rises the producing entity is forced to engage increasingly costly production units, and eventually the production costs exceed the possible sale price of the utility. The only way for the producer to mitigate such a situation is to manipulate consumption in order to lower the demand. The UCP is interconnected with the EDP and uses similar optimization methods.

Each consumer unit is controlled by a consumer agent which is responsible for contributing to achieving the overall DSM strategies while maintaining a sufficient level of local comfort. The amount of deviation from the optimal comfort state is used as currency when a consumer agent participates in an auction process, i.e the more the consumer agent is straying from its desired comfort state, the less likely it will be to win any auction. The consumer agents are cooperative in the sense that they do not lie about their cost for participating in a DSM task, since this could possibly jeopardize their internal comfort levels.

The goal for the agent system is then; for each point in time achieve a total actual consumption as close as possible to the total wanted consumption while keeping all local comfort levels within their individual constraints. In a steady state system this could be seen as a traditional optimization problem, i.e. to find a optimum between two conflicting objective functions. However, since we are dealing with a dynamic system the aspects of adaptation and re-planning becomes important, which requires a more sophisticated solution. Whenever a producer agent needs to implement a DSM action it will distribute this by using a private value first priced, sealed bid auction process. This type of auction based multi agent system has previously been successfully implemented in district heating networks in order to achieve DSM [15]. Strategic decisions are made based on global or local views within the environment, and the specific optimization actions rely on continuous sensor data.

In this study we compare the performance of a fully functional agent system with two other systems displaying increasingly worse availability of sensor data. These three different scenarios are based on the level of system wide knowledge available to the participating agents; global, partial and local. We choose to compare these specific three levels of system wide knowledge because they correspond to infrastructural prerequisites which can normally be found in actual physical systems, and because they display a broad and clear view of the problem discussed.

Global knowledge is the normal operational scenario for the MAS used to operate the DSM strategies. The producer agents are able to continuously supervise the use of production utility and are able to instigate system wide auctions as need arises. The consumer agents are able to uphold their individual QoS by deciding when and how to participate in these auctions, i.e. a DSM task is never forced upon a consumer agent against its will. Partial knowledge means that the producer agents are able to supervise the consumption of production utility, but they are not able to communicate local sensory data with consumer agents or to uphold cooperative behaviour through auctions. A producer agent is, however, still able to instigate uninformed DSM actions. This is normally done by using predefined activation lists, which try to force consumer agents to implement DSM tasks. The local consumer agents might however decide to reject the appointed task without being able to tell the producer about this. In the local scenario the producer agents have little or no knowledge about the continuous consumption of production utility, and they do not have any possibility at all to implement any DSM actions, either by cooperation or force. In such a system the consumer agents are often assigned the task of keeping the local utility use to a minimum while upholding the desired QoS. Depending on the situation such behaviour might or might not be for the good of the global system state, but the consumer agent will never know anything about this.

3 The Experiment

The experiment is based on operational data from an agent based control system operational in a district heating network in the town of Karlshamn in the south of Sweden [14, 15]. This data is used as input when simulating the various scenarios described in the previous sections. District heating networks are good examples of the described problem domain as they display most, if not all, of the mentioned characteristics. The reference data in question is collected during a twenty-four hour period with no DSM strategy active, i.e. no external control is applied to the consumers.

The consumer agents all have different comfort constraints based on a function of size, shape and material of the individual building, i.e. the amount of thermal buffer available [12]. In the operational system each consumer agent has access to sensor and actuator data through an I/O hardware platform, which enables the agent to measure the physical behaviour of the heating system within the building as well as the outdoor temperature.

Each agent has a value of wanted indoor climate, and constantly tries to minimize all deviation from this value. The consumer agent has two basic values to consider, the comfort level and the thermal buffer level. It is possible to adjust the energy buffer during shorter periods of time without the comfort level having the time to react. When a consumer agent responds to an auction it will use its currently available buffer level as the price it is willing pay for implementing a single DSM task. We evaluate the performance of the consumer agents by measuring how they choose to use their individual buffers.

The optimization strategy used in this experiment is that of peak shedding, i.e. at any given moment when the total energy use exceeds a certain threshold the producer agent will try to convince the consumer agents to lower their local energy usage in a coordinated fashion. The success of the system wide optimization is measured by the amount of deviation between the wanted threshold and the resulting actual level.

We use real operational data from the Karlshamn district heating network as input into the simulation model, where actual flow data is used as initial values for the calculations. The implemented agent system is functioning according to the same principles as previously described. In the simulation there are fourteen active agents; one producer agent and thirteen consumer agents. By simulating the described levels of agent knowledge we can evaluate the performance of the agent system during different scenarios.

A simulation run begins by calculating specific solutions to the EDP and the UCP. These solutions yield a wanted system wide consumption level for each time step throughout the day. This wanted consumption level is then used by the producer agent as a decision basis, when deciding when and how to instigate DSM actions throughout each time step. This buffer levels are then adjusted through each time step as the agents perform DSM tasks, which in turn makes it possible to calculate the comfort levels for each time step.

4 Results

The control strategy is evaluated by measuring the flow of hot water into the area. Energy usage in a district heating network is measured by combining the temperature of the water with the flow. Since the supply water temperature in the primary network is more or less stable throughout a single day the flow in itself gives a good estimation of the energy usage within all the buildings. In Figures 1, 2 and 3 we show the flow data achieved during the three different scenarios in relation to the wanted DSM strategy. The straight dashed line is the wanted DSM level. This level of consumption is based on a solution of the Economic Dispatch Problem and the Unit Commitment Problem. The peaks above the dashed line represents peak loads which would need to be satisfied by using financially and environmentally unsound fossil fuel. In other words, the global goal of the agent system is to keep the consumption as close to the straight dashed line as possible.

It is clearly visible that the flow value in the global scenario, Figure 1, most closely resembles the desired DSM strategy, with the partial scenario, Figure 2, being somewhat worse, and finally the local scenario, Figure 3, showing a distinct lack in ability to achieve the desired level of consumption.

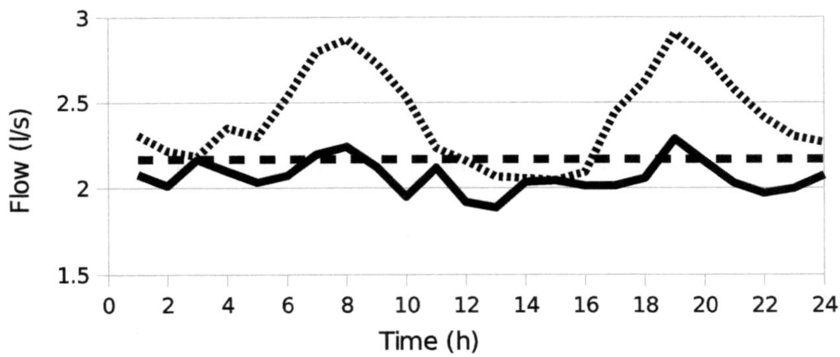

Fig. 1 Global scenario. Agent performance (continuous), reference data (dotted) and wanted DSM level (dashed)

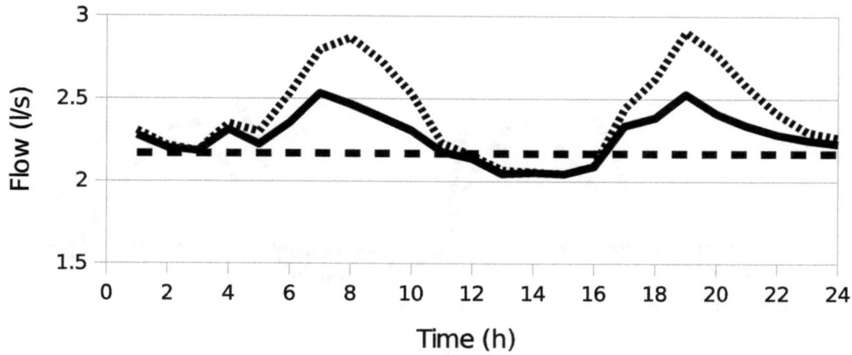

Fig. 2 Partial scenario. Agent performance (continuous), reference data (dotted) and wanted DSM level (dashed)

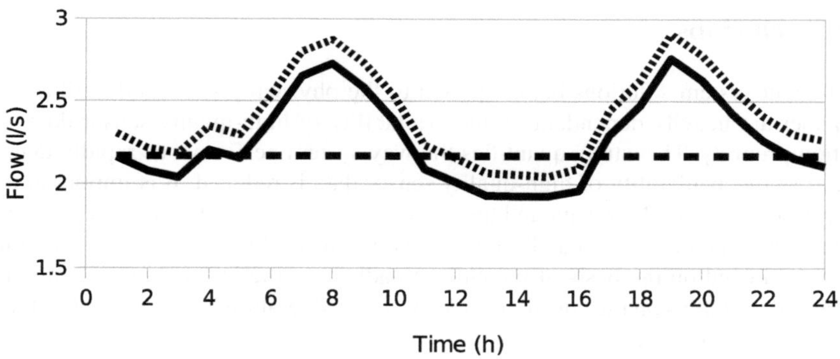

Fig. 3 Local scenario. Agent performance (continuous), reference data (dotted) and wanted DSM level (dashed)

Every agent has an maximum allowed buffer usage of one, with a minimum of zero. The level of comfort will not be negatively effected by a usage between one and zero. If the buffer usage is above one the consumer agent has used more than the allowed buffer and the comfort can be in jeopardy if such a status is allowed to continue for a longer period of time. In other words a consumer agent has an optimal buffer usage of one, i.e. the agent participates in achieving the global goal as much as possible but does this without sacrificing its desired comfort level.

Figure 4 shows the dynamic system wide buffer usage during the whole time period. The range on the y axis is dependent on the amount of consumer agents, since every such agent has a optimal buffer usage of one. In this case study we have thirteen agents, so an optimal usage of the system wide buffer would be thirteen. In the global and partial scenarios the buffer usage clearly follows the reference data as the agents continuously try to counter the varying consumption.

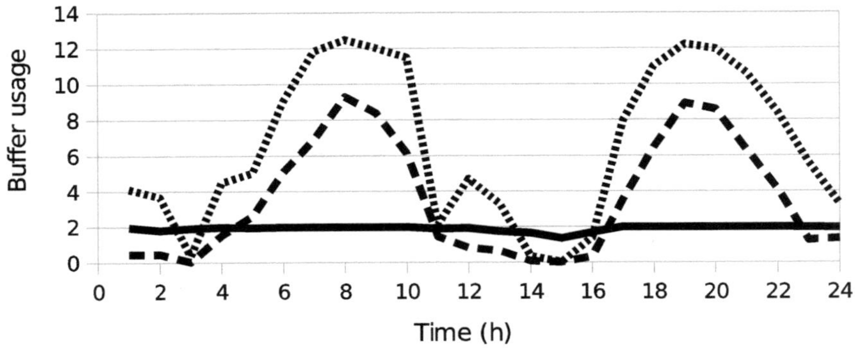

Fig. 4 Buffer usage. Global scenario (dotted), Partial scenario (dashed) and Local scenario (continuous)

5 Conclusions

Multi-agent system solutions being applied to the physical processes described in this paper are heavily dependent on the availability of high-quality sensor data to function properly. This study quantifies the way system performance rapidly deteriorates as the availability of high-quality sensor data is reduced. It is important to factor in both the DSM strategy and the consumer agent comfort value when evaluating an implementation for handling DSM within the problem domain. If a system is only evaluated on the basis of its ability to adhere to the DSM strategy it might give rise to problems on the consumer side as no consideration is given to upholding a sufficient level of QoS.

The local scenario is similar to a type of control system that is often implemented in both electrical grids and district heating networks, as a local uninformed optimization technique. This study indicates that such systems have little global effect in regards to overall production optimization strategies. The reason that the local scenario never goes beyond a certain level in Figure 4 is that the consumer agents are only reacting to their own local peak loads, which are well beyond their own capacity to handle. This is due to the fact that individual peaks are much larger than any individual buffer, so in the local scenario some agents are always maximizing their use of their individual buffer, but without the ability to somehow distribute the load through the producer agents their efforts will always fall short on a system wide scale.

Figure 4 also shows that producer agent knowledge is needed in order to dynamically counter the user demand in regards to the DSM strategy. This is also the buffer usage, which shows that the partial scenario is not able to fully use the available buffer. This is due to the fact that the agents cannot perform cooperative work. The lower use of available buffer of the partial scenario is caused by the fact that although the consumer agent is handed a DSM task, it can choose not to implement the task if the agent considers it to jeopardize its internal QoS level. Since the producer agent never receives any feedback about this, it will not be able to distribute the task to another consumer better suited for the task, and hence the system will on average not utilize the maximum available buffer.

Figure 4 shows that the global scenario is close to using the maximum available buffer on several occasions, while neither the partial or the local scenarios are close to utilizing their full DSM potential.

In this paper we have shown that distributed multi agent systems based on cooperative auctioning are able to achieve the studied DSM strategy, while maintaining an acceptable level of QoS. As the availability and quality of the sensor data diminishes the system performance deteriorates, first into the equivalence of static distributed models and then into the equivalence of simple local optimization models.

This paper is the result of an initial case study in regards to sensor data utilization within industrial multi-agent system applications. In the future we will use this as groundwork while incorporating the financial factors underlying the discussion, in order to further study the economical effects found within such systems.

Acknowledgements. The operational data used in this study was supplied by NODA Intelligent Systems. The project has also been supported by Karlshamns Energi and Karlshamnsbostäder.

References

1. Arvastsson, L.: Stochastic Modeling and Operational Optimization in District Heating Systems. PhD Thesis, Lund Institute of Technology (2001)
2. Aune, M.: Energy Technology and Everyday Life - The domestication of Ebox in Norwegian households. In: Proceedings of ECEEE Summer Study (2001)
3. Dash, R., Rogers, A., Reece, S., Roberts, S., Jennings, N.R.: Constrained Bandwidth Allocation in Multi-Sensor Information Fusion: A Mechanism Design Approach. In: Proceedings of The Eight International Conference on Information Fusion, Philadelphia, PA, USA (2005)
4. Dotzauer, E.: Simple Model for Prediction of Loads in District Heating Systems. Applied Energy 73(3-4), 277–284 (2002)
5. Dutta, P.S., Goldman, C., Jennings, N.R.: Communicating Effectively in Resource Constrained Multi-Agent Systems. In: 20th International Joint Conference on Artificial Intelligence (IJCAI), Hyderabad, India (2007)
6. Goldman, C.V., And Zilberstein, S.: Optimizing Information Exchange in Cooperative Multi-Agent Systems. In: Proceedings of Second International Conference on Autonomous Agents and Multiagent Systems (AAMAS), pp. 137–144 (2003)
7. Jayasima, D.N.: Fault Tolerance in Multisensor Networks. IEEE Transactions on Reliability 45(2), 308–315 (1996)
8. Koay, C.A., Lai, L.L., Lee, K.Y., Lu, H., Park, J.B., Song, Y.H., Srinivasan, D., Vlachogiannis, J.G., Yu, I.K.: Applications to Power System Scheduling. In: Lee, K.Y., El-Sharkawi, M.A. (eds.) Modern Heuristic Optimization Techniques. The Institute of Electrical and Electronics Engineers, Inc. (2008)
9. Lesser, V., Ortiz, C., Tambe, M. (eds.): Distributed Sensor Networks: a multiagent perspective. Kluwer Publishing, Dordrecht (2003)

10. Lin, C.E., Viviani, G.L.: Hierarchical Economic Dispatch for Piecewise Quadratic Cost Functions. IEEE Trans Power Apparatus Systems 103(6), 1170–1175 (1984)
11. Nordvik, H., Lund, P.E.: How to Achieve Energy Efficient Actions as an Alternative to Grid Reinforcement. In: Proceedings of ECEEE Summer Study (2003)
12. Olsson Ingvarsson, L.C., Werner, S.: Building Mass Used as Short Term Heat Storage. In: Proceedings of The 11th International Symposium on District Heating and Cooling, Reykjavik, Iceland (2008)
13. Shen, J., Lesser, V., Carver, N.: Minimizing Communication Cost in a Distributed Bayesian Network Using a Decentralised MDP. In: Proceedings of Second International Conference on Autonomous Agents and Multiagent Systems (AAMAS), pp. 678–685 (2003)
14. Wernstedt, F., Davidsson, P., Johansson, C.: Demand Side Management in District Heating Systems. In: Proceedings of Sixth International Conference on Autonomous Agents and Multiagent Systems (AAMAS), Honolulu, Hawaii (2007)
15. Wernstedt, F., Johansson, C.: Intelligent Distributed Load Control. In: Proceedings of The 11th International Symposium on District Heating and Cooling, Reykjavik, Iceland (2008)

Interoperability, Standards and Metadata for E-Learning

Eugenijus Kurilovas

Abstract. The main research object of the paper is investigation and proposal of interoperability recommendations for e-learning system components – Learning Objects (LOs) and Virtual Learning Environments (VLEs). The main problem in e-learning is not the identification of suitable standards and specifications, but the adoption of these standards and specifications and their application in e-learning practice. Approaches concerning Learning Object Metadata (LOM) standard application profiles are the main topics investigated here because they could provide more quick and convenient LOs search possibilities for the users. Interoperability issues are also analyzed here as significant topics for e-learning systems components quality evaluation.

1 Introduction

Standards and interoperability are the key factors in the success of the introduction of e-learning systems.

The main task under consideration in this paper is to formulate recommendations how to improve e-learning standards, namely to provide recommendations for the improvement of LOM application profiles (APs).

Interoperability issues are also analysed here as the significant part of e-learning systems components (LOs, their repositories (LORs) and VLEs) quality evaluation problems.

LO is referred here as "any digital resource that can be reused to support learning" [18]. VLE is considered here as "specific information system which provides the possibility to create and use different learning scenarios and methods". Metadata is considered here as "structured data about data" [3]. An application profile is referred

Eugenijus Kurilovas
Institute of Mathematics and Informatics, Vilnius, Lithuania
e-mail: eugenijus.kurilovas@itc.smm.lt

G.A. Papadopoulos and C. Badica (Eds.): Intelligent Distributed Computing III, SCI 237, pp. 121–130.
springerlink.com © Springer-Verlag Berlin Heidelberg 2009

here as “an assemblage of metadata elements selected from one or more metadata schemas and combined in a compound schema” [3].

2 E-Learning Interoperability Problems Analysis

The author has analysed the main existing and emerging interoperability standards and specifications for e-learning systems in his PhD thesis and has also formulated the problem of standards and specifications adoption, application and implementation in e-learning practice.

The following interoperability aspects have been analysed:

- Different layers of interoperability.
- Government policies and strategies for interoperability.
- Accessibility interoperability.
- Learning assessment interoperability.
- Learning content and repository interoperability.
- Learning activity interoperability.
- Learner information interoperability.
- Education institution interoperability.

The analysis has shown that while some standards and specifications are only at the beginning stage of adoption, there are already a fair number of standards that have been well adopted, but too many islands exist. The same results were achieved while analysing this topic during the preparation of the ASPECT[1] project proposal. The analysis highlights the need for making old and new standards and specifications work together (i.e., interoperate). The main problem here is not identification of suitable standards and specifications, but the problem how to adopt these standards and specifications and apply and implement them in e-learning practice. E.g., in ASPECT we analyse and compare how SCORM and IMS Common Cartridge packages are used by the teachers in the real pedagogical contexts, and improve their usage.

The main conclusions based on the analysis of the existing and emerging interoperability standards and specifications are:

- The majority of standards and specifications are not adopted and do not conform to the educational practice.
- There exists a problem of complex solutions for the application of standards and specifications in education.
- Standards and specifications often do not cooperate.

First of all, in order to make it easier for educators to discover and use LOs that addresses the needs of their students, to maximise reuse of LOs and minimise the costs associated with their repurposing, the good solutions are lacking for the specific application profiles of IEEE LOM.

[1] eContent*plus* programme’s ASPECT (Adopting Standards and Specifications for Educational Content) project web site. http://aspect-project.org/

3 Learning Objects Reusability Increase

In the author's point of view, one of the main criteria for achieving high LOs effectiveness and efficiency level is LOs reusability [1].

The need for reusability of LOs has at least three elements: (1) Interoperability: LO is interoperable and can be used in different platforms; (2) Flexibility in terms of pedagogic situations: LO can fit into a variety of pedagogic situations; and (3) Modifiability to suit a particular teacher's or student's needs: LO can be made more appropriate to a pedagogic situation by modifying it to suit a particular teacher's or student's needs [14].

There are two main conditions for LOs reusability elsewhere: (1) LOs have to fit different countries national curricula; (2) Different countries' LOM APs have to be oriented towards quick and convenient search of reusable LOs.

The principle of ultimate increase of reusability of LOs is considered by the author as one of the main factors of e-learning systems flexibility [1].

It was analysed that the flexible approach to the system's creation and development should be based on the idea of LOs' partition to two main separate parts, i.e., LOM compliant small pedagogically decontextualised Learning Assets and LOM and IMS Learning Design compliant Units of Learning – UoLs (see, e.g. [2, 11, 12, 13]). VLEs suitable for the flexible system should have a high level of adaptation capabilities [10].

European Learning Resource Exchange (LRE) system's [7] validation in Lithuania performed by the author while implementing FP6 CALIBRATE project[2] has shown that the teachers prefer LOs from national repositories which have the potential to 'travel well' and can be used in different national contexts. These reusable LOs preferred by the teachers are mainly the small decontextualised learning assets. Therefore in order to maximise LOs reusability in Europe LRE should consist mainly of the decontextualised learning assets.

The results of the teachers-experts survey also performed by the author in CALIBRATE show that the teachers would mostly like to find pedagogically decontextualised ultimately reusable LOs and therefore to have a service for quick and convenient search of such LOs. While searching for LOs in CALIBRATE portal the experts have used browsing by subject and advance search services. These advance search services have not contained any services to ease the search of reusable LOs in the portal. The LOs in the portal are described in conformity with the partners' LOM APs, and these APs have not contained any services to simplify the search of reusable LOs. Therefore it took very much time for the experts to find and choose suitable reusable LOs for their UoLs (e.g., lesson plans).

According to [3], the purpose of an AP is to adapt or combine existing schemas into a package that is tailored to the functional requirements of a particular application, while retaining interoperability with the original base schemas. There are several principles described in [3] providing "a guiding framework for the development of practical solutions for semantic and machine interoperability in any domain

[2] FP6 IST CALIBRATE (Calibrating eLearning in Schools) project web site. http://calibrate.eun.org (2008)

using any set of metadata standards": modularity, extensibility, refinement and multilingualism.

One of the mechanisms for APs to achieve modularity is the elements' cardinality enforcement. Cardinality refers to constraints on the appearance of an element. Is it mandatory or recommended or optional? According to [3], "the status of some data elements can be made more stringent in a given context". For instance, an optional data element can be made recommended, and a recommended data can be made mandatory in a particular AP. On the other hand, as an AP must operate within the interoperability constraints defined by the standard, it cannot relax the status of data elements [3].

The author has applied this cardinality enforcement principle in his research. It was analysed that the main LOM elements which vocabulary values could reflect the LOs ultimate reusability deal with structure of LO, its functional granularity (aggregation) level, educational type as well as the kind of relation of this LO with the others (see [11, 13]).

The results of the author's analysis of the last European LOM AP (LRE Metadata AP v3.0) have shown that it would be purposeful to improve it in order to provide more quick and convenient search possibilities for those searching ultimately reusable LOs (i.e., learning assets) by the means of changing (i.e., advancing / enforcing cardinality) the status of a number of LRE AP elements.

These proposals deal with changing the status of the following LOM AP elements from 'optional' to 'recommended' as well as from 'optional' and 'recommended' – to 'mandatory':

- 1.7 General.Structure;
- 1.8 General.Aggregation Level;
- 5.2 Educational. Learning Resource Type; and
- 7.1 Relation.Kind (see Fig. 1).

The author believes that the development of advanced search engine reflecting LOs reusability level based on this research would considerably reduce the time for

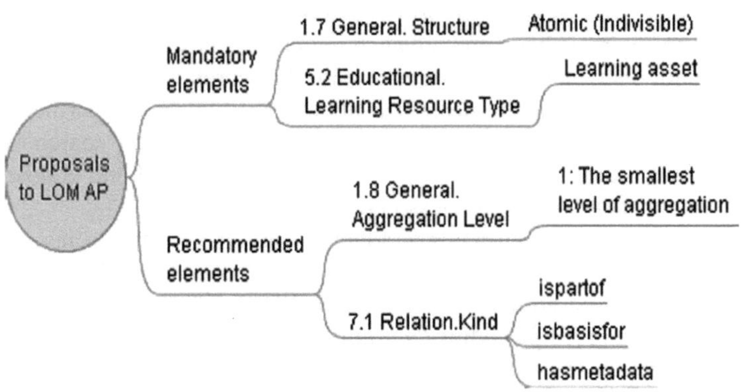

Fig. 1 Proposals to LOM AP [1]

the teachers to find and choose suitable LOs in the repositories. Lithuanian central LOM repository was analysed for this task.

4 Interoperability Issues in E-Learning Components Quality Evaluation Problems

4.1 Multiple Criteria Software Quality Evaluation

In ISO/IEC 14598-1:1999[3] quality evaluation is defined as "the systematic examination of the extent to which an entity (part, product, service or organisation) is capable of meeting specified requirements".

Researchers usually divide e-learning systems components (such as LOs, LORs and VLEs) quality evaluation criteria into technological, pedagogical and organisational evaluation criteria (see [8, 9, 10]).

Different scientific methods are used for the software's quality evaluation.

The section is aimed to consider the problems of expert evaluation of technological quality criteria of LOs and VLEs. The expert evaluation is referred here as the multiple criteria evaluation of e-learning system components aimed at selection of the best alternative based on score-ranking results. According to [4], if the set of decision alternatives is assumed to be predefined, fixed and finite, then the decision problem is to choose the optimal alternative or, maybe, to rank them. But usually the experts (decision makers) have to deal with the problem of optimal decision in the multiple criteria situation where the objectives are often conflicting. In this case, according to [4], "an optimal decision is the one that maximises the decision maker's utility". According to [18], each alternative in multi-criteria decision making problem can be described by a set of criteria. Criteria can be qualitative and quantitative. They usually have different units of measurement and different optimisation direction.

Below in the section the author has analysed LOs and VLEs technological quality criteria suitable for multiple criteria expert evaluation (decision making).

4.2 Interoperability Issues in Learning Objects Quality Evaluation

The author has analysed several LOs technological quality evaluation methods suitable for expert multiple criteria evaluation (see LORI [17], Paulsson and Naeve [15] and scientific projects-based criteria developed in MELT[4] and Q4R[5]).

[3] ISO/IEC 14598-1:1999. Information Technology – Software Product Evaluation – Part 1: General Overview. First edition, 1999-04-15 (1999).

[4] MELT, (eContent*plus* programme) Metadata Ecology for Learning and Teaching project web site. http://melt-project.eun.org (2008).

[5] Q4R, Quality for Reuse project web site. http://www.q4r.org (2008).

It was found out that all the analysed LO evaluation methods have a number of limitations from the technological point of view, e.g.:

- LORI [17], Paulsson and Naeve [15] and MELT criteria do not examine different LO life cycle stages.
- Q4R tool insufficiently examines technological evaluation criteria before LO inclusion in the LOR.
- All tools insufficiently examine LO reusability / interoperability criteria.

It is obvious that a more comprehensive LO technological evaluation method is needed.

The author has proposed the original method / set of LO technological quality evaluation criteria based on the analysis of LO quality evaluation methods listed above.

This set of criteria incorporates LO quality evaluation criteria suitable for the different LO life cycle stages, including criteria before, during and after LO inclusion in the repository as well as LO reusability criteria (see Table 1).

It was also proposed by the author that LO reusability criteria should have the same weight as the other criteria [10].

The LORs quality assurance strategies are further investigated in EdreNe[6] – one of the largest LORs related R&D projects at the moment. The LORs quality assurance strategies priority was ranked the highest by the EdReNe and external experts during the strategic seminar in Lisbon in 2008: 58.8 % experts have ranked it as essential, and 38.3 % – as important [12].

While preparing this set of criteria the author had to exclude all evaluation (e.g., pedagogical, economic etc.) criteria that do not deal directly with the LOs technological quality problems on the one hand, and to estimate interconnected/overlapping criteria on the other. This new author's set of criteria includes LO technological evaluation criteria suitable for different LO life cycle stages (before, during and after LO inclusion in the LOR), as well as LO reusability criteria.

This set of criteria takes into account LORI [17], Paulsson and Naeve [15], MELT, Q4R and the authors' research results published in [1, 8, 10].

We can conclude that interoperability criteria are significant ones for LOs technological quality evaluation, especially while evaluating such sub-criteria as 'Metadata accuracy' and 'Compliance with the main import/export standards' (before LO inclusion in LOR) as well as sub-criteria 'Automatic verification of capability with known protocols' and 'Automatic metadata generation or simplified metadata tagging' (during LO inclusion in LOR).

In the author's point of view the experts-evaluators (decision makers) have to attribute high weights for these criteria while choosing different alternatives (i.e., LOs) suitable for their e-learning system.

---

[6] EU eContent*plus* programme's EdReNe (Educational Repositories Network) project web site. http://edrene.org (2009).

Table 1 Learning Objects technological quality evaluation criteria

Before LO inclusion in LOR	Narrow definition compliance		
	Reusability level: interoperability	Metadata accuracy	
		Compliance with the main import/export standards (e.g., IMS)	
	Reusability level: decontextualisation	Is LO indivisible (atomic)?	LO aggregation (granularity) level
			Is LO modular?
		Does LO have a strong visual element?	
	Reusability level: cultural and learning diversity principles	Is LO flexible (can be modified)?	
		LO suitability for localisation	
		LO internationalisation level	
	Reusability level: accessibility	Is LO designed for all?	
		Compliance with accessibility standards (W3C)	
	LO architecture	Is LO architecture layered in order to separate data, presentation and application logics?	
	Working stability		
	Design and usability	Aesthetics	
		Navigation	
		User-friendly interface	
		Information structurisation	
		Personalisation	
During LO inclusion in LOR	Membership of contribution control strategies	Using LO harvesting	
		Obligatory membership	
	Technical interoperability	Automatic verification of capability with known protocols	
		Automatic metadata generation or simplified metadata tagging	
After LO inclusion in LOR	Retrieval quality	User should be able to retrieve LO in different ways	
	Information quality	Display information strategies Feedback techniques	

4.3 Interoperability Issues in Virtual Learning Environments Quality Evaluation

The author has analysed several VLEs technological quality evaluation methods (see [16] and [5]) suitable for the expert multiple criteria evaluation. It was found out that the analysed VLEs evaluation methods have a number of limitations from the technological point of view, e.g.:

- The method developed in [16] practically does not examine adaptation capabilities criteria.
- The method proposed by [5] insufficiently examines general technological criteria.

Therefore a more comprehensive VLE technological evaluation method / set of criteria is needed. It should include general technological evaluation criteria based on the modular approach and interoperability [9], as well as adaptation capabilities criteria [10]. In the author opinion, VLE adaptation capabilities criteria should have the same weight as the other criteria [10].

Therefore the author has proposed the original comprehensive set of VLE technological evaluation criteria combining both general and adaptation criteria (see Table 2). While preparing this method / set of criteria the author had to exclude all evaluation (e.g., pedagogical, organisational, economic etc.) criteria [9] that do not deal directly with VLEs technological quality problems on the one hand, and to estimate the interconnected / overlapping criteria on the other.

This method includes general technological evaluation criteria based on the modular approach and interoperability, as well as adaptation capabilities criteria.

We can conclude that interoperability criterion is significant one for VLEs technological quality evaluation method.

In the author's point of view the experts-evaluators (decision makers) have to attribute high weight for this criterion (or both sub-criteria) while choosing any alternative (i.e., VLE software package) suitable for their e-learning system.

Table 2 Virtual Learning Environments technological quality evaluation criteria

General criteria	Overall architecture and implementation	Scalability
		Modularity (of the architecture)
		Possibility of multiple installations on a single platform
		Reasonable performance optimisations
		Look and feel is configurable
		Security
		Modular authentication
		Robustness and stability
		Installation, dependencies and portability
	Interoperability	Integration is straightforward
		VLE standard support
	Internationalisation and localisation	Localisable user interface
		Localisation to relevant languages
		Unicode text editing and storage
		Time zones and date localisation
		Alternative language support
	Accessibility	Text only navigation support
		Scalable fonts and graphics
Adaptation criteria	Adaptability (facilities to customise for the educational institution's needs)	Language
		Design
	Personalisation aspects (facilities of each individual user to his/her own view of the platform)	
	Extensibility	Good programming style
		Availability of a documented API
	Adaptivity (all kinds of automatic adaptation to the individual user's needs)	Personal annotations of LOs
		Automatically adapted content

5 Conclusion

Approaches concerning LOM APs are the main while improving LOs usability and creating any metadata strategies. The presented method of improvement of existing LOM APs results in provision of more quick and convenient search possibilities for those searching ultimately reusable LOs. It would be purposeful to improve LRE AP by advancing the status of four of LRE AP elements: 1.7 General.Structure; 1.8 General.Aggregation Level; 5.2 Educational.Learning Resource Type; and 7.1 Relation.Kind. The author proposes to include search service against these elements into extended search service in the repositories.

Interoperability criteria are significant ones for LOs technological quality evaluation, especially while evaluating such sub-criteria as 'Metadata accuracy' and 'Compliance with the main import/export standards' as well as sub-criteria 'Automatic verification of capability with known protocols' and 'Automatic metadata generation or simplified metadata tagging'. Interoperability criteria are significant ones also for VLEs technological quality evaluation.

References

1. Dagienė, V., Kurilovas, E.: Information Technologies in Education: Experience and Analysis. Monograph. – Vilnius: Institute of Mathematics and Informatics, 216 p. (2008) (in Lithuanian)
2. Dagienė, V., Kurilovas, E.: Design of Lithuanian Digital Library of Educational Resources and Services: the Problem of Interoperability. Information Technologies and Control. Kaunas: Technologija 36(4), 402–411 (2007)
3. Duval, E., Hodgins, W., Sutton, S., Weibel, S.L.: Metadata Principles and Practicalities. D–Lib Magazine 8(4) (2002), http://www.dlib.org/dlib/april02/weibel/04weibel.html
4. Dzemyda, G., Šaltenis, V.: Multiple Criteria Decision Support System: Methods, User's Interface and applications. Informatica 5(1-2), 31–42 (1994)
5. Graf, S., List, B. An Evaluation of Open Source E-Learning Platforms Stressing Adaptation Issues. Presented at ICALT 2005 (2005)
6. Institute of Mathematics and Informatics. Computer Learning Tools and Virtual Learning Environments Implementation in Vocational Education. Scientific research report, p. 80 (2005), http://www.emokykla.lt/lt.php/tyrimai/194
7. Jevsikova, T., Kurilovas, E.: European Learning Resource Exchange: Policy and Practice. In: Proceedings of the 2nd International Conference Informatics in Secondary Schools: Evolution and Perspectives (ISSEP 2006), Selected papers, Vilnius, Lithuania, November 7-11, 2006, pp. 670–676 (2006)
8. Kurilovas, E.: Digital Library of Educational Resources and Services: Evaluation of Components. Information Sciences 42-43, 69–77 (2007)
9. Kurilovas, E.: Several aspects of technical and pedagogical evaluation of virtual learning environments. Informatics in Education, Vilnius 4(2), 215–252 (2005)
10. Kurilovas, E., Dagienė, V.: Learning Objects and Virtual Learning Environments Technical Evaluation Tools. In: Proceedings of the 7th European Conference on e-Learning (ECEL 2008), Agia Napa, Cyprus, November 6-7, 2008, vol. 2, pp. 24–33 (2008)

11. Kurilovas, E., Kubilinskienė, S.: Interoperability Framework for Components of Digital Library of Educational Resources and Services. Information Sciences. Vilnius 44, 88–97 (2008)
12. Kurilovas, E., Kubilinskienė, S.: Analysis of Lithuanian LOM Repository Strategies, Standards and Interoperability. In: Proceedings of the 2nd International Workshop on Search and Exchange of e-learning Materials (SE@M 2008) within the 3rd European Conference on Technology Enhanced Learning (EC–TEL 2008), Maastricht, Netherlands, 17-19 September, 2008, vol. 385 (2008), http://sunsite.informatik.rwth-aachen.de/Publications/CEUR-WS/Vol-385/
13. Kurilovas, E., Kubilinskienė, S.: Creation of Lithuanian Digital Library of Educational Resources and Services: the Hypothesis, Contemporary Practice, and Future Objectives. In: Proceedings of the 1st International Workshop on Learning Object Discovery & Exchange (LODE 2007) within the 2nd European Conference on Technology Enhanced Learning (EC–TEL 2007), Sissi, Crete, Greece, vol. 311, pp. 11–15 (2007), http://CEUR-WS.org/Vol-311/
14. McCormick, R., Scrimshaw, P., Li, N., Clifford, C.: CELEBRATE Evaluation report (2004), http://celebrate.eun.org/eun.org2/eun/Include_to_content/celebrate/file/Deliverable7_2EvaluationReport02Dec04.pdf
15. Paulsson, F., Naeve, A.: Establishing technical quality criteria for Learning Objects (2006), http://www.frepa.org/wp/wp-content/files/Paulsson-Establ-Tech-Qual_finalv1.pdf
16. Technical Evaluation of Selected Learning Management Systems (2004), https://eduforge.org/docman/view.php/7/18/LMS%20Technical%20Evaluation%20-%20May04.pdf
17. Vargo, J., Nesbit, J.C., Belfer, K., Archambault, A.: Learning object evaluation: Computer mediated collaboration and inter–rater reliability. International Journal of Computers and Applications 25(3), 198–205 (2003)
18. Wiley. D. A. Connecting Learning Objects to Instructional design Theory: a definition, a Metaphor, and a Taxonomy. Utah State University (2000), http://www.reusability.org/read/
19. Zavadskas, E.K., Turskis, Z.: A New Logarithmic Normalization Method in Games Theory. Informatica 19(2), 303–314 (2008)

Appendix

The work presented in this paper is partially supported by the European Commission under the eContent*plus* programme – as part of the EdReNe project (Project Number ECP-2006-EDU-42002) and ASPECT project (Contract ASPECT-ECP-2007-EDU-417008). The author is solely responsible for the content of this paper. It does not represent the opinion of the European Commission, and the European Commission is not responsible for any use that might be made of data appearing therein.

Studying the Cache Size in a Gossip-Based Evolutionary Algorithm

Juan Luís J. Laredo, Carlos Fernandes, Antonio Mora, Pedro A. Castillo, Pablo García-Sánchez, and Juan Julián Merelo

Abstract. Gossiping is a self-organized and decentralized approach to distribute algorithms through Peer-to-Peer (P2P) networks. Based on such an approach, the Evolvable Agent model is a P2P Evolutionary Algorithm (EA) whose population structure is defined by the gossiping protocol newscast, a protocol that behaves asymptotically as a small-world graph. This paper explores the impact of different cache sizes on the algorithm performance given that cache size is the only tunable parameter in newscast. To this aim, the problem generator wP-PEAKS and the multimodal deceptive problem MMDP have been used as benchmarks. Results show that the quality of the solutions and the run-time of the algorithm are not altered when changing the settings of the cache size. This fact points out that newscast is a robust gossiping protocol for tackling distributed evolutionary computation.

1 Introduction

Evolutionary Algorithms (EAs) are a set of bioinspired meta-heuristics for solving a wide range of problems such as image processing for cancer detection [Odeh et al(2006)], automatic processing of XSLT documents [García-Sánchez et al(2008)] or bankruptcy detection [Alfaro-Cid et al(2008)]; these are only some examples in which EAs have been used to find a solution to the problem. The meta-heuristic workflow begins with a initial population of randomly generated solutions (so-called individuals) and a heuristic that provides a fitness value to every individual. Iteratively, the fittest individuals are preferentially selected and mixed into a new population leading to the problem optimization [Eiben and Smith(2003)].

In spite of tackling a wide range of problems, EAs are hard time consuming tasks with high requirements in computing power which have challenged practitioners to

Juan Luís J. Laredo and co-authors
University of Granada. ATC-ETSIIT. Periodista Daniel Saucedo Aranda s/n 18071, Granada, Spain
e-mail: juanlu@geneura.ugr.es

G.A. Papadopoulos and C. Badica (Eds.): Intelligent Distributed Computing III, SCI 237, pp. 131–140.
springerlink.com © Springer-Verlag Berlin Heidelberg 2009

explore parallel approaches in order to reduce the run-time [Alba and Tomassini(2002)]. Within this context, Peer-to-Peer (P2P) systems [Steinmetz and Wehrle(2005)] appear as powerful parallel infrastructures in which the computational power is built up by a group of interconnected users who share their spare CPU cycles. Such a computation paradigm is devoid of any central server and decentralization arises as a major issue.

Tackling the issue, the Evolvable Agent model (EvAg) presented by the authors in [Laredo et al(2008)] proposes a distributed and decentralized P2P EA in which the population structure of the EA is defined by a gossiping protocol called newscast [Jelasity and van Steen(2002)]. Newscast builds a small-world overlay network connecting every pair of nodes through a short sequence of intermediate nodes from which emerges a good scalability behaviour [Watts and Strogatz(1998)].

Nevertheless, parallelization implies new stochastic components which drastically increase the complexity for the fine-tuning of the algorithm. [Cantú-Paz(1999)] and [Hidalgo and Fernández(2005)] analyse up to seven of these components for one of the most extended models in the parallelization of EAs, the island model. The EvAg model instead, leaves all the parallelization complexity to the underlying P2P protocol. This way, every individual just takes care of the self-scheduling of the evolutionary steps as an agent and the newscast protocol deploys the algorithm in parallel.

In addition, newscast simplifies all the decision making for the parallelization by tuning a single parameter, the cache size. The cache acts as a routing table in which every peer holds a list of neighbours peers. Therefore, the cache size represents the maximum number of connections (edges) that a peer could have. For example, a topology with n peers and a cache size of n will lead to a complete graph topology. Hence, the cache size has to be much smaller than the number of peers in order to get small-world features as a small characteristic path length and a high clustering coefficient [Jelasity et al(2005)].

In this paper, we study the impact of the cache size on the algorithm performance. To this aim, two discrete optimization problems proposed in [Giacobini et al(2006)] have been used. Results show that the success rate of the algorithm remains constant independently of the different cache sizes. This fact points to the robustness of the parallel model and helps decision making. Any cache size guaranteeing a small-world topology will produce equivalent performances.

The rest of the work is structured as follow: Section 2 explains the newscast protocol as population structure for an EA. The overall model architecture of the Evolvable Agent model is summarized in Section 3. Section 4 presents the experiments and analyses the results. Finally, conclusions of the work are drawn in Section 5 and some future lines of work proposed.

2 Newscast as Population Structure for an EA

Newscast is a gossiping protocol for building unstructured P2P overlay networks. Algorithm 1 shows the pseudo-code of the main tasks in the communication process

Algorithm 1. Newscast protocol in node $Node_i$

Active Thread
while $Node_i$ not finished **do**
  sleep ΔT
  $Node_j \Leftarrow$ Random selected node from $Cache_i$
  send $Cache_i$ to $Node_j$
  receive $Cache_j$ from $Node_j$
  $Cache_i \Leftarrow$ Aggregate ($Cache_i$,$Cache_j$)
end while

Passive Thread
while $Node_i$ not finished **do**
  wait $Cache_j$ from $Node_j$
  send $Cache_i$ to $Node_j$
  $Cache_i \Leftarrow$ Aggregate ($Cache_i$,$Cache_j$)
end while

which build the newscast topology. Each node maintains a cache with a maximum of one entry per node in the network.

There are two different tasks that the protocol carries out within each node: The active thread which initiates communications and the passive thread which is waiting for answering.

After ΔT time each $Node_i$ initiates a communication process (active thread). It randomly selects a $Node_j$ from $Cache_i$ with uniform probability. Both $Node_i$ and $Node_j$ exchange their caches and merge them following an aggregation function. In our case, the aggregation consists on picking up the newest item for each cache entry in $Cache_i$, $Cache_j$ and merge them into a single cache that $Node_i$ and $Node_j$ will share.

[Jelasity and van Steen(2002)] show that the influence of the ΔT parameter is just related with the speed of the convergence from the initial random graph to the small-world graph. After converging, the graph properties remain estabilazed on the values influenced by the cache size. Therefore, the cache size plays an important role in newscast. Figure 1 depicts the influence of the cache size (c) in the average path length and the clustering coefficient for different network sizes. A smaller c implies a higher average path length but also a higher clustering coefficient. That means that the distance between any pair of nodes is larger for smaller values of c while neighbour nodes are strongly connected forming clusters.

From $c = 30$ ahead these properties are roughly the same for the considered network sizes (up to 1600 nodes). Hence, we have tested our experiment for cache sizes from 4 to 30.

Finally, as shown in Figure 2, newscast asymptotically creates a small-world graph since keeps the small average path length of random graphs but a much higher clustering coefficient [Jelasity and van Steen(2002), Voulgaris et al(2004)].

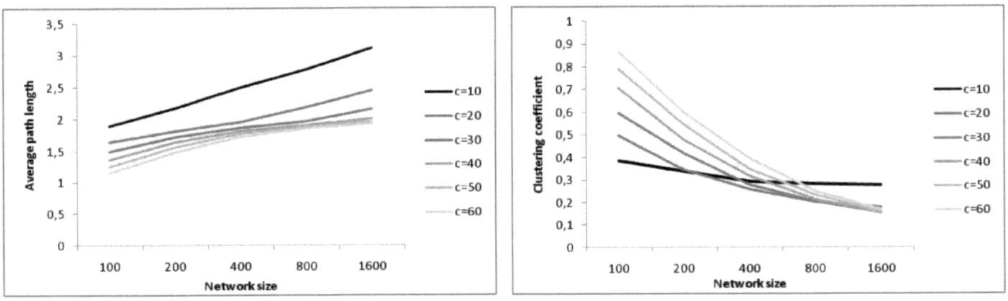

Fig. 1 Average path length (left) and clustering coefficient (right) for different network and cache sizes (c)

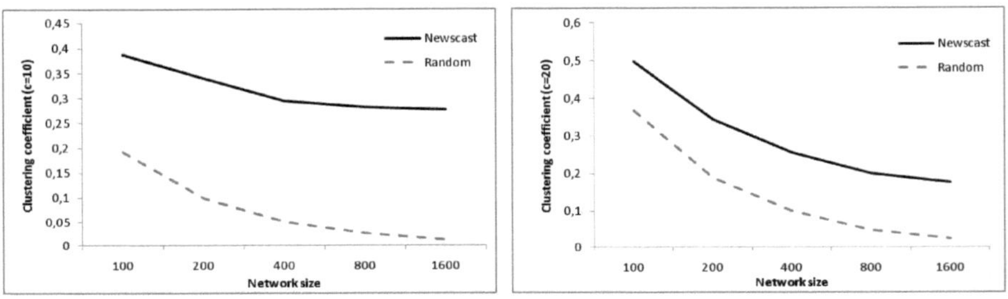

Fig. 2 Clustering coefficients for equivalent random and newscast graphs (i.e. graphs with the same number of edges, ten on the left and twenty on the right). The higher values in the newscast graphs denote a small-world topology [Watts and Strogatz(1998)].

3 Overall Model Description

The overall architecture of our Evolvable Agent model (EvAg) consists of a group of Evolvable Agents (each one running on its own thread) whose main design objective is to carry out the main steps of evolutionary computation: selection, variation (crossover and mutation) and evaluation of individuals. The selection takes place locally within the neighbourhood, that is, the current nodes in the cache of each agent. A decentralized system would be inefficient in supporting a global comparison among all individuals as in the case of roulette wheel or rank-based selection.

Thus, instead of the classical view of a population of solutions managed by an oracle, this model proposes a population of autonomous agents, each one representing a solution which evolves within a neighbourhood defined by the newscast protocol. Despite the model does not follow the canonical sense of population, adjacent EvAgs provide each other with the genetic material that they require to evolve. Therefore, we talk about a population of EvAgs instead of a population of individuals. For the sake of simplicity, we will consider that a peer, node or an EvAg is the same vertex on the newscast graph.

Algorithm 2 shows the pseudo-code of an EvAg where the agent owns an evolving solution (S_t).

Algorithm 2. Evolvable Agent

```
S_t ⇐ Initialize Agent
loop
   Sols ⇐ Local Selection(Neighbourhood)
   S_{t+1} ⇐ Recombination(Sols,P_c)
   S_{t+1} ⇐ Mutate(S_{t+1}, P_m)
   S_{t+1} ⇐ Evaluate(S_{t+1})
   if S_{t+1} better than S_t then
      S_t ⇐ S_{t+1}
   end if
end loop
```

The selection takes place locally into a given neighbourhood where each agent select other agents' current solutions (S_t) Selected solutions are stored in *Sols* ready to be recombined and mutated. Within this process a new solution S_{t+1} is generated. If the newly generated solution S_{t+1} is better than the old one S_t, it replaces the current solution.

4 Experiments

The test-suite to evaluate our proposal is composed of two discrete optimization problems used e.g. in [Giacobini et al(2006)] as benchmarks: The massively multimodal deceptive problem (MMDP) and a version of the problem generator P-PEAKS (wP-PEAKS). They represent a set of difficult problems to be solved by an EA with different features such as epistasis, multimodality, deceptiveness and problem generators. Table 1 shows the settings for the two problem instances.

Table 1 Problem Instances

Problem	Bits	Number of Peaks	Max. eval.	Optimum
MMDP	96	-	30000	16
wP-PEAKS	32	16	30000	1.0

- Massively Multimodal Deceptive Problem (MMDP)
  The MMDP [Goldberg et al(1992)] is a deceptive problem composed of k subproblems of 6 bits each one (s_i). Depending of the number of ones (unitation) s_i takes the values depicted in Table 2.

  The fitness value is defined as the summation of the s_i subproblems with an optimum of k (equation 1). The number of local optima is quite large (22^k),

Table 2 Basic deceptive bipolar function (s_i) for MMDP

Unitation	Subfunction value
0	1.000000
1	0.000000
2	0.360384
3	0.640576
4	0.360384
5	0.000000
6	1.000000

while there are only 2^k global solutions. We consider an instance of intermediate difficulty with $k = 16$.

$$f_{MMDP}(\mathbf{s}) = \sum_{i=1}^{k} fitness_{s_i} \tag{1}$$

- Multimodal Problem Generator (wP-PEAKS)
  The wP-PEAKS problem is a multimodal problem generator. It is the modified version proposed in [Giacobini et al(2006)] of the problem generator P-PEAKS [Jong et al(1997)]. The idea is to generate P random $N-bit$ strings where the fitness value of a string $\mathbf{x}$ is the number of bits that $\mathbf{x}$ has in common with the nearest peak divided by N. The modified version consists in adding weights w_i with only $w_1 = 1.0$ and $w_{[2..P]} < 1.0$. Hence, despite P optima solutions as in the P-PEAKS, there is just one optima and $P-1$ attractors. We consider an instance of $P = 16$ and $N = 32$ with $w_1 = 1.0$ and $w[2..P] = 0.99$ where the optimum fitness is 1.0 (equation 2).

$$f_{wP-PEAKS}(\mathbf{x}) = \frac{1}{N} \max_{1 \le i \le p} \{w_i N - HD(\mathbf{x}, Peak_i)\} \tag{2}$$

4.1 Experimental Setup

Two metrics has been considered to assess the impact of different cache sizes on the algorithm performance: the success rate (SR) and the average evaluations to solution (AES) [Eiben and Smith(2003)]. The SR measures the algorithm quality as the proportion in which the algorithm is able to find the problem optimum out of all the runs. In the case of study, the SR is calculated over 50 independent runs. In order to estimate the algorithm run-time performance, the AES metric measures the average number of evaluations that the algorithm spends in those runs that yield success.

Table 3 summarizes the settings of the experiments. Fourteen different cache sizes for the newscast protocol and two different population/network sizes[1] for the EA have been tested for every problem instance. The rest of the parameters are standard in EAs and a detailed explanation can be found in e.g. [Eiben and Smith(2003)].

[1] Population size referring to an EA is equivalent to network size in terms of a P2P system.

Table 3 Parameters of the algorithms

EA settings	
EA type	Genetic Algorithm
Population size MMDP	200, 440 individuals
Population size wP-PEAKS	400, 800 individuals
Recombination	DPX, $p_c = 1.0$
Selection Parents	Binary Tournament + individual
Newscast settings	
$Cache_{size}$	4, 6, 8, 10,12,14,16,18,20,22,24,26,28,30

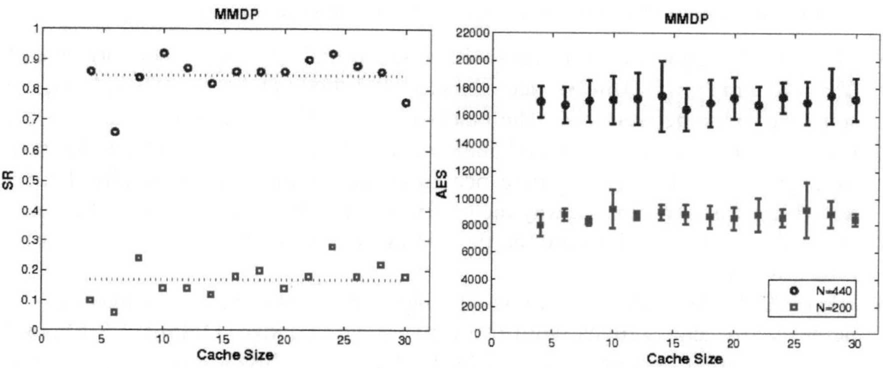

Fig. 3 Success Rate of the EvAg model (*left*) and Average Evaluation to Solution with standard deviation (*right*) for the MMDP. Population/network sizes of $N = 200$ and $N = 440$. Averaged values for the different cache sizes in dotted lines.

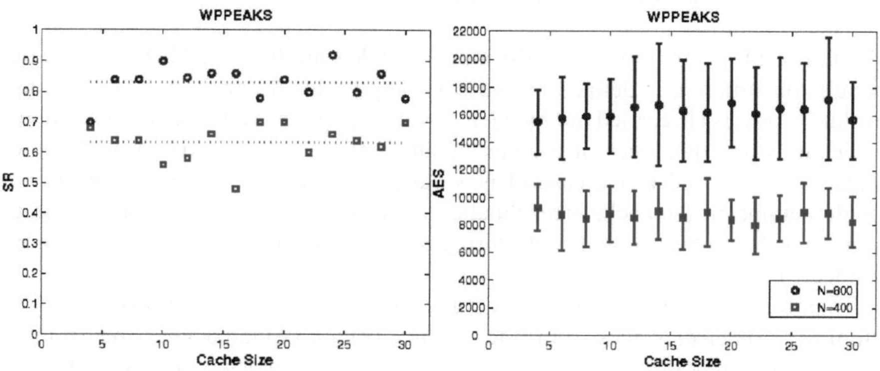

Fig. 4 Success Rate of the EvAg model (*left*) and Average Evaluation to Solution with standard deviation (*right*) for the wP-PEAKS. Population/network sizes of $N = 400$ and $N = 800$. Averaged values for the different cache sizes in dotted lines.

4.2 *Analysis of Results*

Figures 3 and 4 show the SR and AES of the EvAg model when tackling respectively the MMDP and wP-PEAKS problems for different cache and network sizes. In both cases, it can be observed that the network size impacts on the algorithm performance while the cache size does not. Larger network sizes improve the quality of the solutions but increase the computational effort. However, considering different cache sizes, the algorithm does not change either in terms of quality (SR) or run-time (AES).

Results in the two different instances can be analysed as follows:

- MMDP: The algorithm converges with a SR around 0.9 using a network size of $N = 440$. For $N = 200$ the SR decreases to ~0.2 showing that the P2P EA is quite sensitive to the network size when tackling the MMDP. Regarding cache sizes, results around the same averaged quality and run-time show the independence of the algorithm performance with respect to such a parameter. Additionally, there is not a clear trend toward the way the performances vary when changing the cache size values and variations around the average can be explained as an stochastic effect of EAs.
- wP-PEAKS: As in the case above, the algorithm performance is subject to the network size and is robust with respect to the cache size. Despite wP-PEAKS needing a larger network size than MMDP to yield a SR of ~0.9 (i.e. $N = 800$ vs. $N = 440$), an user could set the cache size regardless of the algorithm convergence.

5 Conclusions and Future Work

The Evolvable Agent model is a distributed and decentralized P2P EA in which the population structure is defined by the gossiping protocol newscast. This way, the evolution process is carried out by agents evolving within the P2P neighbourhood.

Features and dynamics of a given population structure are known to have an impact on EAs, and in the case of newscast, the protocol is mainly subject of a single parameter, the cache size. Smaller values for the cache size lead to higher average path lengths and higher clustering coefficients while larger values to a more connected topology.

Therefore, this paper analyses the effects of different cache sizes on the algorithm performance. Values from 4 to 30 have been considered since the protocol induces small-world topologies within that range in networks composed of up to 1600 peers. Results on the problem generator wP-PEAKS and the multimodal deceptive problem MMDP show that the algorithm quality and run-time are not altered when changing the settings of the cache size. This fact translates into the robustness of the EvAg model with respect to the cache size parameter and is specially remarkable from the point of view of tuning the algorithm parameters. Otherwise, setting an adequate cache size could not result a trivial decision when tackling different problem instances and complexities.

As future work, we will try to extend these results in churn scenarios, that is, considering that peers leave and re-join the P2P system it might be that the choice of an adequate cache size were much more determinant.

Acknowledgements. This work has been supported by the Spanish MICYT project TIN2007-68083-C02-01, the Junta de Andalucia CICE project P06-TIC-02025 and the Granada University PIUGR 9/11/06 project.

References

[Alba and Tomassini(2002)] Alba, E., Tomassini, M.: Parallelism and evolutionary algorithms. IEEE Trans. Evolutionary Computation 6(5), 443–462 (2002)

[Alfaro-Cid et al(2008)] Alfaro-Cid, E., Castillo, P.A., Esparcia-Alcázar, A., Sharman, K., Merelo, J.J., Prieto, A., Mora, A.M., Laredo, J.L.J.: Comparing multiobjective evolutionary ensembles for minimizing type i and ii errors for bankruptcy prediction. In: IEEE Congress on Evolutionary Computation, pp. 2902–2908. IEEE, Los Alamitos (2008)

[Cantú-Paz(1999)] Cantú-Paz, E.: Topologies, migration rates, and multi-population parallel genetic algorithms. In: Proceedings of the Genetic and Evolutionary Computation Conference, vol. 1, pp. 91–98. Morgan Kaufmann, Orlando (1999)

[Eiben and Smith(2003)] Eiben, A.E., Smith, J.E.: Introduction to Evolutionary Computing. Springer, Heidelberg (2003)

[García-Sánchez et al(2008)] García-Sánchez, P., Merelo, J.J., Laredo, J.L.J., Mora, A., Castillo, P.A.: Evolving xslt stylesheets for document transformation. In: Rudolph, G., Jansen, T., Lucas, S., Poloni, C., Beume, N. (eds.) PPSN 2008. LNCS, vol. 5199, pp. 1021–1030. Springer, Heidelberg (2008)

[Giacobini et al(2006)] Giacobini, M., Preuß, M., Tomassini, M., et al.: Effects of scale-free and small-world topologies on binary coded self-adaptive CEA. In: Gottlieb, J., Raidl, G.R. (eds.) EvoCOP 2006. LNCS, vol. 3906, pp. 86–98. Springer, Heidelberg (2006)

[Goldberg et al(1992)] Goldberg, D.E., Deb, K., Horn, J.: Massive multimodality, deception, and genetic algorithms. In: Männer, R., Manderick, B. (eds.) Parallel Problem Solving from Nature, vol. 2. Elsevier Science Publishers, B. V., Amsterdam (1992), `http://citeseer.ist.psu.edu/133799.html`

[Hidalgo and Fernández(2005)] Hidalgo, I., Fernández, F.: Balancing the computation effort in genetic algorithms. In: The 2005 IEEE Congress on Evolutionary Computation, 2005, vol. 2, pp. 1645–1652. IEEE Press, Los Alamitos (2005)

[Jelasity and van Steen(2002)] Jelasity, M., van Steen, M.: Large-scale newscast computing on the Internet. Tech. Rep. IR-503, Vrije Universiteit Amsterdam, Department of Computer Science, Amsterdam, The Netherlands (2002), `http://www.cs.vu.nl/pub/papers/globe/IR-503.02.pdf`

[Jelasity et al(2005)] Jelasity, M., Montresor, A., Babaoglu, O.: Gossip-based aggregation in large dynamic networks. ACM Trans. Comput. Syst. 23(3), 219–252 (2005)

[Jong et al(1997)] Jong, K.A.D., Potter, M.A., Spears, W.M.: Using problem generators to explore the effects of epistasis. In: Bäck, T. (ed.) Proceedings of the Seventh International Conference on Genetic Algorithms (ICGA 1997). Morgan Kaufmann, San Francisco (1997), `citeseer.ist.psu.edu/dejong97using.html`

[Laredo et al(2008)] Laredo, J.L.J., Castillo, P.A., Mora, A., Merelo, J.J.: Exploring population structures for locally concurrent and massively parallel evolutionary algorithms. In: IEEE Congress on Evolutionary Computation (CEC2008), WCCI2008 Proceedings, pp. 2610–2617. IEEE Press, Hong Kong (2008)

[Odeh et al(2006)] Odeh, S.M., Ros, E., Rojas, I., Palomares, J.M.: Skin lesion diagnosis using fluorescence images. In: Campilho, A., Kamel, M.S. (eds.) ICIAR 2006. LNCS, vol. 4142, pp. 648–659. Springer, Heidelberg (2006)

[Steinmetz and Wehrle(2005)] Steinmetz, R., Wehrle, K.: What is this peer-to-peer about? In: Steinmetz, R., Wehrle, K. (eds.) Peer-to-Peer Systems and Applications. LNCS, vol. 3485, pp. 9–16. Springer, Heidelberg (2005)

[Voulgaris et al(2004)] Voulgaris, S., Jelasity, M., van Steen, M.: A Robust and Scalable Peer-to-Peer Gossiping Protocol. In: Moro, G., Sartori, C., Singh, M.P. (eds.) AP2PC 2003. LNCS (LNAI), vol. 2872, pp. 47–58. Springer, Heidelberg (2004)

[Watts and Strogatz(1998)] Watts, D., Strogatz, S.: Collective dynamics of 'small-world' networks. Nature 393, 440–442 (1998), http://dx.doi.org/10.1038/30918

A Topic Map for "Subject-Centric" Learning

Gabriel Mihai, Liana Stanescu, Dumitru Burdescu, Marius Brezovan, and Cosmin Stoica Spahiu

Abstract. The paper presents an original algorithm for automated building of a topic map starting from a relational database. This aspect is favorable in e-learning domain because many e-learning systems use a relational database. The process is illustrated on a database used in TESYS e-learning system. The reason for topic map building and topic map graphical viewing is to offer students along with traditional learning modality named "course-centric" a new modality named "subject-centric". This new modality based on topic maps allows learner to navigate through topic map depending on its interest subject or to follow the associations between subjects. Students had the possibility to use these two learning modalities and they emphasized positive aspects of using topic maps, for example the possibility to understand better the relationships of knowledge.

1 Introduction

In the last few years, web-based e-learning has gained an increased popularity in many domains: technical, economic, medical, etc. The e-learning platforms should offer learners interactive and flexible interfaces for learning resources access and should also adapt easily to ones individual needs [1], [2].

At University of Craiova an e-learning system called TESYS has been created, which is used as stand-alone in distance learning form in certain domains (economic), but also in the hybrid learning in medical and engineering domains to complete face-to-face lectures [8].

Over the few years there has been a tendency of passing from "course-centric" learning systems to "subject-centric" learning systems. "Course-centric" learning

Gabriel Mihai · Liana Stanescu · Dumitru Burdescu · Marius Brezovan
Cosmin Stoica Spahiu
Software Engineering Department, University of Craiova, Romania
e-mail: stanescu, burdescu_dumitru@software.ucv.ro,
marius_brezovan, stoica_cosmin@software.ucv.ro

G.A. Papadopoulos and C. Badica (Eds.): Intelligent Distributed Computing III, SCI 237, pp. 141–150.
springerlink.com © Springer-Verlag Berlin Heidelberg 2009

systems are the traditional ones. They assume sequential run-over learning resources along with the lecture time schedules. In this way, learners acquire knowledge step by step in the order established by teacher. It's considered that less motivated students often loose enthusiasm in the middle of the course, having difficulties in knowledge understanding [16].

This is why an e-learning system should offer to the students the possibility of designing themselves the learning modality in order to stay motivated. This feature can be realized with "subject-centric" learning systems based on topic maps. So, the learner can choose himself the subject and topic maps not only permit subjects visualizing but also the relationships between them [16].

The paper presents an original algorithm for automated representation of a relational database with a topic map. This aspect is favorable in e-learning domain because a lot of e-learning systems use a relational database. For example, the Moodle database has around 200 tables. The information about Courses and their organization into categories are stored in the following tables: course, course_categories, course_display, course_meta, course_request. The information about Activities and their arrangement within courses are stored in the next tables: modules, course_modules, course_allowed_modules and course_sections. The database structure is defined, edited and upgraded using the XMLDB system [3]. Also, the Blackboard Learning System uses a relational database for storing necessary data [14].

The proposed algorithm will be illustrated on a database used in TESYS e-learning system.

The paper presents also a topic map graphical view that allows learner navigation for studying the topics and associations that shape in fact the relationships between topics in database. Associations provide the context information necessary to better understand a topic. Associations simulate the way human think and such are essential for knowledge modeling. This tool allows also the learning resources filtering by establishing the search criteria in topic map.

2 Related Work

Topic maps represent a new technology for the structuring and retrieval of information, based on principles used in traditional indexes and thesauri, with inspiration from semantic networks. Topic maps work with topics, the relationships between topics, and links to resources about those topics. Because topic maps are independent of the resources they describe, they enable their use in many different situations. As a result, the topic maps can be used in information access on the Web, in reference book publishing, or in the integration of corporate information repositories [7] [9] [10].

There are some available TM authoring tools, but they are useful to experts in knowledge representation, not to end-users (Ontopia Knowledge Suite [12],

Mondeca Intelligent Topic Manager [13]). Also, there are few specialized education-oriented TM tools that can be used to facilitate the creation, maintenance, search, and visualization of Topic Maps-based learning resources.

We can mention papers that present interesting and modern modalities of using topic maps in e-learning. For example, TM4L is an e-learning environment providing editing and browsing support for developing and using topic maps-based digital course libraries. The TM4L functionality is enhanced by an interactive graphical user interface that combines a hierarchical layout with an animated view, coupled with context sensitive features [4] [5].

Another author proposed topic map ontology, focusing on both students and teachers as active producers of learning resources. Topic maps customize the interface, and the interface should also provide possibilities for online students to share learning resources like "on campus" students do [6].

In [15] we proposed original ways of using topic maps in medical e-learning. The topic map is used for visualizing a thesaurus containing medical terms. The paper presents also how to use the topic map for semantic querying of a multimedia database with medical information and images.

3 TESYS Database Structure

In figure 1 there is a part of the relational database used by the e-learning system called TESYS [8]. This database will be used latter to explain better the topic map automated building and also it's graphical view. The table named Courses stores data about electronic courses, each course being equivalent to a unit of curriculum or an academic subject in traditional learning. Usually, a course contains many chapters, and each chapter a number of topics. Each topic represents one unit of knowledge, being the smallest component. The topic can be a piece of text, a video clip, a picture or a voiced text.

In this database structure the relationships between topics studied at the same course or different courses are important. If a topic uses some knowledge that is presented in other topics, these topics must be linked. As a result, on Topics table a m:m recursive relationship is defined. This special relationship is implemented with Topic_connection table.

4 The Algorithm for Building the Topic Map Starting from a Relational Database

The basic concepts of the topic map (TM) paradigm are [7] [9] [10]:

- Topics that represent subjects and carry names
- Occurrences of a topic that point to relevant resources

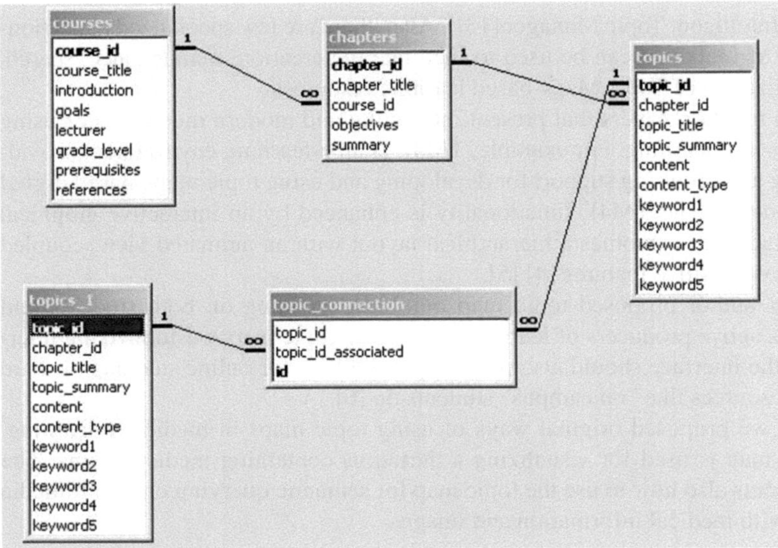

Fig. 1 TESYS database structure

- Associations that connect related topics
- Topic classes, occurrence classes and association classes that help to distinguish different kind of topics, occurrences and associations, respectively

Manual topic map population may require lots of resources: time, money and humans. As a result it is considered the problem of the self-populating the topic maps. The available resources that can act as a source of input to auto-population are identified: ontology, relational or object-oriented database, metadata about resources, index glossary, thesaurus, data dictionary, document structures and link structures or unstructured documents [9].

It can be observed among these resources, the presence of relational database, which can be used for the design and self- population of topic maps, solution adopted by us also.

A relational database consists of tables, columns, rows, keys and foreign keys. The transformation will probably not map all tables into the topic map, but selected ones. The obvious mapping is [9]:

Table → topic class; Row → topic instance of corresponding class; Column → name or occurrence; Key → topic id; Foreign key → association The topic map generation process is a general one. It can be used for any relational database. The result is a topic map that offers an intuitive view of the content from the learner point of view. Topic map content is stored in an xml file that respects xtm syntax [11].

Initially, the topic map generation process builds a topic types array: "database", "table", "row", "column", "relationship", "part", "whole", "primary", "foreign" that will be used as basic types for types and associations. From database meta-data the algorithm needs the following:

- The array with table names
- The relationships between tables

Topic map generation process has the following steps:

1. Topic generation for database and tables The algorithm produces a topic that is an instance of the topic database. This topic has an id that contains the database name. Example: For ELearning database it will be generated the topic with the id "ELearning", being an instance of the topic database.

```
<topic id="ELearning">
<instanceOf>
        topicRef xlink:href="#database␣" />
      </instanceOf>
<baseName><baseNameString>ELearning </baseNameString>
</baseName>
</topic>
```

In the same way, the algorithm creates a topic for each table in the database. The topic id is given by table's name, because this is unique in the database. For the database in figure 1, it is generated such a topic for tables: courses, chapters, topics, topics1, topic_connection.

Example: For table courses it is generated the topic with id "courses" that is an instance of the topic table.

2. Topic generation for table columns and records The algorithm will generate a topic for each column in a table. The topic id is given by the next syntax: table name. column name. This is our choice because each topic must have a unique id.

Example: For column course_title in table courses the generated topic has the id "courses. course_title". This topic is an instance of the topic column.

```
<topic id="courses.␣course_title">
   <instanceOf>
      <topicRef xlink:href="#column" />
   </instanceOf>
  <baseName>
     <baseNameString>course_title </baseNameString>
  </baseName>
 </topic>
```

The method generates a topic for each table record. Topic unique id has the next syntax: table nume. Row. Primary key value. The record content is considered as topic occurrence.

Example: The record with primary key value 3 in table chapter will be represented by a topic with the id "chapters.Row.3" that is an instance of topic row. One of the columns in table chapters is chapter_title. For this column content it is created an occurrence that is an instance of topic chapters.chapter_title and having the text value "B Trees".

```
<topic id="chapters.Row.3">
  <instanceOf>
        <topicRef xlink:href="#row" />
  </instanceOf>
  <baseName>
     <baseNameString> chapters.Row.3 ></baseNameString>
  </baseName>
  <occurrence>
     <instanceOf>
      <topicRef xlink:href="#chapters.chapter_title" />
      </instanceOf>
        <resourceData>B Trees </resourceData>
  </occurrence>
</topic>
```

3. Associations generation process 3.1 associations corresponding to relationships between tables For each relationship in the database it is generated an association of type "relationship". The association identification it is generated intuitively using the tables' names, the primary key and the foreign key. This development mode takes into consideration to offer information about the database structure for facilitating the learning process.

Example: For the relationship 1: m between tables courses and chapters it is generated an association with the next id: "courses.course_id-chapters.course_id". This association is an instance of the topic relationship. In this association, the table courses contains the primary key and plays the role "primary" and the table chapters containing the foreign key plays the role "foreign".

```
<association id="courses.course_id-chapters.course_id">
  <instanceOf>
    <topicRef xlink:href="#relationship" />
  </instanceOf>
  <member>
    <roleSpec>
      <topicRef xlink:href="#primary" />
     </roleSpec>
<topicRef xlink:href="#␣courses␣" />
  </member>
  <member>
    <roleSpec>
 <topicRef xlink:href="#foreign" />
     </roleSpec>
 <topicRef xlink:href="#␣chapters" />
  </member>
</association>
```

3.2 associations between database and tables The association between database and its tables is of type "part-whole". The topic representing the database plays the role "whole" and every topic representing a table plays the role "part".

Example: The association between the topic representing the database ELearning and topics representing the tables (courses, chapters, topics, topics1, topic_connection) has the id "Database:ELearning.Tables", being an instance of the topic part-whole.

3.3 associations between table and records The fact that a table contains records it is represented by an association of type "part-whole" between table and its records.

Example: For table courses it is generated an association with the next id: "Table:courses.Rows", that is an instance of the topic part-whole. In this association the topic representing the table courses plays the role "whole" and every topic representing a record plays the role "part".

3.4 associations between records involved in a 1:m relationship

This association is of type "related-to". In order to be generated, for every value of the primary key, the records that contain the same value in the foreign key column must be founded. As a result, this association is established between the topics already generated for every these record.

Example: Tables courses and chapters are involved in a 1:m relationship. The course entitled "Databases" contains 3 chapters stored in the table chapters. This fact is represented by an association of type "related-to" between the topic representing the corresponding record in the table courses and the topics representing connected records from the table chapters. Every topic plays the role "related".

5 Topic Map Graphical View

Topic map content is explored using a graphical interface with multiple views. The viewing interface for topic maps is organized in two windows: the left window displays a list of all topics, topic types, associations, association types, occurrence types and member types. As a topic type we have: column, database, row, table, etc.

In topic map there are 3 association types: "part-whole" (defined between the topics that represent database and tables or between topics representing a table and its records), "relationship" (defined between the topics representing tables implied in a 1:m relationship), and "related-to" (defined between the topics representing the table records bound with a 1:m relationship).

The learner can select an item from the list displayed in the left window and he will see in the right window the item's graphical representation. Topic map viewing tool intends to offer learner many information about the selected item. Unlike TM4L, our viewing tool displays for each topic involved in an association its occurrence content also. Example: for the topic that represents a record in Courses table, the learner will see information like: lecturer, grade_level, introduction, etc. Topic content can be visualized separately by selecting it in the left window.

Another original element in this graphical window is that the learner can see directly the record content involved in 1:m relationship implemented in topic map by "related-to" association. Beside these associations viewing that offer the topic better understanding, the learner can go directly to study the associated topic.

In figure 2 there are presented details of an association of type "related-to" defined between topics representing records in the table topics. Between these records there is a semantic relationship. Every topic in this association plays the role "related". In order to offer details, for every topic it is presented the occurrence content: content, content_type, topic_title, keyword1, keyword2, keyword3, etc.

The users can use the topic map as a navigation tool. They can navigate through topic map depending on their interest subject, having in this way big advantages. They don't have to be familiar with the logic of the database, they will learn about the semantic context, in which a collection and its single items are embedded and they may find useful items that they would not have expected to find them in the beginning.

At this moment, the graphical tool allows only a simple search based on topic types. The learner can specify a topic type and the application will display a list with all the topics of the selected type. Selecting an item in this list, the graphical window will display automatically its details.

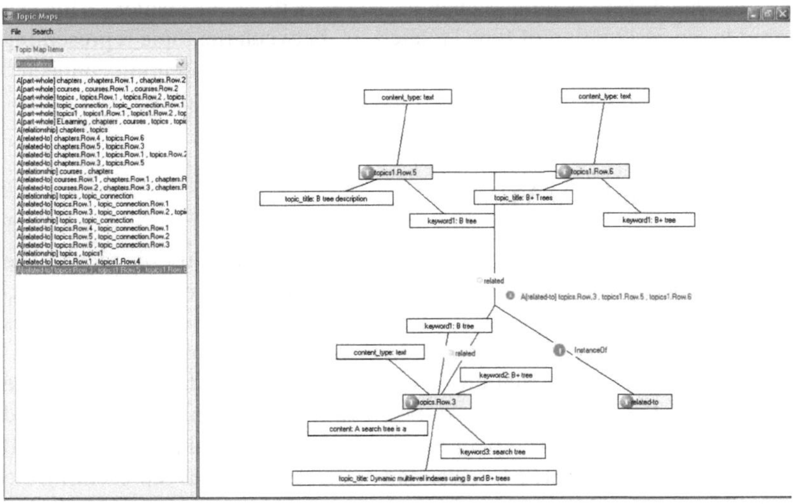

Fig. 2 An association of type "related-to"

6 Users' Feedback

A number of 60 students participated to the following experiment: they were asked to study the discipline Database using TESYS system, an on-line e-learning platform that uses a tree structure for displaying the learning content: the learner chooses the course, then a chapter, and finally a lesson. The existing relationships between learning objects are implemented as hyperlinks. The student can also use some search criteria. After that they had to study the same discipline using the topic map created with this software tool. The students' opinion over this two learning modalities was tested with a number of question presented in table 1.

The students' answers emphasized the fact that using topic maps in the e-learning field presents positive aspects: they are easy to use, the student can easy to pick a subject and see the relationships between subjects.

The students consider that viewing a large number of subjects in topic map can be a negative aspect. In this case, the student can feel "lost" in the middle of a large amount of information.

As a conclusion the students prefer the new modality based on topic maps.

Table 1 Questionnaire

Question	The number of "Yes" answers
Is it easy to find necessary subject in TM?	42
Is it easy to find associated subjects in TM?	50
Does TM enables the study in a free way?	33
Does TM keeps you motivated?	48
Is TM easy to use?	53
Is it difficult to understand the subjects' order in the course using TM?	49
Is it easy to understand the whole structure of knowledge using TM?	56
Are too many subjects displayed in TM?	32
What kind of method do you prefer between "course-centric" and "subject-centric"?	46 – subject-centric

7 Conclusions

The e-learning systems must have powerful and intuitive tools for viewing the learning resources, for browsing the lessons or topics and relationships between them, and also for searching the relevant information. An important feature of an e-learning system is the presentation way of the semantic relationships between topics using an appropriate navigational structure.

This aim can be achieved using a modern concept - topic map. Topic Maps are an emerging Semantic Web technology that can be used as a means to organize and retrieve information in e-learning repositories in a more efficient and meaningful way.

As a result, the paper presents two important aspects:

- The algorithm for topic map automated building starting from a relational database. The existing topic maps software doesn't allow this thing. This aspect is useful because there are many e-learning systems that store the educational content in a database.

- A topic map graphical view with important facilities for learner: topic map navigation useful in studying topics that represent in fact learning objects and associations between them. This window allows learner to filter the information based on his interest

The students have found useful this new modality of knowledge visualization.

Acknowledgements. This research was partially supported by the Romanian National University Research Council under the PCE Grant No. 597.

References

1. Rosenberg, M.: E-Learning: Strategies for Delivering Knowledge in the Digital Age. McGraw-Hill, New York (2001)
2. Wentling, T., Waight, C., Gallaher, J., La Fleur, J., Wang, C., Kanfer, A.: E-Learning: A Review of Literature (2000), `http://learning.ncsa.uiuc.edu/papers/elearnlit.pdf`
3. Moodle (2009), `http://docs.moodle.org/en/Development:Database_schema_introduction`
4. Dicheva, D., Dichev, C., Dandan, W.: Visualizing topic maps for e-learning. In: Advanced Learning Technologies, Proc. ICALT 2005, pp. 950–951. IEEE Computer Society Press, Los Alamitos (2005)
5. Dandan, W., Dicheva, D., Dichev, C., Akouala, J.: Retrieving information in topic maps: the case of TM4L. In: ACM Southeast Regional Conference, pp. 88–93 (2007)
6. Kolas, L.: Topic Maps in E-learning: An Ontology Ensuring an Active Student Role as Producer. In: Proceedings of World Conference on E-Learning in Corporate, Government, Healthcare, and Higher Education, Ed/ITLib Digital Library, Association for the Advancement of Computing in Education (AACE), pp. 2107–2113 (2006)
7. Garshol, L.M.: Metadata? Thesauri? Taxonomies? Topic Maps! Journal of Information Science 30(4) (2004) ISSN 0165-5515
8. Stanescu, L., Mihaescu, M.C., Burdescu, D.D., Georgescu, E., Florea, L.: An Improved Platform for Medical E-Learning. In: Leung, H., Li, F., Lau, R., Li, Q. (eds.) ICWL 2007. LNCS, vol. 4823, pp. 392–403. Springer, Heidelberg (2008)
9. Rath, H.: The Topic Maps Handbook. Empolis GmbH, Empolis GmbH, Gutersloh, Germany (2003)
10. TopicMaps, `http://www.topicmaps.org/`
11. XML Topic Maps (XTM) 1.0, `http://topicmaps.org/xtm/1.0/index.html`
12. Ontopia (2009), `http://www.ontopia.net/`
13. Mondec (2009), `http://www.mondeca.com/`
14. Blackboard (2009), `http://www.blackboard.com/Teaching-Learning/Learn-Platform.aspx`
15. Stanescu, L., Burdescu, D., Mihai, G., Ion, A., Stoica, C.: Topic Map for Medical E-Learning. Studies in Computational Intelligence, vol. 162, pp. 305–310. Springer, Heidelberg (2008)
16. Matsuura, S., Naito, M.: Creating a Topic Maps based e-Learning System on Introductory Physics. Leipziger Beitrage zur Informatik XII, 247–260 (2008)

Emergent Properties for Data Distribution in a Cognitive MAS

Andrei Olaru, Cristian Gratie, and Adina Magda Florea

Abstract. Emergence is a key element in the research of multi-agent systems. Emergent properties provide higher level features of a system formed by simpler individuals. So far, emergence has been studied mostly by means of systems formed by reactive agents – that are more simple to design and implement. As computational power increases and even small devices become more capable, cognitive agents become a promising solution for the solution of more complex problems. This paper presents a simple example of a multi-agent system inspired from the Beliefs-Desires-Intentions model, designed to manifest emergent properties: a data storage system that assures distribution and replication of data without the need for these features to be controlled from the exterior.

1 Introduction

Emergence is a key concept in the recent development of multi-agent systems. Several definitions of emergence exist [3] but none is yet generally accepted. Many times emergence is defined by its effect – the formation of patterns in the structure or behaviour of certain systems.

In the context of a large number of agents, forming a complex system [1], emergence offers the possibility of obtaining a higher level function or property by using agents with lower level implementations. The use of emergence allows a considerable reduction in the complexity needed for the individual agents, therefore they can run on simpler and smaller devices.

Many recent papers on emergence and multi-agent systems deal with emergent properties and behaviour in systems formed of a large number of reactive agents. Reactive agents are used because they are simple and need very limited computational

Andrei Olaru · Cristian Gratie · Adina Magda Florea
University Politehnica of Bucharest, Splaiul Independentei 313, 060042 Bucharest, Romania
e-mail: `cs@andreiolaru.ro`, `cgratie@yahoo.com`, `adina@cs.pub.ro`

G.A. Papadopoulos and C. Badica (Eds.): Intelligent Distributed Computing III, SCI 237, pp. 151–159.
springerlink.com © Springer-Verlag Berlin Heidelberg 2009

capacity, but emergent functions may be more complex and the structure is robust [14, 8, 2].

Nowadays, the capabilities of even very basic computing devices have considerably increased, allowing for a much more complex internal structure for agents – the possibility to hold reasonable amounts of data and to have a more nuanced behaviour. Cognitive agents have knowledge about the surrounding environment, have goals they desire to fulfill, make plans and take action in order to fulfill them.

The purpose of this paper is to study emergent properties in a multi-agent system formed of cognitive agents. A system of cognitive agents used for the storage of data has been designed and implemented. This system is, at this time, very simple and limited, but it makes a good framework for the study of emergent properties.

The system was designed with the purpose of manifesting emergent properties like uniform distribution and availability of data. Individual agents were given local goals that do not directly imply that the data is uniformly distributed or that data is not completely lost. No agent has the capability to have a global image on the system or even on a significant part of it. Agents' knowledge is limited to beliefs about its immediate or almost immediate neighbours. Yet the system as a whole keeps data well replicated and distributed.

The paper is organised as follows. Section 2 is dedicated to related work in the field of emergence and multi agent systems. Section 3 presents the main topic of this paper: a cognitive multi-agent system designed to manifest emergence. Section 4 describes the results obtained from experiments. The last section is dedicated to the conclusions.

2 Related Work

What we know is that emergence appears in the context of complex systems [1] – systems composed of a large number of interacting individual entities. Emergence needs two levels of perspective: the inferior, or micro level of the individual entities and the superior, or macro level of the whole system. A simple definition is that "emergence is the concept of some new phenomenon arising in a system that wasn't in the system's specification to start with" [12]. A more elaborated definition is that "a system exhibits emergence when there are coherent emergents at the macro-level that dynamically arise from the interactions between the parts at the micro-level. Such emergents are novel with respect to the individual parts of the system" [5]. An "emergent" is a notion that can represent a property, a structure or a behaviour that results from emergence.

The essence of emergence is not actually the novelty or the unexpectedness of the emergent – as these will fade at later experiments although the emergents will stay the same – but the difference between the description of the individual and the description of the emergent [12]. If the minimal description of the individual is taken, it cannot be used to describe the emergents resulting from the system, therefore the emergent is considered as novel and, potentially, unexpected.

More formally, a better definition of emergence [2] would be that an emergent is, in the context of an interacting set of agents whose dynamics are expressed in a vocabulary D, a global phenomenon – static or dynamic, but nevertheless invariant – that is observed by the agents or by an external observer and can only be interpreted in a vocabulary D' that is different from D.

There are a few important features of emergence [5, 6, 7]:

- emergence occurs out of the interactions between the parts of a complex system.
- emergence is defined in relation with two levels – it is manifested at the higher level, arising from interactions at the lower level.
- the representation of the emergents cannot be reduced to the specifications of the individuals.
- emergents only arise after a certain time in which the system has evolved.
- once they occurred, emergents will maintain identity over time.
- emergents arise without any centralised or exterior control.
- the emergent phenomena is robust and flexible, i.e. it is not influenced by damage on the system (even if the emergent is temporarily not observable, it will arise again as the system evolves).

In most examples of emergence developed so far in the field of multi-agent systems, reactive agents have been used. They are easier to implement and control, and they are suitable for small devices with low computational power.

Notable examples of emergents in reactive agent systems relate to the formation of a certain geometrical or geometry-related structure or behaviour: arrangement of agents in a circular or almost circular shape [2, 9, 14]; detection of areas in an image [4]; gathering of resources in a single area [11]; foraging of food or transportation of loads [13]; emergence of a traffic direction [10].

Although reactive agent systems may be very useful, there are many advantages that a cognitive agent has over a reactive agent. First, it is proactive. Even if there are no signals, perceptions or stimuli from the environment, a cognitive agent may act by itself, taking action according to its objectives. Second, a cognitive agent is aware of its situation and may reason about it. It is aware of what it is suppose to fulfil as final goal and is capable of making plans and taking action towards the realisation of its goal. The cognitive agent can use its experience and information about the environment and the consequences of past actions to develop a better plan every time a similar situation occurs.

3 A Cognitive Multi-agent System for the Distribution of Data

A simple application has been developed and experiments have been performed in order to demonstrate how emergents may arise in a system formed of cognitive agents. The system models a data storage network. Its purpose is to store pieces of data (of equal size) injected from the environment and to provide the environment with the data that is requested. There is a limited set of pieces of data, with certain similarities between them. The desired emergents are the replication, distribution and availability of the data across the system.

The experiments were performed with a set of agents placed on a rectangular grid. Each agent can communicate only with its 8 neighbours. Each agent has a limited capacity of 4 data chunks (6 different data chunks are used in the experiments). The agents must always have some capacity left, ready for data chunks that might be injected from the environment. However, agents must try to store as much data as possible, in order to not waste the capacity of the system.

The desired emergent properties of the system are replication of data, distribution (possibly uniform) and availability. That means that, after letting the system evolve for some time, the following properties should exist: there should be more than one agent holding one piece of data; in any area of the grid there should exist copies of all pieces of data; if requested to, any agent should be able to get hold of any piece of data.

Following the ideas in [14, 9, 3, 2], in order to obtain the desired emergents the individual agents were designed with selfish objectives that reflect the spirit of the global objectives. In order to obtain data replication and distribution, an agent is "curious" and, if capacity is available, it will be interested in and it will request a copy of any piece of data it does not hold and that is similar to what it already holds. Using similarity is good because it is more likely for subsequent external requests to be made for similar content. In order to obtain variation and uniformity in the data distribution pattern, an agent will "lose interest" in pieces of data that are already held by most of the neighbour agents. In order for an agent to be able to get hold of pieces of data not held by itself or any of the neighbours, agents will be able to collaborate for the common goal of finding a certain piece of data that, if data is well distributed, cannot be very far away.

Agents were implemented using the Beliefs-Desires-Intentions model. Beliefs of an agent are associations of the type $\langle DataID, AgentID \rangle$ specifying the pieces of data held by itself and by each agent in the vicinity. Knowledge about other agents is also associated with the *ID* of the neighbour agent that provided that knowledge. Beliefs that have not been refreshed or revised for a long time will be discarded, or "forgotten", as the probability for them to still be true decreases.

According to the ideas above, the goals of an agent are:

- In case there is an external request for data, provide that data or try to find it in the vicinity.
- Maintain 25% of the capacity free, ready for potential data coming from the environment.
- In case there is available capacity (over 25%), request interesting data from a neighbour.
- Process and respond to messages from neighbour agents.
- If all other objectives are complete (capacity 75%, no messages), discard some data that is already contained by most of the surrounding agents.

In order to fulfill its objectives, an agent has the following available actions:

- Send a request for data to a neighbour.

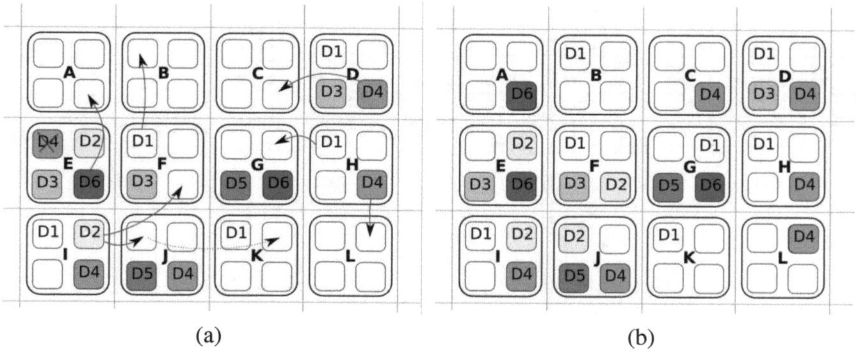

Fig. 1 Group of agents, before (a) and after (b) the data transfers indicated by the arrows. There are 6 data pieces in the system: D1 to D6.

- Send a piece of data to a neighbour (as reply to a request only).
- Receive data or knowledge from a neighbour.
- Discard a piece of data.
- Broadcast to the all neighbours own knowledge and intentions.

All agents in the system work simultaneously. At each step, one agent chooses a goal that is neither already completed (like already having 25% of the capacity free) nor impossible to attain (the agent cannot retrieve specified data from any of the neighbours). After choosing a goal, the agent makes a plan formed of the actions it needs to take to fulfill the goal, and starts performing those actions.

The example in Figure 1 presents a group of agents in a system where data did not get to spread very much. Consider every agent has sufficient knowledge about his vicinity. For example, agent **I** has the following beliefs:

$\langle I : D1 \rangle$ $\langle I : D2 \rangle$ $\langle I : D3 \rangle$
$\langle J : D4 \rangle$ $\langle J : D5 \rangle$
$\langle E : D4 \rangle$ $\langle E : D2 \rangle$ $\langle E : D3 \rangle$ $\langle E : D6 \rangle$
$\langle F : D1 \rangle$ $\langle F : D3 \rangle$
$\langle G : D5 \rangle$ (knows that from F)

Each agent that has less than 75% of its capacity full will ask for some piece of data from one of its neighbours. Agent **E** will discard data $D4$, as both **I** and **J** also have it. Supposing that agent **K** needs D2 (most likely because there is an external request for it) it will communicate its intentions to its neighbours and **J** will collaborate with it and retrieve D2.

There are several tweaks that had to be performed on the system in order for it to function well and for the agents not to be overwhelmed by messages and/or by actions to perform. These tweaks were implemented in the agents' policies. First, there must exist a good balance between how fast the agent can learn new knowledge and revise its beliefs and how often an agent broadcasts its own knowledge and

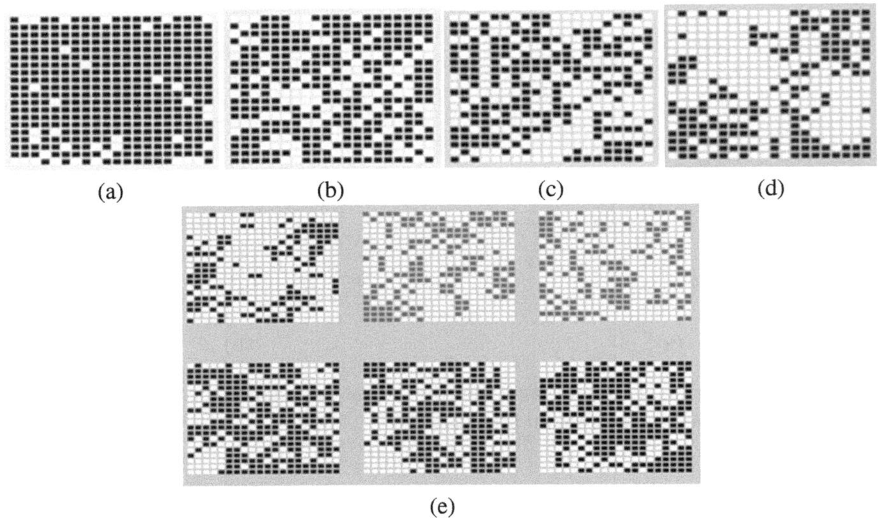

Fig. 2 The distribution of one piece of data in a system with agents of capacity 4 in the context of a total number of pieces of data of: (a) 3, (b) 4, (c) 6, (d) 8. The simultaneous distributions of 6 pieces of data (e).

intentions to its neighbours. If an agent broadcasts all its knowledge each time it leans something new, all agents will be flood with messages they don't have time to process. If the agents broadcast their knowledge too rarely, this might result in the agents having outdated beliefs about their neighbours. The settings used for the experiments were to broadcast all new knowledge about every 10 to 20 steps.

Another element that needs good balance is the rate at which an agent forgets old beliefs. If an agent forgets too quickly, it might have already forgotten some important information at the time it is supposed to be using it. If an agent forgets too slowly, it will have a lot of information that is outdated and of no use anymore.

4 Results and Discussion

Many experiments were carried out throughout the development of the system, first for assuring that the agents were not overwhelmed by messages and actions to perform, second for the actual study of the evolution of the system. The presented experiments involved a square grid of 400 agents.

Several cases were studied, according to the number ND of pieces of data injected into the system: $ND = 75\% \cdot agent\_capacity$, $ND = 100\% \cdot agent\_capacity$, $ND = 150\% \cdot agent\_capacity$ and $ND = 8(number\ of\ neighbours)$.

Case 1 (75%) The system acted as expected: every agents contains approximately all data (see the distribution for one piece of data in Figure 2 (a)). Here,

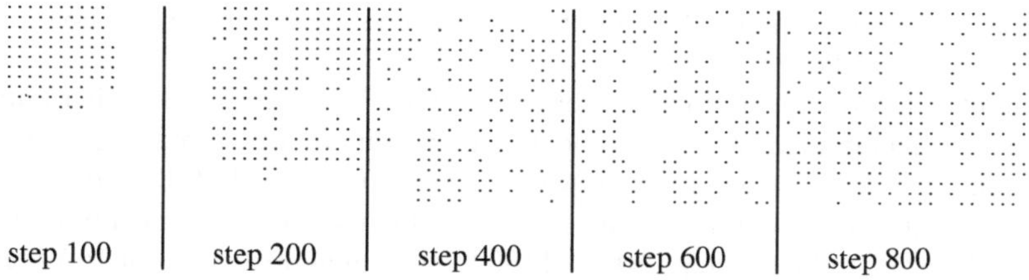

Fig. 3 The distribution of one piece of data injected at moment 0.

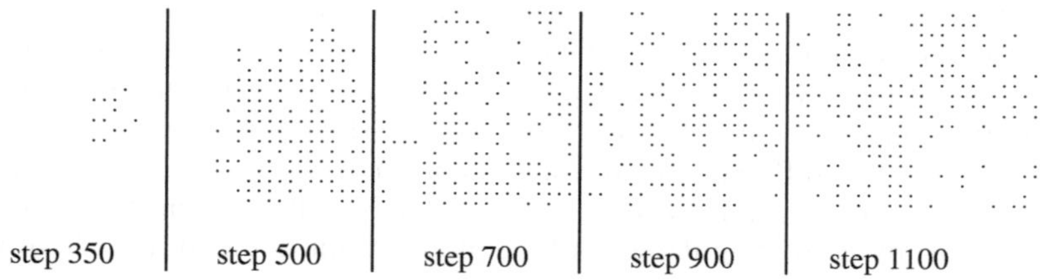

Fig. 4 The distribution of one piece of data injected at moment 300.

"approximately" means that, due to the "loss of interest" in certain pieces of data, they are discarded and, later, acquired again. Data is almost perfectly distributed. The "holes" that are present (agents not holding that data) "move" around the system, however there is always a neighbour that has that data.

Cases 2 and 3 (100% and 150%) After the system stabilises, data remains evenly distributed throughout the system (Figure 2 (b, c)). The "holes" present in the distribution of a certain piece of data are larger as more data is stored in the system, however the uniform distribution makes it easy for agents to quickly obtain data, if necessary.

Case 4 (200%) Closer to reality, the case when more data is present in the system yields a less even distribution (Figure 2 (d)). However, no data is farther than two neighbours away. In this case, when the pressure on agents is high (each agent only holds 30% to 50% of the data available in the system), similarity between data becomes important and observable. Figure 2 (e) shows the data distributions for 6 of the 8 pieces of data, of which the first three have a high similarity between them, as well as the last three. It is easy to observe that the distributions of similar data are resemblant, showing that similar data tends to be found in the same areas.

It is also interesting to follow how the distribution of one piece of data evolves as the system runs. Figure 3 presents the evolution of data injected at moment 0 – when the system was started – in the top left corner. At first the distribution is solid (step 100). When most of the agents start to hold that data, agents in the central part start discarding it (step 200). Next, the phenomenon spreads throughout the system and the distribution becomes poor (steps 400, 600). Finally (step 800), the distribution stabilises at a normal value of 179 out of 400 agents, with the system holding 5 different pieces of data.

A slightly different evolution is observed in Figure 4, that shows the distribution of a piece of data injected at a later time, together with other two (making a total of 8). Initially, the distribution grows slower (the system is already full) and no solid distributions are found. An interesting fact is that the number of agents holding the data stabilises much faster, steps 500 to 1100 having an almost constant 160 agents holding the data.

Although all the examples presented here involve the same number of agents, this number can be changed easily and the system scales with no problem, as every agents interacts only with and has knowledge only about its close vicinity.

What is important to point out is that very good distributions are obtained although none of the agents has that as objective. Moreover, agents are not capable of measuring in any way the distribution of the data. Their objectives are local and mostly selfish, but a different global result is obtained. As shown in the previous subsection, the system and the agents have been specifically designed to produce the emergents, by translating the idea of the global result to the local scale of the individual.

The developed system makes a good platform for the study of emergent properties in a simple cognitive agent system. Although the designed system is still very simple, there are important differences between the implemented agents and reactive agents. It must be pointed out the the agents reason permanently about what data they should get and what they should keep. A simple form of collaboration has been implemented – agents are able to share intentions. Reactive agents would not have been able to have beliefs about their neighbours and would not have been able to reason about what action would be best in the context.

5 Conclusion

Emergence is an essential notion in the field of multi-agent systems. It provides the possibility of obtaining a more complex outcome from a system formed of individuals of lower complexity. Although there are many recent studies concerning emergence, there is yet no clear methodology on how to specifically design a system so that it would manifest emergence. Moreover, most implementations use reactive agents, that have limited cognitive and planning capacities.

The paper presents a cognitive multi-agent system in which the agents' interactions allow the emergence of specific properties required for solving the problem, by giving the agent local goals that naturally lead to the global, desired, goal of the system.

As future work, the system will be improved, so that the emergents will be more nuanced. Agent interests will have a stronger influence and collaboration will be used in a greater measure.

References

1. Amaral, L., Ottino, J.: Complex networks: Augmenting the framework for the study of complex systems. The European Physical Journal B-Condensed Matter 38(2), 147–162 (2004)
2. Beurier, G., Simonin, O., Ferber, J.: Model and simulation of multi-level emergence. In: Proceedings of IEEE ISSPIT, pp. 231–236 (2002)
3. Boschetti, F., Prokopenko, M., Macreadie, I., Grisogono, A.: Defining and detecting emergence in complex networks. In: Khosla, R., Howlett, R.J., Jain, L.C. (eds.) KES 2005. LNCS (LNAI), vol. 3684, pp. 573–580. Springer, Heidelberg (2005)
4. Bourjot, C., Chevrier, V., Thomas, V.: A new swarm mechanism based on social spiders colonies: From web weaving to region detection. Web Intelligence and Agent Systems 1(1), 47–64 (2003)
5. De Wolf, T., Holvoet, T.: Emergence versus self-organisation: Different concepts but promising when combined. In: Brueckner, S.A., Di Marzo Serugendo, G., Karageorgos, A., Nagpal, R. (eds.) ESOA 2005. LNCS (LNAI), vol. 3464, pp. 1–15. Springer, Heidelberg (2005)
6. Goldstein, J.: Emergence as a construct: History and issues. Emergence 1(1), 49–72 (1999)
7. Heylighen, F.: The science of self-organization and adaptivity. In: The Encyclopedia of Life Support Systems, pp. 1–26 (2002)
8. Mamei, M., Vasirani, M., Zambonelli, F.: Selforganising spatial shapes in mobile particles: the TOTA approach. In: Engineering Self-Organising System, pp. 138–153 (2004)
9. Mamei, M., Zambonelli, F.: Spatial computing: the TOTA approach. In: Babaoğlu, Ö., Jelasity, M., Montresor, A., Fetzer, C., Leonardi, S., van Moorsel, A., van Steen, M. (eds.) SELF-STAR 2004. LNCS, vol. 3460, pp. 307–324. Springer, Heidelberg (2005)
10. Picard, G., Toulouse, F.: Cooperative agent model instantiation to collective robotics. In: Gleizes, M.-P., Omicini, A., Zambonelli, F. (eds.) ESAW 2004. LNCS (LNAI), vol. 3451, pp. 209–221. Springer, Heidelberg (2005)
11. Randles, M., Zhu, H., Taleb-Bendiab, A.: A formal approach to the engineering of emergence and its recurrence. In: EEDAS-ICAC, pp. 1–10 (2007)
12. Standish, R.: On complexity and emergence, pp. 1–6 (2001) Arxiv preprint nlin.AO/0101006
13. Unsal, C., Bay, J.: Spatial self-organization in large populations of mobile robots. In: Proceedings of the 1994 IEEE International Symposium on Intelligent Control, pp. 249–254 (1994)
14. Zambonelli, F., Gleizes, M., Mamei, M., Tolksdorf, R.: Spray computers: Frontiers of self-organization for pervasive computing. In: Proceedings of the 13th IEEE Int'l Workshops on Enabling Technologies, WETICE, pp. 403–408 (2004)

Dynamic Process Integration Framework: Toward Efficient Information Processing in Complex Distributed Systems

Gregor Pavlin, Michiel Kamermans, and Mihnea Scafeş

Abstract. The Dynamic Process Integration Framework (DPIF) supports efficient creation of distributed systems for collaborative reasoning, which is relevant for contemporary problems, such as situation assessment, prediction, planning and evaluation of alternatives in a decision making process. In particular, the DPIF is a multi agent systems approach supporting (i) a systematic encapsulation of heterogeneous processes and (ii) negotiation-based self configuration mechanisms which automate creation of meaningful work flows implementing complex collaborative reasoning processes.

1 Introduction

This paper introduces a service oriented architecture supporting complex collaborative processing in distributed systems. The presented approach is relevant for many contemporary applications that require reasoning about complex processes and phenomena in real world domains. For example, in crisis management advanced information processing is required for (i) identification of critical situations, (ii) impact assessment which takes into account possible evolution of physical processes, (iii) planning and evaluation of countermeasures and (iv) decision making. This can be achieved only through adequate processing of large quantities of very heterogeneous

Gregor Pavlin
Thales Nederland B.V. , D-CIS Lab Delft
e-mail: gregor.pavlin@icis.decis.nl

Michiel Kamermans
Thales Nederland B.V. , D-CIS Lab Delft
e-mail: michiel.kamermans@icis.decis.nl

Mihnea Scafes
University of Craiova
e-mail: scafes_mihnea@software.ucv.ro

G.A. Papadopoulos and C. Badica (Eds.): Intelligent Distributed Computing III, SCI 237, pp. 161–174.
springerlink.com © Springer-Verlag Berlin Heidelberg 2009

information, based on rich expertise about different aspects of the physical world, which exceeds the cognitive capabilities of a single human expert; a human does not have knowledge of all the relevant mechanisms in the domain and cannot process the huge amounts of available information. On the other hand, full automation of decision making processes in such settings is not feasible, since the creation of the required domain models as well as inference are intractable problems. Namely, automated inference processes involve many variables and relations with accompanying representation and specific inference mechanisms.

In such settings the solutions must support collaborative processing based on a combination of automated reasoning processes and cognitive capabilities of multiple human experts, each contributing specific expertise and processing resources. Key to effective combination of human-based expertise and automated reasoning processes is a framework which allows each piece of the relevant information is adequately considered in the final processing outcome. The main elements of such a framework are:

- Standardized data formats that facilitate sharing of heterogeneous information.
- Filtering services which provide stakeholders in a decision making process with the information they can process in the context of their role. In principle, filtering services must (i) transform very heterogeneous data/observations to more abstract information types, i.e. extract high-level information through interpretation of heterogeneous cues, and (ii) route the interpretation results to the consumers that can make use of the extracted information.

In this paper we focus on the latter challenge, which is tackled with the help of the Dynamic Process Integration Framework (DPIF). The DPIF supports seamless integration of heterogeneous domain knowledge and processing capabilities into coherent collaborative processes. Processes are encapsulated by modules, each using identical collaboration functionality. The DPIF architecture supports efficient construction of information processing systems; collaboration between the processes is based on the specification of the processing capabilities (i.e. services).

The presented approach exploits service oriented architecture in new ways. In contrast to traditional MAS approaches [13], the DPIF facilitates integration of human cognitive capabilities right into the problem solving processes in work flows; humans are not mere users, but contribute processing resources. Furthermore, we exploit MAS and service oriented architectures to implement hybrid collaborative reasoning systems with emergent problem solving capabilities. In general, a key to efficient collaborative processing in complex domains are work flows, in which peers with different processing capabilities exchange relevant information. The presented paper focuses on efficient distributed configuration of meaningful processing work flows by using local knowledge of relations between different variables (i.e. phenomena). In this way centralized ontologies and centralized composition of work flows, typical for traditional approaches [12], are avoided.

The paper is organized as follows: in section 2 a rationale for decentralized collaborative reasoning in work flows is provided and the basic features of the DPIF are introduced; aection 3 explains how meaningful work flows between heterogeneous processes can be dynamically implemented, without centralized knowledge of relations between the variables in the reasoning processes. In particular, we emphasize a combination of service composition, decentralized validation methods and advanced negotiation mechanisms, which allow a systematic incorporation of various criteria into the work flow creation processes; section 4 introduces basic DPIF architecture principles.

The presented work is partially funded by the European Union under the Information and Communication Technologies (ICT) theme of the 7th Framework Programme for R&D, ref. no: 224318.

2 Collaborative Processing

Reasoning about domains requires knowledge about typical dependencies (i.e. relations) between relevant phenomena in the physical world. By using (i) domain models capturing the relations between relevant phenomena and (ii) evidence based on observations of certain phenomena, we can assess (i.e. estimate) the states of the domain that cannot be observed directly. In addition, with the help of models, the future evolution in the domains can be predicted. However, in complex domains reliable reasoning can be achieved only by relating large quantities of information of very heterogeneous types with very different semantics. Such dependencies can be explained only through complex models.

Irrespectively of the used models, it is unlikely that a single model designer or an expert understands all the relevant phenomena and relations between them in complex domains. Instead, a set of relatively simple domain models will exist, with each model capturing a small subset of the relevant variables and the corresponding relations. Thus, reasoning based on the relations between the entire available evidence can be achieved only by combining simpler processes, each using a limited domain model. The outputs of simple processes are used as inputs of other simple processes. In other words, the reasoning is based on work flows established between heterogeneous processes. In such data-driven work flows, difficult problems can be solved through collaboration of heterogeneous processes, each focusing on a relatively small subset of relevant aspects in the targeted domain.

We illustrate such processing by using an example from the environmental management domain. We assume a chemical incident in which the impact is mitigated through a collaboration of experts captured by Fig. 1. We assume that the factory staff (FS) at the incident have an overview of the current state of the damaged system; FS can estimate the quantity of the escaping chemical and its type. This information can be used by a chemical expert at the incident location (CE1) to estimate the type and quantity of toxic fumes resulting from the fire. By knowing the location of the fire, the meteorological conditions as well as the quantity and type of the produced fumes, chemical expert (CE2) can (i) estimate the zones in which the concentration of the toxic gases have exceeded critical levels and (ii) identify areas which are likely to be critical after a certain period of time. The CE2 makes use of the domain knowledge about the physical properties of the gases and their propagation

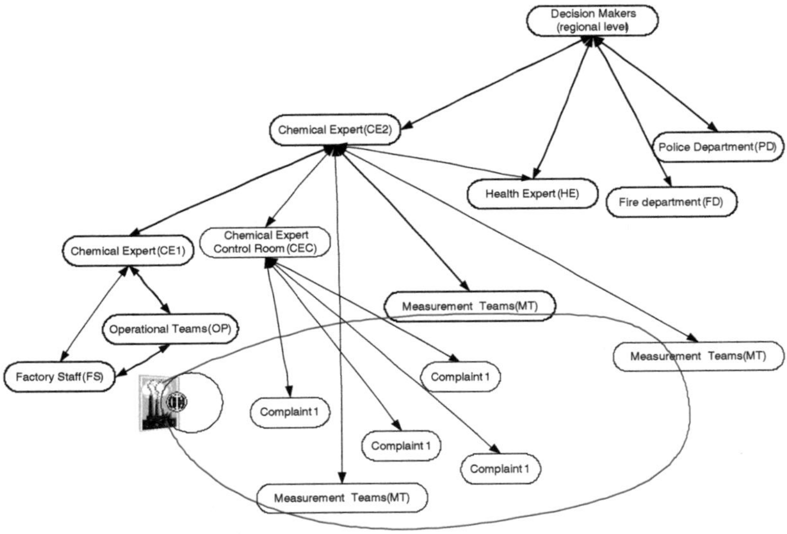

Fig. 1 A workflow in a decision making process. Arrows denote information flow between different experts, each processing relevant information of different types. The circled region denotes the initial estimate of the area where concentration is likely to be critical.

mechanisms. In addition, it guides fire fighter teams (MT) which can measure gas concentrations at specific locations in order to provide feedback on a more accurate estimation of the critical area. A map showing the critical area is supplied to a health expert (HE) who uses the information on population obtained from the municipality to estimate the impact of the toxic fumes on the humans in case of exposure. Finally, the estimated impact on the population is supplied to decision makers, who choose between no action, evacuation and sheltering. This decision also considers estimated time and costs in case of the evacuation from the danger zone as well as the estimated costs and duration of the preventive evacuation. The former estimate is provided by the fire brigade representatives while the later estimate is supplied by the police department. In other words, in such a system, each expert can be viewed as a module providing predefined services which in turn require services from other experts. Thus, the situation analysis in the presented example can be viewed as a work flow between different, weakly coupled processing services, each specialized for specific aspects of the domain (e.g. types and quantities of toxic fumes produced by burning the chemical, propagation of fumes, measurements with chemical probes, etc.). Moreover, a processing service can be provided by a human (e.g. a chemical expert analyzing the extent of the contamination) or by an automated reasoning process (e.g. detection of gases based on automatic fusion of sensor data). Note that, for the sake of clarity, the used example is a significant abstraction of real crisis management processes.

Moreover, each incident is a unique combination of events, requiring a specific work flow consisting of a particular combination of processing nodes. Because sequences of events in incidents are unpredictable, it is impossible to specify an adequate work flow a priori. For example, given the wind direction, experts for the evacuation of hospitals and schools might be needed. Instead, if the toxic gas is blown to the open sea, no evacuation is required. Thus, in the latter case experts for the evacuation are not needed. However, a crisis can be viewed as an arbitrary combination of known types of events/processes, each understood by a human expert or modeled by an artificial agent. For example, the way chemicals burn and react, the effects of exposure to toxic fumes, evacuation approaches in hospitals and schools, etc. are independent of the location and time. Therefore, we can obtain general knowledge about such processes/phenomena which can be used for the analysis in any situation involving such phenomena. In other words, mapping between experts and event types can be made a priori, which is the case in the domains we are focusing on.

Clearly, a major challenge is creation of adequate work flows which correctly integrate the relevant processes and support globally coherent processing in decentralized collaborative systems. In the following text we explain how this can be achieved in an efficient and sound way.

2.1 Dynamic Process Integration Framework

The Dynamic Process Integration Framework (DPIF) supports decentralized creation of work flows that facilitate collaborative problem solving. The DPIF is a service-oriented approach (SOA) which supports efficient composition of very heterogeneous processing services provided by different experts and automated reasoning processes. In the context of the DPIF, the information processing is abstracted from human or machine instances; a reasoning process is either provided by a human expert or an automated system implemented by a software agent. Each process provides a well defined service in form of an estimate, prediction, etc. The inputs for each of such processes are provided by other processes or by direct observations (i.e. sensor measurements and reports from humans).

A human expert or an automated inference process is represented in the system by a functional module which (i) supports standardized collaboration protocols and (ii) allows incorporation of arbitrary reasoning approaches. In other words, the modules provide a uniform communication/collaboration infrastructure allowing seamless combination of heterogeneous processes provided by human experts or implemented through AI techniques. Each module registers in the DPIF-based system (i) the services supported by its local processing capabilities and (ii) the required inputs, i.e. types of services that should be provided by other modules in the system.

By using the registered services, modules distributed throughout different networked devices can autonomously form work flows in which heterogeneous processes introduce collaborative reasoning. The configuration of work flows is based on the relations between variables captured by local models; each module *knows* what service it can provide and what it needs to do this. This local knowledge is captured by the relations between the variables in partial domain models. Thus, *no centralized ontology* describing relations between different services of various modules is required, whose creation is likely be intractable.

In other words, globally coherent collaborative processing is possible by combining local processes, without any global description of relations between inputs and outputs.

In the following discussion we focus on (i) principles for the creation of valid work-flows based on the local processing capabilities of different modules and (ii) describe the basic elements of the DPIF architecture.

3 Processing Work Flows

A basic work flow element in the DPIF is a local process. Moreover, in the following discussion the term local process refers to a reasoning process provided either by a human expert or an automated system implemented by a software agent. Each local process corresponds to a function $F : \{X_1,...,X_n\} \to Y$, mapping values in a domain $\{X_1,...,X_n\}$ to values of some variable of interest Y. The value of Y for particular values of arguments is given by $y = f_y(x_1,...,x_n)$. The output of this function corresponds to an interpretation of the values of the variables representing relevant states in the domain.

Such functions can be either explicit, based on some rigorous theory, or implicit, when they are provided by humans or sub-symbolic processes, such as for example neural networks. An example of a mathematically rigorous mapping is the function $x_{ce1} = f_{x_{ce1}}(x_{fs})$, an explicit formula describing the relations between the fume volume per time unit represented by X_{CE1} and the escape rate of chemicals denoted by X_{FS}. This function is used by the Chemical Expert CE1 in figure 1. An implicit mapping, on the other hand, is performed by the health expert (HE) who estimates the critical regions with respect to the impact on the residents (see figure 1). HE interprets information about critical concentration X_{CE2} in combination with information on population distribution X_{pop} by using an implicit function $x_{HE} = f_{x_{HE}}(x_{CE2}, x_{pop})$.

3.1 From Local to Global Processes

As was pointed out in the preceding text, an expert or an artificial agent often cannot observe values of certain variables; i.e. variables cannot be instantiated. Instead, the inputs to the local function are supplied by other processes forming a collaborative work flow (see section 2). Thus, the inputs to one function are outputs of other functions used by the information suppliers. From a global perspective this can be seen as a function composition; in a function, each variable which cannot be instantiated is replaced by a function. This process continues until a function is obtained in which all variables are instantiated, i.e. all free variables in the resulting nested function have been reduced to direct observations. In this way, a global function emerges as different processes are connected in a work flow. The resulting function is a composite mapping between directly observable variable states and hidden variables of interest.

In other words, a work flow in a DPIF system corresponds to a full composition of functions, in which each variable replaced by a function corresponds to a required service. This yields the value of the variable of interest. As an example, if we have six service suppliers shown in figure 2, using the following functions:

$$x_a = f_a(x_b, x_c), x_b = f_b(x_d), x_c = f_c(x_e, x_f),$$
$$x_d = f_d(x_g), x_e = f_e(x_h), x_f = f_f(x_i).$$

then the work flow supporting collaborative computation of the value for x_a (see figure 2) corresponds to the composite function

$$f_a(f_b(f_d(x_g)), f_c(f_e(x_h), f_f(x_i))) \tag{1}$$

It is important to bear in mind that in DPIF no explicit composition takes place in any of the agents. Instead, the sharing of function outputs in a work flow corresponds to such a composite function; i.e. a work flow models a (globally emergent) function, mapping all relevant directly observable evidence to a description of some unknown state of interest.

Each work flow is a system of systems, in which exclusively local processing leads to a globally emergent behavior equivalent to processing the fully composed mapping from direct observations to the state of the variable of interest.

3.2 Decentralized Validation of Work Flows

The mapping in a work flow can be described through different ad-hoc or theoretically sound frameworks (such as Bayesian reasoning). However, the relation between work flows and function composition has several interesting properties regardless of the theory used to describe the exact functions. Work flows, in general, can be represented by directed graphs. In most applications the processing requires directed acyclic graphs, as *directed cycles* lead to the unwanted situation, where the state estimate of some variable is based (in part) on other estimates of the same state, without using new evidence; i.e. an update of a state estimate results in a new update of the same state estimate. The resulting estimate is likely to be misleading, since it is assuming that each state estimate update was based on new evidence.

In the context of function composition, cyclic work flows would lead to infinite composition sequences: if some function is directly or indirectly dependent upon itself, then this would lead to an infinitely recursive composition of the full function.

In other words, irrespective of the used processing method, cycles in work flows must be prevented. Therefore, an integral part of the DPIF is a cycle detection mechanism based on the cycle detection principles introduced in [2]. The used approach

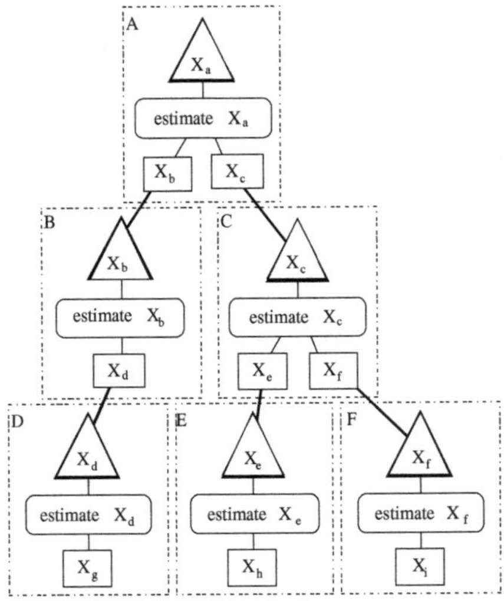

Fig. 2 A self-organized collection of agents. Each agent supplies information concerning a particular variable of interest in the domain, and does so based on other inferred or directly observed variables states.

allows a fully decentralized detection of cycles based on peer to peer communication. These principles were initially used in modular systems using Bayesian networks [2], but the technique discussed is easily carried over to a more generic domain.

In order to effect cycle-free work flows, each DPIF agent must be aware of the outputs of other agents that influence the inputs of this agent. In [2] it was shown that this information can be obtained during the creation of work flows by passing simple information between the peers in a work flow; as agents link up to form work flows, the composition of the network influencing the inputs of each agent can be recorded. By using such knowledge, agents can locally determine whether extension of a DPIF work flow with a new peer would introduce cycles; i.e. each agent can find out whether or not it is setting up an (in)direct dependency on itself, without any central authority.

The cycle detection approach described in [2] can cope with very complex work flows corresponding to arbitrary graphs. While in this paper we cannot describe all interesting aspects of this approach, we illustrate the basic cycle detection principles from [2] with the help of an example. In figure 3(a), agent A supplying x_a forms a work flow with agent B supplying x_b. Agent A knows the output variable set $S_A = \{x_a, x_b\}$, and agent B knows the output variable set $S_B = \{x_b\}$. In figure 3(b), an agent C, able to supply x_c, joins the work flow. Agent C knows its current set of output variables; $S_C = \{x_c\}$. Before joining, agent A verifies whether the proposed connection does not introduce cycles. This is the case if a simple intersection test yields $S_A \cap S_C = \emptyset$. This holds, so the connection is allowed and Agent C joins the work flow. As it does so, Agent A updates its variable set to $S_A = \{x_a, x_b, x_c\}$; i.e. A obtains all the relevant information about the services that influence its function by peer to peer sharing of local information.

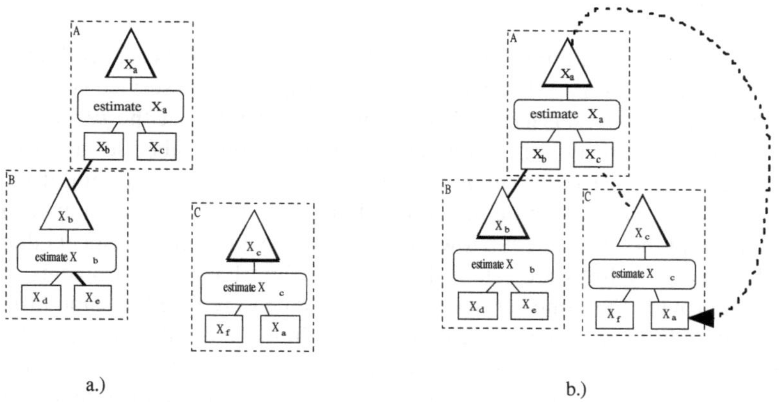

Fig. 3 a.) a partially formed collaboration, allowing the top agent to perform inference, leading to information about the state of variable x_a. b.) a potential connection which, if allowed (right), would lead to a cycle forming in the work flow.

However, when agent C looks for suppliers, the only available agent supplying x_a is A, the one to which C is already connected to. As C conducts a verification step, in which the variable sets $S_C = \{x_c\}$ and $S_A = \{x_a, x_b, x_c\}$ are tested for empty set intersection, the intersection $S_A \cap S_C \neq \emptyset$, and so C knows that a cycle would be introduced if the service x_A were supplied.

In fact, in [2] it was shown that cycles can be detected in work flows in completely decentralized manner by collaboration of peers exchanging asynchronous messages. Peers check the intersections of dynamically assembled variable sets at different levels of the work flow, and as new agents join the work flow, the new network layout needs to be reflected in all agents whose downstream network has changed by new connections. Thus, we can view the task of cycle detection as a combination of (i) checks which *travel* upstream (i.e. toward the top agent) until the top agent of the network is reached, (ii) messages conveying the updated topology, and (iii) control messages which lock/unlock the agents for local checks.

In general, this approach allows for an intelligent handling of loops and cycles in work flows, where the choice on whether to allow a connection or not is dependent on the function performed by an agent that is responsible for expanding a work flow. There exist functions which require that all inputs are provided, in order to yield an output. In such cases, an agent modeling such a function may decide to abandon a work flow when one or more of its inputs would lead to a cycle (or loop). On the other hand, there are also functions which yield output even when some inputs are left unknown, such as for example marginal conditional probabilities expressed with the help of Bayesian networks. In these cases, an agent modeling such a function may keep participating in the work flow, provided it can ensure that the critical inputs otherwise responsible for introducing cycles are kept unsupplied; i.e. they are ignored in the evaluation of the function.

3.3 Negotiation

In the DPIF, communication links between local processes in agents are facilitated firstly using service discovery: whenever an agent supplying some service (we will call this service the *parent service*, and the agent implementing it the *parent*, or *manager agent*) in a work flow requires data relating to some other service (we will call this required service the *child service*, and the agent implementing it the *child*, or *contractor agent*), a communication link needs to be established between the parent agent and the child agent. However, there are two important aspects that affect whether and why links are established: i) we might have several agents in the system that provide the same service, i.e. that are able to realize the same task, and ii) we cannot always assume that a service providing agent will automatically agree to supply the service asked for by a requesting agent. For example, the providing agent might be overloaded, or it might even consider that establishing a link is inappropriate, given the current context.

In addition, on its own service discovery can only offer links between agents based on a broad level of service matching, while for the system to solve a particular problem, a finer level of control is required to match services on however many additional parameters may be of importance to particular links. For this we use negotiation. Rather than performing perfect matching at the service discovery level, negotiation allows us to filter potential links based on additional service parameters.

Negotiation,in general, consists of three elements:

- protocols, i.e. sets of rules that describe the steps of negotiation processes, such as Contract Net (CNET), monotonic concession protocol (MCP), Rubinstein's alternating offers and auctions [11] [9] [13].
- subject - the item being negotiated about. In service negotiation, the negotiation subject is the service with its parameters.
- strategies - the set of decisions that agents will make during the negotiation in order to reach a preferred agreement.

Establishing links is based on one-to-many negotiation [1]; i.e. one agent (the manager) negotiates with multiple agents (possible contractors) about a service, with an arbitrary set of parameters (multi-issue subject) [6], [7]. We illustrate how negotiation takes place with such multi-issue subjects, by look at the example described in figure 1. We emphasize the step when the CE2 decides to guide MTs to measure gas concentrations at a location *X*. In addition, CE2 decides that measurement devices *DEV_X* are the most appropriate for the measurements. Other devices can be used as well but with less precision. CE2 initiates a negotiation over the multi-issue subject (*Gas measurement, location, device*) with all MTs that are able to provide the service *Gas measurement*. MTs propose various deals for this negotiation: the location they are currently placed at and the device they have. Ideally, MTs would propose the best deal (*Gas measurement, X, DEV_X*). CE2 must decide what MTs to choose by taking into account various parameters: the distance between location *X* and locations where MTs are placed, the differences in precision between device *DEV_X* and devices MTs have.

The default negotiation in DPIF is an implementation of CNET. However, the system allows for the use of arbitrary negotiation schemes, supporting domain specific one-to-many generic negotiation approaches where the protocol, the subject type and strategies are specified in the configuration phase.

4 Processing Modules: Architecture

Each local process (human or machine-based) is encapsulated by a module which is implemented through a software agent. In the DPIF, agents provide a uniform interface between collaborating local processes. A key feature of DPIF agents is asynchronous, data-driven processing in complex work flows. This is achieved through a combination of weakly coupled processes. Each module consists of at least two basic processes implemented through asynchronous threads communicating via a local blackboard (see Figure 4). The Communication Engine is a thread that provides inter-module communication, collaboration and negotiation capabilities. Communication Engines in different agents establish work flows between local processes in different agents by executing service discovery and negotiation (see section 3.3). The Processing Engine, on the other hand, is a thread which encapsulates arbitrary automated or human based inference. The Processing Engine, implements transformation between different types of information based on interpretation of complex cues. Moreover, each Processing Engine can keep track of one or more local processes simultaneously. The Communication Engine supplies the Processing Engine with inputs and communicates the results of local inference process to other agents.

The Communication and Processing engines must be able to execute simultaneously. Reasoning can be computationally expensive which requires certain time, but at the same time a module should be able to negotiate about possible collaboration with other modules, collect outputs of other processes which are obtained asynchronously from other collaborating modules, etc. Both threads communicate through a limited set of messages via a local blackboard. Newly received inputs are transformed and placed by the communication engine on the internal blackboard, which triggers callback of adequate functions in the reasoning thread. Through cascaded processing in the resulting system of collaborating modules, each piece of information influences the outcome of the reasoning processes, such as estimates/predictions or evaluations (i.e. reasoning result); with each process, new observations and a particular expertise are incorporated into the overall process.

Note that, irrespective of the type of the local processes, the Communication Engine and the Processing Engine in each agent use the same mechanisms for the creation and maintenance of correct work flows between local processes in different agents. The uniformity of configuration mechanisms can be used for a seamless integration of human experts into complex processing work flows. A DPIF agent representing an expert has a very simple Processing Engine which delivers the required inputs to a human expert via a suitable GUI. The expert's conclusions, i.e. his/her service, are also formulated with the help of the GUI and routed to interested

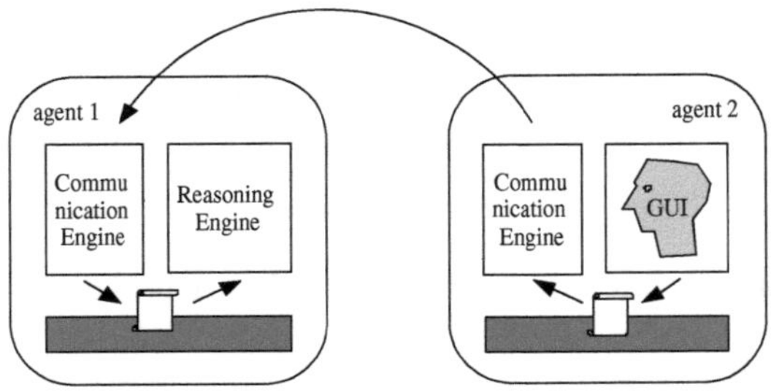

Fig. 4 Interaction between agents providing heterogeneous processing services. Both agents use identical communication engine. However, agent 1 encapsulates automated processing while agent 2 introduces human-based processing.

agents via the Communication Engine. Thus, agents in such a case merely provide automated routing of information between the experts and automated creation of connections between the relevant experts.

5 Conclusions and Future Work

The DPIF supports uniform encapsulation and combination of heterogeneous processing capabilities which are required for collaborative reasoning about complex domains. The processing capabilities can be provided by human experts or automated reasoning processes. In the DPIF context, human-based and automated processes are abstracted to functions with well defined outputs and inputs; each function provides a particular reasoning service given certain inputs.

The DPIF provides *function wrappers*, software components which standardize interfaces between the functions. The interfaces are based on standardized service descriptions as well as uniform self-configuration, negotiation and logical routing protocols. With the help of the DPIF wrapping components, processing modules can be built, such that very heterogeneous functions can easily be combined.

The processing modules form work flows in which heterogeneous functions correspond to suppliers and consumers; outputs of some functions are inputs to other functions and so on. In other words, a work-flow corresponds to a set of nested functions that capture dependencies between very heterogeneous variables.

Creation of work flows and routing of information is based on the relations between different types of information. These relations are captured by *local functions* wrapped by different modules. The DPIF approach assumes that each expert or an automated agent can declare the inputs and outputs of the contributed local functions, which is sufficient for automated creation of globally meaningful work flows by using service discovery. Thus, in contrast to traditional approaches to processing in work flows [12], *neither centralized configuration of work flows nor centralized knowledge of the combination or routing rules are needed.*

Similarly to the OpenKnowledge framework [8], the DPIF approach avoids creation of centralized heavy weight ontologies describing all aspects of the domain. Both approaches create work flows on the fly by establishing peer to peer communication between appropriate providers based on the matching of the provider capabilities. While the DPIF requires mere specification of the provided and supplied services, the OpenKnowledge framework also requires specification of *interaction models* shared by the collaborating peers. Such interaction models define work flows for each processing task a priory and the OpenKnowledge approach assumes that collaborating peers understand interaction protocols. This can introduce additional complexity to the system configuration in which services and processes are specified. Since the DPIF is targeting professional bureaucracy systems [10], it is assumed that experts do not have to share knowledge about their local processes.

However, the success of the system depends critically on the efficiency of the service specification. We assume that services (i.e. capabilities) are described by the experts themselves, as they join a DPIF-based system; i.e. no central configuration authority exists. In order to facilitate configuration, we have been developing a service configuration tool which supports alignment of service descriptions provided by different experts. With the help of such a tool, domain experts create rigorous service ontologies by using human readable keywords and free text (see [4] for the explanation of the basic principles).

In principle, arbitrary automated reasoning techniques can be integrated in the DPIF. However, globally coherent reasoning in such work flows can be achieved only by using rigorous approaches to designing local models and combining partial processing results. An example of such a system is the Distributed Perception Networks (DPN), a modular approach to Bayesian inference [3]. The DPN is a fully automated DPIF variant that supports exact decentralized inference through sharing of partial inference results obtained by running inference processes on local Bayesian networks [5] in different collaborating DPN agents. If the local Bayesian networks are designed according to the rules introduced in [3], it can be shown that collaboratively computed posterior distribution for any variable in the distributed system correctly captures all evidence.

A basic version of the DPIF as well as a prototype of the service configuration tool have been implemented and are currently being enhanced in the context of the FP7 DIADEM project. In this project we are investigating incorporation of advanced negotiation techniques as well as integration of Multi Criteria Decision Analysis and Scenario Based Reasoning methods facilitating human-based processing in work flows.

References

1. Jennings, N.R., Faratin, P., Lomuscio, A.R., Parsons, S., Sierra, C., Wooldridge, M.: Automated negotiation: Prospects, methods and challenges. International Journal of Group Decision and Negotiation 10(2), 199–215 (2001)
2. Kamermans, M.: Distributed perception networks: Effecting consistent agent organisation and optimising communication volume in a distributed multi-agent network setting. Masters Thesis, Informatics Institute, University of Amsterdam (2008)
3. Pavlin, G., de Oude, P., Maris, M., Nunnink, J., Hood, T.: A multi agent systems approach to distributed Bayesian information fusion. International Journal on Information Fusion (2008) (to appear)

4. Pavlin, G., Wijngaards, N., Nieuwenhuis, K.: Towards a single information space for environmental management through self-configuration of distributed information processing systems. In: Proceedings of the European conference TOWARDS eENVIRONMENT, Prague (2009)
5. Pearl, J.: Probabilistic Reasoning in Intelligent Systems: Networks of Plausible Inference. Morgan Kaufmann, San Francisco (1988)
6. Scafeş, M., Bădică, C.: Preliminary design of an agent-based system for human collaboration in chemical incidents response. In: Ultes-Nitsche, U., Moldt, D., Augusto, J.C. (eds.) Proc.of MSVVEIS 2009 – 7th Int.Workshop on Modelling, Simulation, Verification and Validation of Enterprise Information Systems. INSTICC Press (2009)
7. Scafeş, M., Bădică, C.: Service negotiation mechanisms in collaborative processes for disaster management. In: Proceedings of the 4th South East European Doctoral Student Conference (DSC 2009), Research Track 2: Information and Communication Technologies, Thessaloniki, Greece, SEERC (2009)
8. Siebes, R., Dupplaw, D., Kotoulas, S., Perreau de Pinninck, A., van Harmelen, F., Robertson, D.: The openKnowledge system: An interaction-centered approach to knowledge sharing. In: Meersman, R., Tari, Z. (eds.) OTM 2007, Part I. LNCS, vol. 4803, pp. 381–390. Springer, Heidelberg (2007)
9. Smith, R.G.: The contract net protocol: High-level communication and control in a distributed problem solver. IEEE Trans. Comput. 29(12), 1104–1113 (1980)
10. van Aart, C.J., Wielinga, B., Schreiber, G.: Organizational building blocks for design of distributed intelligent system. International Journal of Human-Computer Studies 61(5), 567–599 (2004)
11. Vidal, J.M.: Fundamentals of Multiagent Systems: Using NetLogo Models (unpublished) (2006), `http://www.multiagent.com/fmas`
12. Kiepuszewski, B., van der Aalst, W.M.P., ter Hofstede, A.H.M., Barros, A.P.: Workflow patterns. Distributed and Parallel Databases, 5–51 (2003)
13. Wooldridge, M.: An Introduction to MultiAgent Systems. John Wiley & Sons, Chichester (2002)

WELSA: An Intelligent and Adaptive Web-Based Educational System

Elvira Popescu, Costin Bădică, and Lucian Moraret

Abstract. This paper deals with an intelligent application in the e-learning area (WELSA), aimed at adapting the courses to the learning preferences of each student. The technical and pedagogical principles behind WELSA are presented, outlining the intelligent features of the system. The learner modeling and adaptation methods are also briefly introduced, together with their realization in WELSA. Finally, the platform is validated experimentally, proving its efficiency and effectiveness on the learning process, as well as the high degree of learner satisfaction with the system.

1 Introduction

An important class of intelligent applications in e-learning are the adaptive ones, namely those that aim at individualizing the learning experience to the real needs of each student. The rationale behind them is that accommodating the individual differences of the learners (in terms of knowledge level, goals, learning style, cognitive abilities etc) is beneficial for the student, leading to an increased learning performance and/or learner satisfaction. A common feature of these systems is that they build a model of learner characteristics and use that model throughout the interaction with the learner [2]. An adaptive system must be capable of managing learning paths adapted to each user, monitoring user activities, interpreting them using specific models, inferring user needs and preferences and exploiting user and domain knowledge to dynamically facilitate the learning process [3].

The idea dates back to 1995-1996, when the first intelligent and adaptive Web-based educational systems (AIWBES) were developed [2]. Since then, both the intelligent techniques employed evolved and the range of learner characteristics that

Elvira Popescu · Costin Bădică · Lucian Moraret
Software Engineering Department, University of Craiova,
A.I. Cuza 13, 200585 Craiova, Romania
e-mail: popescu_elvira@software.ucv.ro,
badica_costin@software.ucv.ro, dnlac@yahoo.com

G.A. Papadopoulos and C. Badica (Eds.): Intelligent Distributed Computing III, SCI 237, pp. 175–185.
springerlink.com © Springer-Verlag Berlin Heidelberg 2009

the systems adapt to expanded. A relatively recent characteristic that has started to be taken into account is the learning style of the student, i.e. the individual manner in which a person approaches a learning task, the learning strategies activated in order to fulfill that task. More formally, learning styles represent a combination of cognitive, affective and other psychological characteristics that serve as relatively stable indicators of the way a learner perceives, interacts with and responds to the learning environment [8].

This paper deals with an intelligent e-learning platform that adapts to the learning style of the students, as its name suggests: **W**eb-based **E**ducational system with **L**earning **S**tyle **A**daptation (WELSA). The next section gives an overview of the system, briefly introducing its main features: i) an implicit and dynamic learner modeling method; ii) a dynamic adaptation approach; and iii) an intelligent way of indexing and organizing the learning material. The system architecture is also presented, as well as an example of the platform at work. The following 2 sections describe in more detail the main components responsible with the system's functionality: modeling component (section 3) and adaptation component (section 4). Finally, the system usability is evaluated in section 5 and some conclusions are drawn in section 6.

2 WELSA Overview

2.1 Main Features

The main pedagogical goal of WELSA is to provide an educational experience that best suits the learning style of each student. In order to do that, the system first needs to identify these learning preferences. The learning style model employed in WELSA is that described in [14], called ULSM (Unified Learning Style Model). ULSM is actually a collection of learning preferences extracted from the main learning style models proposed in the literature, which was conceived to cover a wide range of characteristics, while at the same time aiming for independence between the learning preferences and the least possible overlap. More specifically, ULSM integrates learning preferences related to: perception modality, way of processing and organizing information, as well as motivational and social aspects. The advantages of ULSM versus traditional learning style models in Web-based learning settings have been outlined in [14].

Regarding the method used for the identification of the students' preferences, an implicit approach was chosen, based on the analysis of the students' interactions with the system. This overcomes the psychometric flaws of the traditional measuring instruments, while at the same time is more user friendly, not requiring any additional work from the part of the student. Furthermore, with this intelligent modeling mechanism, the learner model may be continuously updated (i.e. dynamic modeling method) (see section 3).

As far as the adaptation technologies are concerned, WELSA makes use of both adaptive presentation and adaptive navigation support technologies [2], providing the student with an individualized path through the learning material. The process is fully automated, based on a set of built-in adaptation rules: the course pages are dynamically generated by the system for each student, according to her/his learner model (see section 4).

In order to achieve the modeling and adaptation objectives, the system uses a fine grained representation of the learning content: the most complex learning object (with the coarsest granularity) is the course, while the finest granularity learning object (LO) is the elementary educational resource. This fine-grained representation facilitates the automatic combination of LOs as well as their reuse in different contexts. Each such elementary LO corresponds to a particular educational resource (e.g. a .txt file containing a definition or a .jpg file containing an example), which has a metadata file associated to it. The set of metadata that we propose describe the learning object from the point of view of instructional role, media type, level of abstractness and formality, type of competence etc. These metadata were created by enhancing core parts of Dublin Core [6] and Ullrich's instructional ontology [19] with some aspects specific to learning styles. It is worth pointing out that these metadata are independent of any particular learning style model, thus ensuring the independence between the domain and adaptation models.

The internal representation of the course is XML-based (more details about the content organization can be found in [12]), with the structure of the course (chapters, sections, subsections...) being defined and stored separately from the actual content (the elementary LOs). The Web pages for each student are generated dynamically by the course player, starting from the structure defined in the XML chapter file and filling it with the corresponding LOs.

2.2 Main Functionalities

WELSA's functionalities are primarily addressed at the students, who can learn by browsing through the course and performing the instructional activities suggested (play simulations, solve exercises etc). They can also communicate and collaborate with their peers by means of the forum and chat. Students' actions are logged and analyzed by the system, in order to create accurate learner models. Based on the identified learning preferences and the built-in adaptation rules, the system offers students individualized courses.

WELSA provides also functionalities for the teachers, who can create courses by means of the dedicated authoring tool; they can also set certain parameters of the modeling process, so that it fits the particularities of their course.

Figure 1 shows how WELSA appears for a learner who is studying a course on Artificial Intelligence (more specifically the chapter on "Constraint satisfaction problems", based on the classical textbook of Poole, Mackworth and Goebel [11]).

A few notes should be made regarding the course pages: the first resource (LO) on the page is entirely visible (expanded form), while for the rest of LOs only the title is shown (collapsed form). Of course, the student may choose to expand or collapse any resource, as well as lock them in an expanded state by clicking the

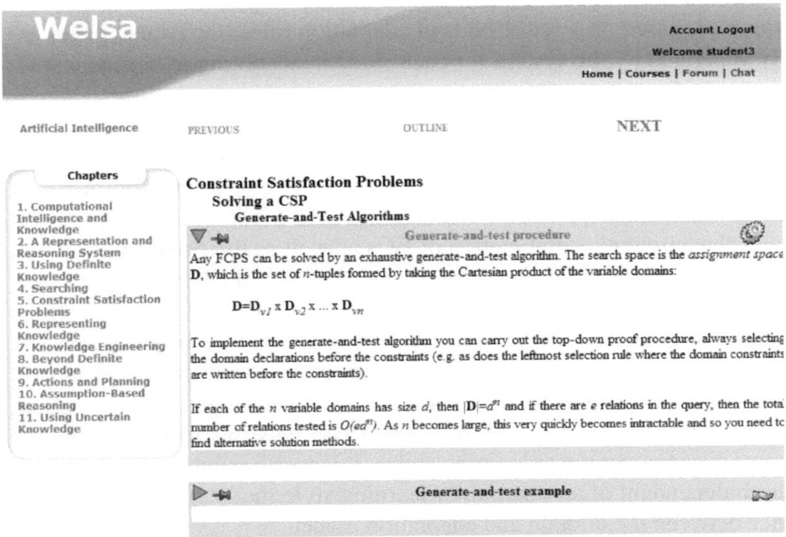

Fig. 1 A snapshot of WELSA (student view)

corresponding icons. Also, there are specific icons associated to each LO, depending on its instructional role and its media type, in order to help the learner browse more effectively through the resources. Finally, navigation can be done by means of the Next and Previous buttons, the course outline or the left panel with the chapter list.

2.3 Architecture

The overall architecture of WELSA is illustrated in Fig. 2. As can be seen in the figure, WELSA is composed of three main modules:

- an authoring tool for the teachers, allowing them to create courses conforming to the internal WELSA format (XML-based representation)
- a data analysis tool, which is responsible for interpreting the behavior of the students and consequently building and updating the learner model, as well as providing various aggregated information about the learners
- a course player (basic learning management system) for the students, enhanced with two special capabilities: i) learner tracking functionality (monitoring the student interaction with the system); ii) adaptation functionality (incorporating adaptation logic and offering individualized course pages).

The first module was presented in detail in [13]. The other two modules will be described in the following two sections respectively.

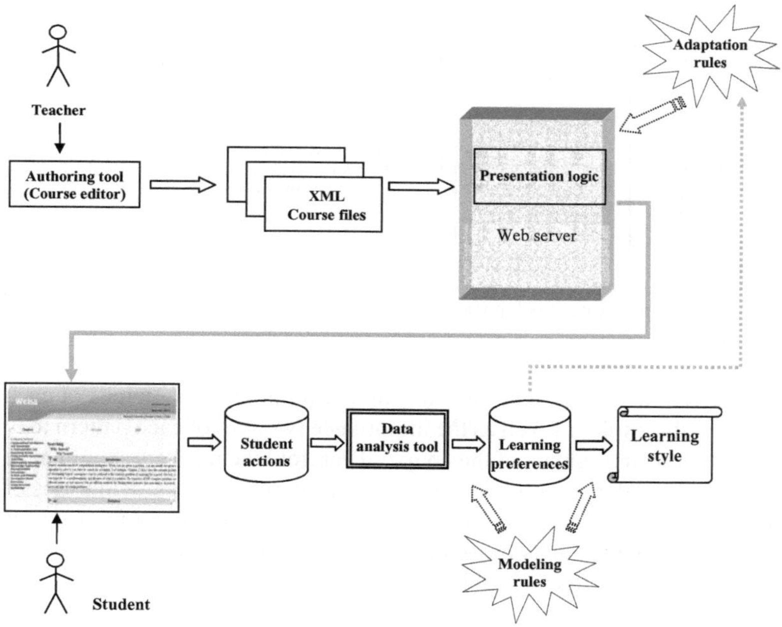

Fig. 2 Schematic WELSA architecture

As far as the implementation is concerned, Java-based and XML technologies are employed for all WELSA components. Apache Tomcat 6.0 is used as HTTP Web server and servlet container and MySQL 5.0 is used as DBMS.

3 WELSA Analysis Tool

Once the learner actions are recorded by the course player, they have to be processed by the Analysis tool, in order to yield the learning preferences of the students. The modeling mechanism is depicted in Fig. 3

In order to determine the pattern values, a pre-processing phase of the raw data (i.e. the student actions and the associated timestamps) is necessary, for computing LO access times and filtering spurious values. The data are then aggregated to obtain the pattern values for each student (e.g. total time spent on the course, total number of actions performed while logged in, time spent on each type of LO, number of hits on each category of LOs, the order of accessing the LOs, the number of navigation actions of a specific type, the number of messages in chat / forum etc). The reliability levels of these patterns are calculated as well (i.e. the larger the number of available relevant actions, the more reliable the resulted pattern). Next the Analysis tool infers the ULSM preferences of each student, using modeling rules based on the pattern values, their reliability levels and their weights.

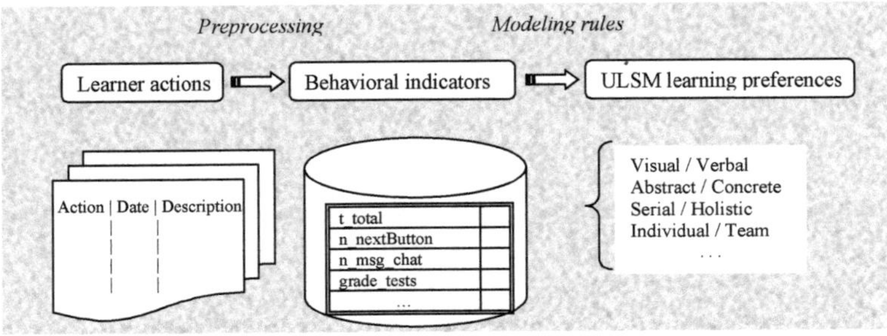

Fig. 3 WELSA learner modeling mechanism

It should be noted that these rules also take into account the specificities of each course: the pattern thresholds as well as the importance (weight) of each pattern may vary with the structure and subject of the course. Therefore the teachers should have the possibility to adjust the predefined values to correspond to the particularities of her/his course or even to eliminate some of the patterns, which are not relevant for that course. This is why the Analysis tool has a configuration option, which allows the teacher to modify the pattern weight and threshold values.

4 WELSA Course Player

WELSA doesn't store the course Web pages but instead generates them on the fly, by means of the course player module. The schematic representation of this component's architecture is illustrated in Fig. 4.

The main function of the course player is to generate the Web pages so that they can be visualized by the students. These Web pages are dynamically composed from the elementary learning objects, following the structure indicated in the XML course and chapter files. An example of such a Web page resulted from composing several LOs is included in Fig. 1.

Another function of the course player is to track student actions (down to click level) and record them in a database for further processing. This is done with the help of JavaScript code added to the HTML page, coupled with Ajax technology. Thus the application can communicate with the Web server asynchronously in the background, without interfering with the display and behavior of the existing page.

Apart from the two specific functionalities (Web page generation and learner monitoring), WELSA course player also incorporates some basic LMS (learning management system) functions, such as: administrative support (registration and authentication) and communication and collaboration tools (discussion forum, chat).

This variant of the course player presented so far is used for the non-adaptive version of WELSA (employed for experimental purposes). For the final system however, the player was enhanced with an adaptation mechanism, allowing the dynamic generation of individualized courses for each student. More specifically, the

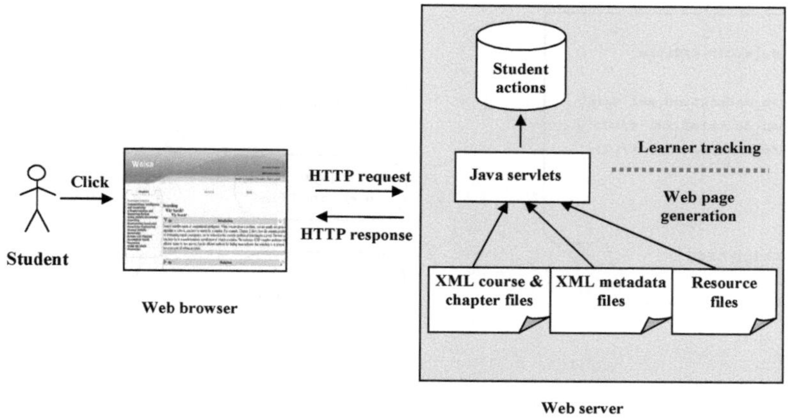

Fig. 4 Course player schematic architecture

adaptation sub-component queries the learner model database, in order to find the ULSM preferences of the current student. Based on these preferences, the component applies the corresponding adaptation rules and generates the new Web page. As explained in section 2, these adaptation rules involve the use of LO metadata, which are independent of any learning style; however, they convey enough information to allow for the adaptation decision making (i.e. they include essential information related to the media type, the level of abstractness, the instructional role etc).

The conception of these adaptation rules was a delicate task, since it involved interpretation of the learning style literature (which has a rather descriptive nature) in order to extract the prescriptive instructional guidelines. Our pedagogical goal was to offer students recommendations regarding the most suited learning objects and learning path, but let the students decide whether they want to follow our guidelines or not. We therefore decided to rely on sorting and adaptive annotation techniques rather than direct guidance or hiding/removing fragments. We also decided to use the popular "traffic light metaphor", to differentiate between recommended LOs (with a highlighted green title), standard LOs (with a black title, as in case of the non-adaptive version of WELSA) and not recommended LOs (with a dimmed light grey title) [16].

The adaptation mechanism is illustrated in Fig. 5, with a fragment of a Web page from the AI course generated for a student with a preference towards *Concrete, practical examples* rather than *Abstract concepts and generalizations*. The page is dynamically composed by selecting the appropriate LOs (mainly of type *Example*), each with its own status (highlighted in case of the LOs of type *Example* and standard in case of LOs of type *Definition*) and ordered correspondingly (first the notion of "Constraint satisfaction problem" is illustrated by means of two examples and only then a definition is provided).

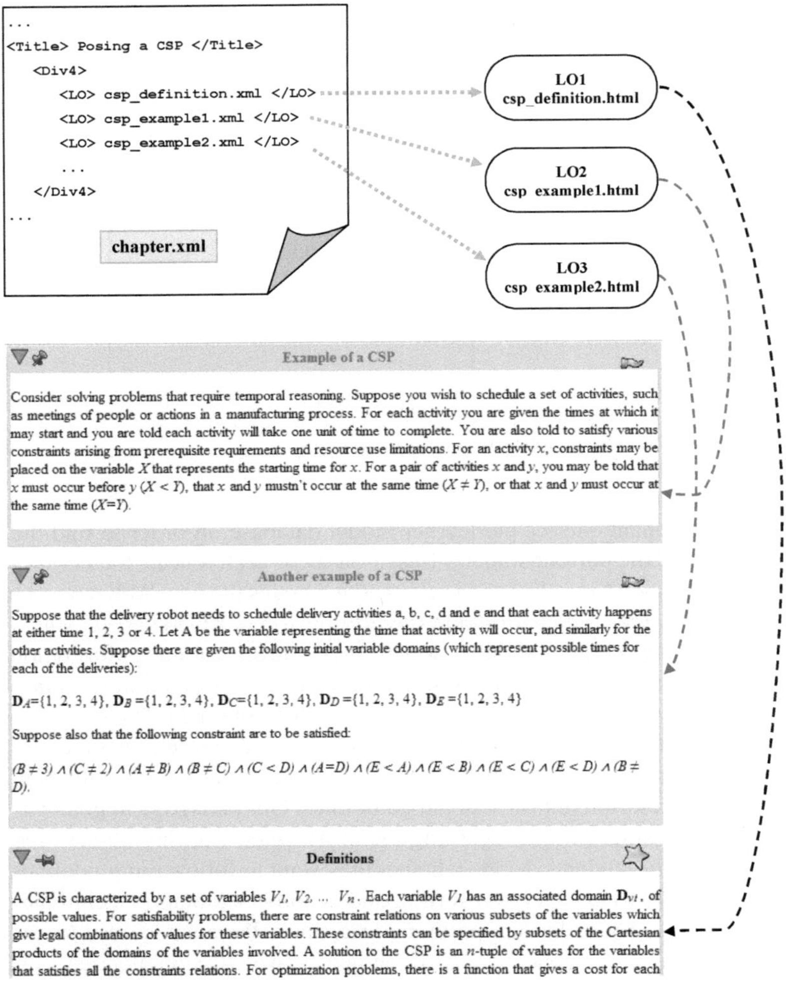

Fig. 5 Composing a page from elementary LOs for a student with *Concrete* preference

5 System Validation

Taking into account the structure and mission of the WELSA system, its validation had to be performed on three directions: i) the precision of the modeling method; ii) the efficiency and effectiveness of the adaptation approach; iii) the usability and acceptability of the platform as a whole.

As far as the modeling method is concerned, an experiment involving 71 undergraduate students was realized. The learners studied an AI course module on "Search strategies and solving problems by search" and all of their interactions with WELSA were recorded by the course player. Next, the Analysis tool computed the values of the behavioral patterns and applied the modeling rules, inferring the ULSM learning preferences of each student. In order to evaluate the validity of our modeling method, the results obtained by the Analysis tool (implicit modeling method) were compared with the reference results obtained using the ULSM questionnaire (explicit modeling method); good precision results were obtained, with an average accuracy of 75.70%, as reported in [14].

In order to assess the effect of adaptation on the learners, we performed another experiment in which the students had to interact with the adaptive version of WELSA. After studying another AI course module on "Constraint satisfaction problems", the students were asked to fill in an opinion questionnaire, comparing their experiences in the adaptive versus non-adaptive sessions. The results obtained are very encouraging [15], with a perceived improvement in terms of enjoyment and overall satisfaction (for 65.63% of the students), as well as motivation and learning effort (for 56.25% of the students).

The final step of our research was the global evaluation of WELSA system. After following the course sessions, the students were asked to assess various aspects of their learning experience with WELSA, on a 1 to 10 scale (e.g. course content, presentation, platform interface, navigation options, expand/collapse functionality for the resources, communication tools, the course as a whole). All in all, very good marks were assigned to most of the features, with only one feature (the communication tools) receiving lower (but still satisfactory) marks. We can therefore conclude that students had a very positive learning experience with WELSA. These findings are reflected also in the readiness of the students to adopt WELSA system for large scale use, with 87.50% of the students willing to do so and only 6.25% reluctant.

6 Conclusions

The WELSA system described in this paper is an intelligent e-learning platform, aimed at adapting the course to the learning preferences of each student. Unlike similar systems ([1], [4], [5], [7], [9], [10], [17], [18], [20]), WELSA is based not on a single learning style model, but on a distilled complex of features extracted from several such learning style models (ULSM). Furthermore, the identification of the student's learning style is realized using an implicit modeling method, which only a small number of related systems attempt to use ([5], [7], [17]). Finally, WELSA was thoroughly tested and experimental data is available regarding the efficiency and effectiveness of the adaptation on the learning process.

As future work, the system could be extended by adding more tools and functionalities borrowed from LMSs, such as: more advanced communication and collaboration tools (as the student surveys suggested), student involvement tools (student portfolio, bookmarks, calendar/schedule, searching facilities, context sensitive help etc). Another possible extension could be made to the adaptation component, by incorporating a wider variety of adaptation actions, including also collaboration level adaptation.

References

1. Bajraktarevic, N., Hall, W., Fullick, P.: Incorporating learning styles in hypermedia environment: Empirical evaluation. In: Proc. Workshop on Adaptive Hypermedia and Adaptive Web-Based Systems, pp. 41–52 (2003)
2. Brusilovsky, P., Peylo, C.: Adaptive and Intelligent Web-based Educational Systems. International Journal of Artificial Intelligence in Education 13(2-4), 159–172 (2003)
3. Boticario, J.G., Santos, O.C., van Rosmalen, P.: Issues in Developing Standard-based Adaptive Learning Management Systems. In: Proc. EADTU 2005 Working Conference: Towards Lisbon 2010: Collaboration for Innovative Content in Lifelong Open and Flexible Learning (2005)
4. Carver, C.A., Howard, R.A., Lane, W.D.: Enhancing student learning through hypermedia courseware and incorporation of student learning styles. IEEE Transactions on Education 42, 33–38 (1999)
5. Cha, H.J., Kim, Y.S., Lee, J.H., Yoon, T.B.: An Adaptive Learning System with Learning Style Diagnosis based on Interface Behaviors. In: Workshop Proceedings of Intl. Conf. E-learning and Games (Edutainment 2006), Hangzhou, China (2006)
6. Dublin Core Metadata Initiative (Accessed, April 2009), `http://dublincore.org/`
7. Graf, S.: Adaptivity in Learning Management Systems Focussing on Learning Styles. PhD Thesis, Vienna University of Technology, Austria (2007)
8. Keefe, J.W.: Learning style: an overview. NASSP's Student Learning Styles: Diagnosing and Prescribing Programs, 1–17 (1979)
9. Lee, C.H.M., Cheng, Y.W., Rai, S., Depickere, A.: What Affect Student Cognitive Style in the Development of Hypermedia Learning System? Computers & Education 45, 1–19 (2005)
10. Papanikolaou, K.A., Grigoriadou, M., Kornilakis, H., Magoulas, G.D.: Personalizing the interaction in a Web-based educational hypermedia system: the case of INSPIRE. User-Modeling and User-Adapted Interaction 13, 213–267 (2003)
11. Poole, D., Mackworth, A., Goebel, R.: Computational Intelligence: A Logical Approach. Oxford University Press, Oxford (1998)
12. Popescu, E., Badica, C., Trigano, P.: Description and organization of instructional resources in an adaptive educational system focused on learning styles. In: Studies in Computational Intelligence, vol. 78, pp. 177–186. Springer, Heidelberg (2008)
13. Popescu, E., Trigano, P., Badica, C., Butoi, B., Duica, M.: A Course Authoring Tool for WELSA Adaptive Educational System. In: Proc. ICCC 2008, pp. 531–534 (2008)
14. Popescu, E.: Diagnosing Students' Learning Style in an Educational Hypermedia System. In: Cognitive and Emotional Processes in Web-based Education: Integrating Human Factors and Personalization, Advances in Web-Based Learning Book Series, IGI Global, pp. 187–208 (2009)
15. Popescu, E.: Evaluating the Impact of Adaptation to Learning Styles in a Web-based Educational System. In: Spaniol, M., et al. (eds.) ICWL 2009. LNCS, vol. 5686, pp. 343–352. Springer, Heidelberg (2009)
16. Popescu, E., Badica, C.: Providing Personalized Courses in a Web-Supported Learning Environment. In: Proc. WI-IAT 2009 (Workshop SPeL), IEEE Computer Society Press, Los Alamitos (in press, 2009)
17. Sangineto, E., Capuano, N., Gaeta, M., Micarelli, A.: Adaptive course generation through learning styles representation. Journal of Universal Access in the Information Society 7(1), 1–23 (2008)

18. Triantafillou, E., Pomportsis, A., Demetriadis, S.: The design and the formative evaluation of an adaptive educational system based on cognitive styles. Computers & Education 41, 87–103 (2003)
19. Ullrich, C.: The Learning-Resource-Type is Dead, Long Live the Learning-Resource-Type! Learning Objects and Learning Designs 1, 7–15 (2005)
20. Wang, T., Wang, K., Huang, Y.: Using a Style-based Ant Colony System for Adaptive Learning. Expert Systems with Applications 34(4), 2449–2464 (2008)

Autonomous Execution of Tasks by Swarm Carrier Agents in Swarm-Array Computing

Blesson Varghese and Gerard McKee

Abstract. How can a bridge be built between autonomic computing approaches and parallel computing system? The work reported in this paper is motivated towards bridging this gap by proposing swarm-array computing, a novel technique to achieve autonomy for distributed parallel computing systems. Among three proposed approaches, the second approach, namely 'Intelligent Agents' is of focus in this paper. The task to be executed on parallel computing cores is considered as a swarm of autonomous agents. A task is carried to a computing core by carrier agents and can be seamlessly transferred between cores in the event of a pre-dicted failure, thereby achieving self-ware objectives of autonomic computing. The feasibility of the proposed approach is validated on a multi-agent simulator.

1 Introduction

Inspirations from nature have led computing scientists to focus on biologically inspired computing paradigms. Amorphous computing [1], evolutionary computing [2] and organic computing [3] are such areas that focus on abstracting designs from nature. Lately, autonomic computing inspired by the autonomic human nervous system [4] is the emphasis of distributed computing researchers which is considered in this paper.

With the aim of building large scale systems [5], reducing cost of ownership [6, 7] and reallocating management responsibilities from administrators to the com-puting

Blesson Varghese
Active Robotics Laboratory, School of Systems Engineering, University of Reading, Whiteknights Campus, Reading, Berkshire, UK, RG6 6AY
e-mail: b.varghese@student.reading.ac.uk

Gerard McKee
School of Systems Engineering, University of Reading, Whiteknights Campus, Reading, Berkshire, UK, RG6 6AY
e-mail: g.t.mckee@reading.ac.uk

G.A. Papadopoulos and C. Badica (Eds.): Intelligent Distributed Computing III, SCI 237, pp. 187–196.
springerlink.com © Springer-Verlag Berlin Heidelberg 2009

system itself [8, 9, 10], autonomic computing principles have paved necessary foundations towards self-managing systems. Self-managing systems are characterized by four objectives, namely self-configuration, self-healing, self-optimizing and self-protecting and four attributes, namely self-awareness, self-situated, self-monitoring and self-adjusting. [4, 11, 12].

Autonomic computing researchers have adopted six different approaches, namely emergence-based, component/service-based, control theoretic based, artificial intelligence, swarm intelligence and agent based approaches to achieve self-managing systems.

The emergence based approach for distributed systems considers complex beaviours of simple entities with simple behaviours without global knowledge [13]. Autonomic computing research on emergence based approaches is reported in [13, 14].

The component/service based approach for distributed systems employ service-oriented architectures, and is implemented in many web based services. These approaches are being developed for large scale networked systems including grids. Autonomic computing research on component/service based approaches is reported in [15, 16, 17].

The control theoretic based approach aims to apply control theory for developing autonomic computing systems. The building blocks of control theory are used to model computing systems and further used to study system properties. Using a defined set of control theory methodologies, the objectives of a control system can be achieved. Research on control theoretic based approaches applied to autonomic computing is reported in [18, 19].

The artificial intelligence based approaches aim for automated decision making and the design of rational agents. The concept of autonomy is realized by maximizing an agent's objective based on perception and action in the agent's environment with the aid of information from sensors and in-built knowledge. Work on artificial intelligence approaches for autonomic computing is reported in [20, 21].

The swarm intelligence based approaches focus on designing algorithms based on collective behaviour of swarm units that arise from local interactions with their environment. The algorithms considered are population-based stochastic methods executed on distributed processors. Autonomic computing research on swarm intelligence approaches is reported in [22, 23].

The agent based approaches for distributed systems is a generic technique adopted to implement emergence, component/service, artificial intelligence or swarm intelligence based approaches. The agents act as autonomic elements or entities that perform distributed task. The domain of software engineering considers agents to facilitate autonomy and hence have a profound impact on achieving the objectives of autonomic computing. Research work based on multi-agents supporting autonomic computing are reported in [9, 24, 25].

However, though all of the autonomic computing approaches above aims to-wards the objectives of autonomic computing, few researchers have applied auto-nomic computing concepts to parallel computing systems. This is surprising since most distributed computing systems are closely associated with the parallel computing paradigm. The benefits of autonomy in computing systems, namely reducing cost of ownership and reallocating management responsibilities to the system itself are also relevant to parallel computing systems.

How can a bridge be built between autonomic computing approaches and parallel computing system? The work reported in this paper is motivated towards bridging this gap by proposing swarm-array computing, a novel technique to achieve autonomy for distributed parallel computing systems.

Swarm-array computing is biologically inspired by the theory of autonomous agents in natural swarms, abstracted and implemented in swarm robotics. This technique considers the computational resource as a swarm of resources and the task to be executed as a swarm of sub-tasks. Hence, the approach considers complex interactions between swarms of sub-tasks and swarms of resources. The interactions between swarm agents bring about the notion of intelligent agents or swarm agents carrying the sub-tasks and intelligent cores or swarm of cores executing the sub-task. The interactions between different swarms give rise to the notion of landscapes. In other words, the approach can be viewed as a computational approach emerging from the interaction of multi-dimensional arrays of swarm agents.

In this paper, a swarm-array computing approach is proposed as a solution that aims to apply autonomic concepts to parallel computing systems and in effect achieve the objectives and attributes of self-managing systems. Unlike another swarm-array computing approach proposed earlier [26], the task to be executed on parallel computing cores is considered to be a swarm of autonomous agents.

The remainder of the paper is organized as follows. Section 2 considers three approaches of swarm-array computing. The second approach is of focus in this paper. Section 3 investigates the feasibility of the proposed approach by simulations. Section 4 concludes the paper with a discussion on potential applications of swarm-array computing and future work.

2 Swarm-Array Computing Approach

Swarm-array computing is a swarm robotics inspired computing approach that is proposed as a path to achieve autonomy in parallel computing systems. The foundations of swarm-array computing are the existing computing paradigms of parallel and autonomic computing. There are three approaches based on intelligent cores, intelligent agents and swarm of intelligent cores and intelligent agents that bind the swarm-array computing constituents together.

In the first approach, the cores of a parallel computing system are considered to be intelligent. These intelligent cores are implemented as autonomous swarm agents and form the landscape representing the computing space. A parallel task to be executed resides within a queue and is scheduled onto the cores by a scheduler. The intelligent cores interact with each other to transfer tasks from one core to another at the event of a hardware failure.

The cores of a parallel computing system can be considered as a set of autonomous agents, interacting with each other and coordinating the execution of tasks. In this case, a processing core is similar to an organism whose function is to exe-cute a task. The focus towards autonomy is laid on the parallel computing cores abstracted onto intelligent cores. The set of intelligent cores hence transform the parallel computing system into an intelligent swarm. The intelligent cores hence form a swarm-array. A parallel task to be executed resides within a queue and is scheduled onto different cores by the scheduler. The swarm of cores collectively executes the task.

In the second approach of swarm-array computing, only the task to be executed is considered to be intelligent. The intelligent swarm agents are autonomous and form the landscape. A parallel task to be executed resides in a queue, which is mapped onto carrier swarm agents by the scheduler. The carrier swarm displace through the cores to find an appropriate area to cluster and execute the task. The intelligent agents interact with each other to achieve mobility and successful execution of a task.

The task to be executed on the parallel computing cores can be considered as a swarm of autonomous agents. To achieve this, a single task needs to be decomposed and the sub tasks need to be mapped onto swarm agents. The agent and the sub-problems are independent of each other; in other words, the swarm agents are only carriers of the sub-tasks or are a wrapper around the sub-tasks.

The swarm displaces itself across the parallel computing cores or the environment. The goal would be to find an area accessible to resources required for executing the sub tasks within the environment. In this case, a swarm agent is similar to an organism whose function is to execute on a core. The focus towards autonomy is laid on the executing task abstracted onto intelligent agents. The intelligent agents hence form a swarm-array.

The intelligent agents described above are an abstract view of the sub-tasks to be executed on the hardware cores. Intelligence of the carrier agents is demonstrated in two ways. Firstly, the capabilities of the carrier swarm agents to identify and move to the right location to execute a task. In this case, the agents need to be aware of their environments and which cores can execute the task. Secondly, the prediction of some type of core failures can be inferred by consistent monitoring of power consumption and heat dissipation of the cores. If the core on which a sub-task being executed is predicted to fail, then the carrier agents shift from one core to another gracefully without causing an interruption to execution, hence making the system more fault-tolerant and reliable. An agent can shift from one core to another by being aware of which cores in the nearest vicinity of the currently executing core are available.

In the third approach of swarm-array computing, both the intelligent cores and intelligent agents are considered to form the landscape. Hence, the approach is called a combinative approach. A parallel task to be executed resides in a queue, which is mapped onto swarm agents by a scheduler. The swarm agents can shift through the landscape utilizing their own intelligence, or the swarm of cores could transfer tasks from core to core in the landscape. The landscape is affected by the mobility of intelligent agents on the cores and intelligent cores collectively executing a task by accommodating the intelligent agent.

A combination of the intelligent cores and intelligent swarm agents leads to intelligent swarms. The intelligent cores and intelligent agents form a multi-dimensional swarm-array. The arena in which the swarms interact is termed as a landscape.

The landscape is a representation of the arena of cores and agents that are interacting with each other in the parallel computing system. At any given instance, the landscape can define the current state of the computing system. Computing cores that have failed and are predicted to fail are holes in the environment and obstacles to be avoided by the swarms.

In this paper, the focus is on the second approach based on intelligent agents. The feasibility and experimental studies based on the first approach is reported in [26]. The third approach will be reported elsewhere.

3 Experimental Studies

Simulation studies were pursued to validate and visualize the proposed approach in swarm-array Computing. Various simulation platforms were considered, namely network simulators, which could predict behaviours of data packets in networks, and multi-agent simulators, that could model agents and their behaviours in an environment. Since FPGA cores are considered in this paper, network simulators were not an appropriate choice. The approach proposed in this paper considers executing cores as agents; hence a multi-agent simulator is employed. This section is organized into describing the experimental environment, modelling the experiment and experimental results.

The computing systems available for parallel computing are multi-core processors, clusters, grids, field programmable gate arrays (FPGA), general purpose graphics processing units (GPGPU), application-specific integrated circuit (ASIC) and vector processors. With the objective of exploring swarm-array computing, FPGAs are selected as an experimental platform for the proposed approaches.

FPGAs are a technology under investigation in which the cores of the computing system are not geographically distributed. The cores in close proximity can be configured to achieve a regular grid or a two dimensional lattice structure. Another reason of choice to look into FPGAs is its flexibility for implementing re-configurable computing.

The feasibility of the proposed swarm-array computing approach was validated on the SeSAm (Shell for Simulated Agent Systems) simulator. The SeSAm simulator environment supports the modelling of complex agent-based models and their visualization [28, 29].

The environment has provisions for modelling agents, the world and simulation runs. Agents are characterized implemented in the form of an activity diagram by a reasoning engine and a set of state variables. The state variables of the agent specify the state of an agent. The world provides knowledge about the surroundings the agent is thriving. A world is also characterized by variables and behaviours and defines the external influences that can affect the global behaviour of the agent.

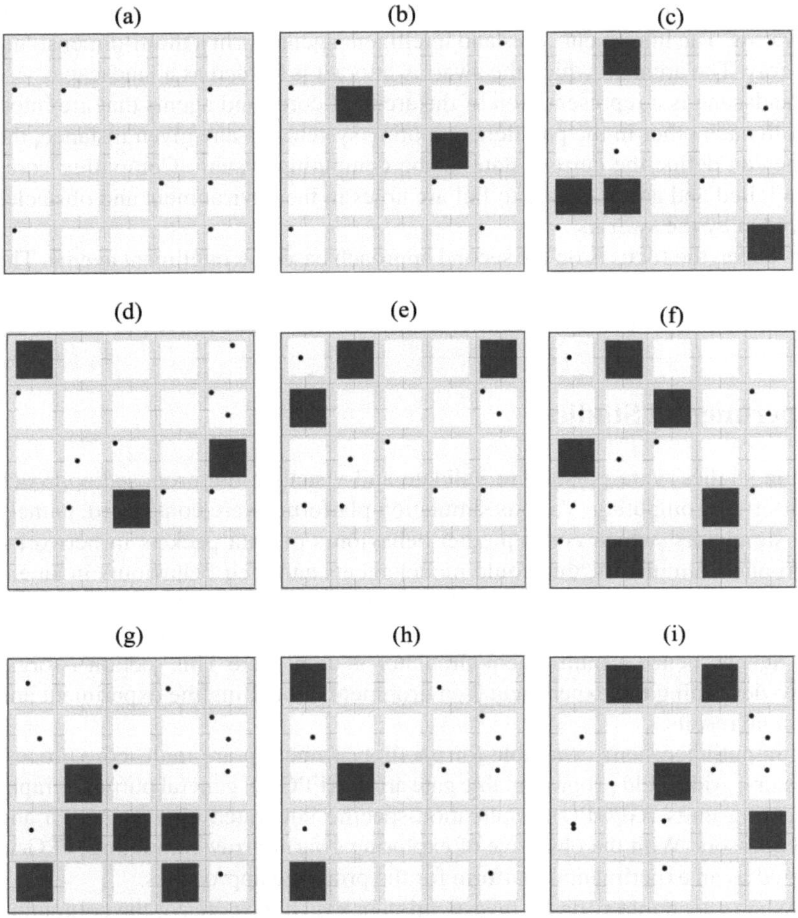

Fig. 1 Sequence of nine simulation screenshots (a) - (i) of a simulation run from initialization on the SeSAm multi-agent simulator. Figure shows how the carrier agents carrying sub-tasks are seamlessly transferred to a new core when executing cores fail.

Simulation runs are defined by simulation elements, namely situations, analysis lists, simulations and experiments that contribute to the agent-based model being constructed.

As considered in Section 2, the swarm-array computing approach considers the computing platform, the problem/task and the landscapes. The parallel computing platform considered in the studies reported in this paper is FPGAs and is modelled in SeSAm. The hardware cores are arranged in a 5 X 5 regular grid structure. The model assumes serial bus connectivity between individual cores. Hence, a task scheduled on a core can be transferred onto any other core in the regular grid.

The breakdown of any given task to subtasks is not considered within the problem domain of swarm-array computing. The simulation is initialized with sub-tasks scheduled to a few cores in the grid. Each subtask carrying agent consistently monitors the hardware cores. This is possible by sensory information (in our model, temperature is sensed consistently) passed onto the carrier agent. In the event of a predicted failure, the carrier agent displaces itself to another core in the computing system. The behaviour of the individual cores varies randomly in the simulation. For example, the temperature of the FPGA core changes during simulation. If the temperature of a core exceeds a predefined threshold, the subtask being executed on the core is transferred by the carrier agent to another available core that is not predicted to fail. During the event of a transfer or reassignment, a record of the status of execution of the subtask maintained by the carrier agent also gets transferred to the new core. If more than one sub-task is executed on a core predicted to fail, each sub-task may be transferred to different cores.

Figure 1 is a series of screenshots of a random simulation run developed on SeSAm for nine consecutive time steps from initialization. The figure shows the executing cores as rectangular blocks in pale yellow colour. When a core is predicted to fail, i.e., temperature increases beyond a threshold, the core is displayed in red. The subtasks wrapped by the carrier agents are shown as blue filled circles that occupy a random position on a core. As discussed above, when a core is predicted to fail, the subtask executing on the core predicted to fail gets seamlessly transferred to a core capable of processing at that instant.

The simulation studies are in accordance with the expectation and hence are a preliminary confirmation of the feasibility of the proposed approach in swarm-array computing. Though some assumptions and minor approximations are made, the approach is an opening for applying autonomic concepts to parallel computing platforms.

4 Discussion and Conclusion

The impact that swarm-array computing can bring about can be foreseen by taking into account the industrial or business perspective and research perspective. From the industrial viewpoint, achieving autonomy in parallel computing systems is productive. The path towards autonomy can be equated to increasing reliability of geographically dispersed systems and hence reduction in total cost for maintenance.

From the research perspective, achieving mobility of swarm agents in a heterogeneous parallel computing environment opens a new avenue to be explored. De-centralized checkpointing accomplished by the aid of swarm carrier agents is an-other interesting area of research. With these objectives, swarm-array computing can hence be proposed as a new approach for closer examination and investigation.

From an application oriented point of view, swarm-array computing can be more assuring for applications that demand reliability. One potential application that can be influenced includes space applications. Space crafts employ FPGAs, a special purpose parallel computing system that are subject to malfunctioning or failures of hardware due to 'Single Event Upsets' (SEUs), caused by radiation on moving out of the protection of the atmosphere [30] - [32]. One solution to over-come this problem is to employ reconfigurable FPGAs. However, there are many overheads in using such technology and hardware reconfiguration is challenging in space environments. In other words, replacement or servicing of hardware is an extremely limited option in space environments. On the other hand software changes can be accomplished. In such cases, the swarm-array computing approach can provide solutions based on agent mobility within the abstracted landscape and hence minimize overheads in software uploading and exclude requirement to reconfigure hardware.

In this paper, a swarm-array computing approach based on intelligent agents that act as carriers of tasks has been explored. Foundational concepts that define swarm-array computing and associated elements are also introduced. The feasibility of the proposed approach is validated on a multi-agent simulator. Though only preliminary results are produced in this paper, the approach gives ground for expectation that autonomic computing concepts can be applied to parallel computing systems and hence open a new avenue of research in the scientific community.

Future work will aim to study the third proposed approach or the combinative approach in swarm-array computing. Efforts will be made towards implementing the approaches in real time and exploring in depth the fundamental concepts associated with the constituents of swarm-array computing.

References

1. Abelson, H., Allen, D., et al.: Amorphous computing. Communications of the ACM 43(5) (2000)
2. Hedberg, S.R.: Evolutionary Computing: the spawning of a new generation. IEEE Intelligent Systems and their Applications 13(3), 79–81 (2008)
3. Schmeck, H.: Organic Computing - A New Vision for Distributed Em-bedded Systems. In: Proceedings of the 8th IEEE Symposium on Object-Oriented Real-Time Distributed Computing, pp. 201–203 (2005)
4. Hinchey, M.G., Sterritt, R.: 99% (Biological) Inspiration. In: Proceedings of the 4th IEEE International Workshop on Engineering of Autonomic and Autonomous Systems, pp. 187–195 (2007)
5. Lin, P., MacArthur, A., et al.: Defining Autonomic Computing: A Software Engineering Perspective. In: Proceedings of the Australian Software Engineering Conference, pp. 88–97 (2005)
6. Sterritt, R., Hinchey, M.: Autonomic Computing - Panacea or Poppy-cock? In: Proceedings of the 12th IEEE International Conference and Workshops on the Engineering of Computer-Based Systems, pp. 535–539 (2005)
7. Sterritt, R., Bustard, D.: Autonomic Computing - a Means of Achieving Dependability? In: Proceedings of the 10th IEEE International Conference and Workshop on the Engineering of Computer-Based Systems, pp. 247–251 (2003)
8. Nami, M.R., Sharifi, M.: Autonomic Computing a New Approach. In: Proceedings of the First Asia International Conference on Modelling and Simulation, pp. 352–357 (2007)

9. Jarrett, M., Seviora, R.: Constructing an Autonomic Computing Infrastructure using Cougaar. In: Proceedings of the 3rd IEEE International Workshop on Engineering of Autonomic and Autonomous Systems, pp. 119–128 (2006)
10. Lightstone, S.: Foundations of Autonomic Computing Development. In: Proceedings of the 4th IEEE Workshop on Engineering of Autonomic and Autonomous Systems (2007)
11. Marshall, T., Dai, Y.S.: Reliability Improvement and Models in Autonomic Computing. In: Proceedings of the 11th International Conference on Parallel and Distributed Systems, pp. 468–472 (2005)
12. King, T.M., Babich, D., et al.: Towards Self-Testing in Autonomic Computing Systems. In: Proceedings of the 8th International Symposium on Autonomous Decentralized Systems, pp. 51–58 (2007)
13. Anthony, R.J.: Emergence: a Paradigm for Robust and Scalable distributed applications. In: Proceedings of the International Conference on Autonomic Computing, pp. 132–139 (2004)
14. Saffre, F., Halloy, J., et al.: Self-Organized Service Orchestration Through Collective Differentiation. IEEE Transactions on Systems, Man and Cybernetics, Part B, 1237–1246 (2006)
15. Zeid, A., Gurguis, S.: Towards Autonomic Web Services. In: Proceedings of the 3rd ACS/IEEE International Conference on Computer Systems and Applications (2005)
16. Almeida, J., Almeida, V., et al.: Resource Management in the Autonomic Service-Oriented Architecture. In: Proceedings of the IEEE International Conference on Autonomic Computing, pp. 84–92 (2006)
17. Parashar, M., Li, Z., et al.: Enabling Autonomic Grid Applications: Requirements, Models and Infrastructure. In: Babaoğlu, Ö., Jelasity, M., Montresor, A., Fetzer, C., Leonardi, S., van Moorsel, A., van Steen, M. (eds.) SELF-STAR 2004. LNCS, vol. 3460, pp. 273–290. Springer, Heidelberg (2005)
18. Diao, Y., Hellerstein, J.L., et al.: Self-Managing Systems: A Control Theory Foundation. In: Proceedings of the 12th IEEE International Conference and Work-shops on the Engineering of Computer-Based Systems, pp. 441–448 (2005)
19. Zhu, Q., Lin, L., et al.: Characterizing Maintainability concerns in Auto-nomic Element Design. In: Proceedings of the IEEE International Conference on Software Maintenance, pp. 197–206 (2008)
20. Kephart, J.O., Walsh, W.E.: An Artificial Intelligence Perspective on Auto-nomic Computing Policies. In: Proceedings of the 5th IEEE International Workshop on Policies for Distributed Systems and Networks, pp. 3–12 (2004)
21. Peddemors, A., Niemegeers, I., et al.: A System Perspective on Cognition for Autonomic Computing and Communication. In: Proceedings of the 16th Inter-national Workshop on Database and Expert Systems Application, pp. 181–185 (2005)
22. Hinchey, M., Dai, Y.S., et al.: Modeling for NASA Autonomous Nano-Technology Swarm Missions and Model Driven Autonomic Computing. In: Proceedings of the 21st International Conference on Advanced Information Net-working and Applications, pp. 250–257 (2007)
23. Carrasco, L.M.F., Marin, H.T., et al.: On the Path Towards Autonomic Com-puting: Combining Swarm Intelligence and Excitable Media Models. In: Proceedings of the 7th Mexican International Conference on Artificial Intelligence, pp. 192–198 (2008)
24. Guo, H., Gao, J., et al.: A Self-Organized Model of Agent-Enabling Autonomic Computing for Grid Environment. In: Proceedings of the 6th World Congress on Intelligent Control and Automation, pp. 2623–2627 (2006)

25. Hu, J., Gao, J., et al.: Multi-Agent System based Autonomic Computing Environment. In: Proceedings of the International Conference on Machine Learning and Cybernetics, pp. 105–110 (2004)
26. Varghese, B., McKee, G.T.: Towards Self-ware via Swarm-Array Computing. Accepted for publication in the International Conference on Computational Intelligence and Cognitive Informatics, Paris, France (2009)
27. Bacon, J.: Concurrent Systems Operating Systems, Database and Distributed Systems: An Integrated Approach. Addison-Wesley, Reading (1997)
28. Klugl, F., Herrler, R., et al.: SeSAm: Implementation of Agent-Based Simulation Using Visual Programming. In: Proceedings of the Fifth International Joint Conference on Autonomous Agents and Multi-Agent Systems, Japan, pp. 1439–1440 (2006)
29. SeSAm website, `http://www.simsesam.de`
30. O'Bryan, M.V., Poivey, C., et al.: Compendium of Single Event Effects Results for Candidate Spacecraft Electronics for NASA. In: Proceedings of the IEEE Radiation Effects Data Workshop, pp. 19–25 (2006)
31. Johnson, E., Wirthlin, M.J., et al.: Single-Event Upset Simulation on an FPGA. In: Proceedings of the International Conference on Engineering of Reconfigurable Systems and Algorithms, USA (2002)
32. Habinc, S.: Suitability of Reprogrammable FPGAs in Space Applications. Report for the European Space Agency by Gaisler Research under ESA contract No. 15102/01/NL/FM(SC) CCN-3 (2002)

A Privacy Preserving E-Payment Scheme

Giannakis Antoniou, Lynn Batten, Shivaramakrishnan Narayan,
and Udaya Parampalli

Abstract. Secure electronic payment is an integral part of e-commerce. The e-payment systems in current use require the revealing of the purchaser's payment information to the seller, while purchasers are not always happy to subject the details of their payment information to possible misuse by sellers. As a result, the uptake of e-commerce is hampered from reaching its full potential. In this paper, we propose an online payment scheme which uses the traditional e-payment infrastructure but which reveals no payment information to the seller. This is done with only an incremental increase in computational power.

1 Introduction

A fundamental part of e-commerce is the e-payment system, prompting the customer for possibly the most sensitive information in the transaction. The customer is rightly concerned about who gains access to this information as the transaction proceeds. Our thesis is that, if access to the payment information is limited to the relevant financial organization, then the purchaser's fear of a violation of privacy

Giannakis Antoniou
University of Melbourne, Department of CSSE
e-mail: gant@csse.unimelb.edu.au

Lynn Batten
Deakin University, School of Information Technology
e-mail: lynn.batten@deakin.edu.au

Shivaramakrishnan Narayan
University of Calgary, Department of Computer Science
e-mail: snarayan@ucalgary.ca

Udaya Parampalli
University of Melbourne, Department of CSSE
e-mail: udaya@csse.unimelb.edu.au

G.A. Papadopoulos and C. Badica (Eds.): Intelligent Distributed Computing III, SCI 237, pp. 197–202.
springerlink.com © Springer-Verlag Berlin Heidelberg 2009

(misusing purchaser's payment information by others), theft from his bank account and so on, will be significantly reduced, and his willingness to increase his exposure to e-commerce will rise.

In this paper we assume that a purchaser is using a credit card, issued by a bank or similar financial organization, and that the payment information required for online transactions is the Credit Card Details (CCD) comprising the full name of the owner, a credit card number, the expiry date of the credit card and a security number (known as CVV2).

Most of the papers proposing e-payment schemes (such as in [1, 2, 3, 5]) claim that their solutions can be applied to the e-commerce environment where the purchaser pays and the seller delivers the product to the delivery address. However, in these schemes, the delivery address known to the seller is a proxy of that of the purchaser; the seller does not know the actual address of the purchaser. The lack of appropriate use of the delivery address, as we shall show in this paper, can result in an attack which we identify and call the seller-in-the-middle attack.

In [2], the authors present the Virtual Credit Card (VCC) scheme which involves generation of dynamic credit card numbers during the e-payment transaction with the aim of avoiding theft of credit card details. In restricting financial information of the purchaser to the financial organization alone, we introduce an online payment scheme which reveals no CCD to the seller. We refer to this new scheme as the PCCP scheme. The one fundamental difference between this scheme and current online payment schemes is that it is secure against both a replay attack and a seller-in-the-middle attack. The PCCP scheme has an incremental increase in computational complexity over that of existing e-payment systems, caused by the addition of a hash function in the generation of a ticket. Our proposed solution is applicable only for on-line transactions.

The paper is organised as follows: Section 2 describes traditional e-payment systems; Section 3 proposes the new e-payment scheme. Section 4 presents the results of the implementation. Section 5 concludes the paper.

2 Traditional E-Payment Systems

While existing e-payment schemes are implemented on various computing infrastructures run by proprietary software belonging to a variety of organizations, they all essentially follow the uniform payment method. In the e-payment procedure, at least three entities are involved: the purchaser who wants to buy goods, the seller who wants to sell goods, and the bank which transfers money from the purchaser's credit card account to the seller's bank account. However, other entities are normally included in the procedure, such as the Payment Gateway, the Seller Processor and Credit Card Interchange (CCI). The communication between the seller and the bank is handled by the Payment Gateway, Seller Processor and CCI. Hence payment details of the transactions are transferred to multiple points leading to greater risk of losing the privacy of the purchaser's CCD. In the scheme we propose in Section 3, we retain the above communication relationship without risking this loss of privacy to the purchaser.

2.1 The E-Commerce Environment - The Traditional Protocol

SSL [4] is the de facto communication protocol for traditional e-payment systems allowing two parties to exchange confidential information. When the purchaser selects the goods and offers payment, the purchaser and the seller create an SSL connection over which the purchaser receives the digital certificate of the seller, verifies the authenticity of the digital certificate by using the public key of the appropriate Certified Authority (CA) and exchanges a secret key. Both the purchaser and seller use that secret key to encrypt and decrypt the messages exchanged during the communication session.

The protocol requires a set of personal information belonging to the purchaser. We assume that the list below is required and is complete:

- *CCD*: This is the Credit Card Details
- *Credit Card Company*: This is the credit card company, such as VISA or MasterCard. This information identifies the correct CCI.
- *Amount*: This is the amount that the seller asks the bank to charge to the credit card.
- *SellerAccount*: This is the bank account of the seller to which the seller wants to deposit the amount from the purchaser.

The purchaser encrypts and sends the CCD along with the identity of the credit card company to the seller.

P→S: CCD, Credit Card Company

The seller, who usually stores the CCD, forwards (through the Payment Gateway) the purchaser's CCD along with the bank account details and amount in order to get the agreed payment.

S→B: CCD, SellerAccount, Amount

The SSL protocol is unable to prevent the Seller-in-the-middle attack which is described below.

Seller-in-the-middle attack: In this attack, a malicious seller (S1) impersonates the real seller (S2) and accepts CCD from a purchaser (P). He then contacts S2 and, impersonating P, forwards P's CCD and receives the goods from S2. In this attack S2 thinks that the CCD belongs to S1. If the attack is successful, while money gets deducted from P's account the goods will be delivered to S1. S1 may have been lucky enough to read leaked conversations between P and S2 or may in fact advertise goods from a website hoping to attract purchasers to it. The difference between this attack and a standard man-in-the-middle attack is that in the latter, S1 simply taps into communications between P and S2 and reads and forwards communications to each; in the former, S1 intercepts communications from P to S2 and changes a delivery address so that S2 will deliver goods to S1. Thus, the seller, bank and purchaser face several problems because of the traditional protocol's inability to retain confidentiality of the purchaser's CCD.

For the purpose of this paper, we consider a desirable e-payment system to be one which allows a purchaser to buy from a seller without allowing the seller (or any other entity apart from the bank) to be able to misuse the CCD of the purchaser. In the next section we propose a new e-payment system, the PCCP which can prevent the *Seller-in-the-middle attack*.

3 The Proposed Privacy-Preserving E-Payment Scheme

The scheme we propose can be built over the existing traditional protocol for e-payment described in Section 2.1. The scheme takes advantage of the CCD that the bank and the purchaser share during the issuance of the credit card. As mentioned in Section 1, our aim is to keep the CCD from the seller.

Our scheme requires actions from all three entities, the purchaser, the seller and the bank. In addition to the personal information (eg.email address, delivery address) and CCD required in the existing e-payment protocol, our scheme uses the information below:

- *Cost*: This is the amount that the purchaser wants to be charged
- *Nonce*: This is a number generated by the seller and the purchaser. The role of the nonce is to prevent a seller (or any other entity) from re-using a ticket which is issued during the transaction.
- *BankID*: This is the identification of the bank which issues the credit card.
- *SellerAccount*: This is the bank account of the seller to which the seller wants to deposit the amount from the purchaser.
- *CCI*: This is the Credit Card Interchange.

The scheme is described below beginning with the issuance of the credit card to the purchaser.

1. The purchaser gives personal information to the bank and the bank issues and delivers the credit card which contains the CCD, including a shared password (r) to the purchaser and computes:
   X1= $\mathscr{H}$(CCD||r) where $\mathscr{H}$ is a cryptographic hash function. The symbol '||' concatenates the CCD and the r.
   The bank stores CCD, r and X1. The bank will use X1 in a later stage (step e) in order to identify the credit card account of a transaction.
2. The purchaser and the seller agree on the amount for the purchase and the nonce. Optionally, the purchaser and the seller may agree on the start and the expiration date (fromDate and untilDate) during which the request to the bank can take place. In addition, the seller gives its bank account information to the purchaser (SellerAccount), in order for the purchaser to generate a ticket which is linked with that account. The purchaser announces to the seller the delivery address (Delivery Address) for receipt of the product.

σ' = { @cost, @Delivery Address, @nonce, [@fromDate, @untilDate], @SellerAccount }
The symbol @ before the variables indicates the variables that the seller will forward to the bank in step 4.

3. The purchaser calculates the ticket and sends it to the seller.
   σ = { cost, Delivery Address, nonce, [fromDate, untilDate], SellerAccount }
   X1= $\mathcal{H}$(CCD ||r)
   Z={ CCD, r }
   X2= F(Z, σ) where F is an HMAC and accepts the two parameters, Z and σ.
   Ticket= { CCI, BankID, X1, X2 }
4. The seller forwards the ticket as well as the agreed information (σ') to the bank.
5. The bank finds the credit card account (including the CCD) related to the ticket, based on the value X1 (rather than the credit card number).
   CCD = FindAccount(X1)
6. The bank uses CCD and the information received from the seller (σ') in order to verify the validity of the ticket. The verification process follows:
   Z'={ CCD, r }
   X2'=F(Z', σ')
   If X2 = X2' AND NOT @nonce already used Then
   If today is between FromDate and UntilDate Then
   VALID REQUEST . PERFORMS THE TRANSACTION
   Else OUT OF DATE TICKET End If
   Else INVALID REQUEST End If
7. The bank generates and sends a response R to the seller for every transaction request, whether successful or unsuccessful. In case the transaction was unsuccessful and the seller requests the purchaser to generate a new ticket, the seller will forward R to the purchaser.
   R=F(Z, msg), msg

4 Implementation

We implemented a prototype of the purchaser's calculations on a cell phone (Sony Ericsson K610i) which supports Java Platform 7. We used SHA-256 to implement the function $\mathcal{H}$ and SHA-256 with a 20 alphanumeric digit key to implement the HMAC function F.

The result (Table 1) of the prototype implementation verifies our expectations on the low computational complexity of our ticket generation scheme - essential for the scheme to be useful in practice. We executed the application on the cell phone ten times and each execution took between 198 ms and 224 ms. We expect that the same results can be achieved on any cell phone which supports Java Platform 7 and higher. The bank only computes X1 and so its execution time is expected to be on a high power machine with lower execution time than that of the purchaser.

Table 1 Results of the implementation

Calculation	**Execution Time**		
	Minumum	Maximum	Average
X1= $\mathscr{H}$(CCD ||r)	46 ms	62 ms	56 ms
X2= F(Z, σ)	148 ms	166 ms	155 ms
Total	**198 ms**	**224 ms**	**211 ms**

5 Conclusion and Future Work

In this paper, we proposed the PCCP scheme which ensures that financial information of a purchaser is known only to the purchaser's financial organization. It allows a purchaser to pay on-line by using a credit card without allowing the seller to gain access to or misuse the CCD given by the purchaser.

The potential of schemes such as the proposed PCCP and the VCC scheme in [2] in encouraging the use of e-commerce is high as they do not require any new infrastructure and they are easily implemented. In particular, in the PCCP scheme, purchasers do not need to change their e-commerce habits, sellers do not need to adopt any new technology and the banks need only to implement a simple validation of the ticket. In future work, we will develop and implement a complete communication protocol.

References

1. Ashrafi, M.Z., Ng, S.K.: Enabling Privacy-Preserving e-Payment Processing. In: Haritsa, J.R., Kotagiri, R., Pudi, V. (eds.) DASFAA 2008. LNCS, vol. 4947, pp. 596–603. Springer, Heidelberg (2008)
2. Molloy, I., Li, J., Li, N.: Dynamic Virtual Credit Card Numbers. In: Dietrich, S., Dhamija, R. (eds.) FC 2007 and USEC 2007. LNCS, vol. 4886, pp. 208–223. Springer, Heidelberg (2007)
3. Rubin, A.D., Wright, R.N.: Off-line generation of limited-use credit card numbers. In: Syverson, P.F. (ed.) FC 2001. LNCS, vol. 2339, p. 187. Springer, Heidelberg (2002)
4. Schneier, B., Wagner, D.: Analysis of the SSL 3.0 protocol. In: The Second USENIX Workshop on Electronic Commerce Proceedings, 29–40 (1996)
5. Shamir, A.: SecureClick: A Web Payment System with Disposable Credit Card Numbers. In: Syverson, P.F. (ed.) FC 2001. LNCS, vol. 2339, p. 223. Springer, Heidelberg (2002)

Monitoring a BitTorrent Tracker for Peer-to-Peer System Analysis

Mircea Bardac, George Milescu, and Răzvan Deaconescu

Abstract. This paper introduces a BitTorrent tracker monitoring architecture and implementation. The tracker monitoring architecture is designed with robustness, scalability and extensibility in mind, making it suitable for large scale BitTorrent tracker monitor analysis. The solution is implemented as part of a BitTorrent experiment for studying the behavior of a real-life swarm. The peer-to-peer network can be analyzed considering both statistic data variation and geographical distribution of data, leaving room for other analysis methods to be implemented if needed. The paper presents challenges and solutions found while designing, implementing and integrating the components of the tracker monitoring solution.

1 Introduction

This paper presents the problems encountered and the solutions applied as part of a BitTorrent tracker monitoring solution used in the Ubuntu Torrent Experiment - experiment started on the launch of Ubuntu 9.04 (April 23) and is currently being run by the University Politehnica of Bucharest, Computer Science Department.

The collected data in the experiment covers all the torrent transfers that have been reported to the tracker we have been monitoring. The geographic reports generated during the experiment are limited to Romania, as the experiment was primarily

Mircea Bardac
University Politehnica of Bucharest, Splaiul Independenţei 313
e-mail: mircea.bardac@cs.pub.ro

George Milescu
University Politehnica of Bucharest, Splaiul Independenţei 313
e-mail: george.milescu@cs.pub.ro

Răzvan Deaconescu
University Politehnica of Bucharest, Splaiul Independenţei 313
e-mail: razvan.deaconescu@cs.pub.ro

G.A. Papadopoulos and C. Badica (Eds.): Intelligent Distributed Computing III, SCI 237, pp. 203–208.
springerlink.com © Springer-Verlag Berlin Heidelberg 2009

advertised to the Romanian user communities. The monitoring architecture described as part of this article allows further customization and extension in order to provide reports on any selections of data subsets.

Tracker monitoring has not been intensively researched, previous work only existing as a monolithic Perl script last updated in 2003, used for log analyzing [2]. It has limited logging capabilities (only graph generation) and does not provide an extensible interface for plugging or exchanging components.

Tracker monitoring offers an overview on the peer-to-peer network based on the declared status of the peers. Direct interactions between the peers can not be monitored on the tracker, requiring either modified clients or analyzing the log on each client, which are both unfeasible.

2 Tracker Monitor Infrastructure

The underlying system configuration used for both the experiment frontend and the tracker monitoring backend was an OpenVZ container with Debian Stable (Lenny) running on an Intel Core(tm) 2 Quad Processor, having a guaranteed 512 MB RAM (burstable to 1280 MB) and 200 GB of HDD space.

The experiment uses a BitTornado 0.3.18 tracker configured to log everything to a plain-text log file. 12 torrent files of the (K)Ubuntu 9.04 ISO images were configured to send announce messages to the tracker every minute. In order to prevent an increased system load on message processing from occurring, a monitoring architecture was designed having as primary goals robustness and scalability.

The monitoring solution was built in Python 2.5 using a multi-layer architecture, where sub-components can eventually be executed on separate machines, allowing the entire monitoring system to scale as described in section 3.

A MySQL database was setup with the IP geolocation information available from `http://iplocationtools.com/`. This geolocation IP information was used for both the experiment frontend[1] and the monitoring backend, as presented in section 4.

3 Tracker Monitor Architecture

Possible crashes of the monitoring system required having a system that could easily recover to any previous state, thus saving the resources for re-processing previously logged tracker messages. This was a high priority requirement of the entire monitoring system and was implemented using special objects (which can be serialized) designed for storing the status of both the system and each torrent processor.

The tracker log is constantly being monitored by the Log Parsing Component. The log file is updated by the tracker with the announce messages received from the peers downloading a file from the swarm using the provided torrent files.

[1] `http://torrent.cs.pub.ro/`

Each torrent file has an associated Torrent Processor instance, responsible for processing torrent-related messages arrived from the Log Parsing Component.

The Torrent Processors have Logging Engines attached to them. The Logging Engines are responsible for producing the results in the monitoring system. These results can be either graphs, structured data or even new messages. In the last case, the new messages are being delivered by the Logging Engine acting as a Proxy to another monitoring entity.

In order to use the data from the monitoring engine, two timed events are being used: data_dump and data_publish. Time passing is measured using the timestamps in the tracker log. Each Torrent Processor can determine the time passed since the last dump or publish based on the current processed log entry. This also has the advantage of allowing delayed log processing (in the case of recovery from crash or distributed processing, as described in section 4). Both the data_dump event and the data_publish event can be interpreted differently by the Logging Engines, depending on their structure or purpose.

When the *data_dump* event is triggered, the torrent processor executes in order: 1. call the data_dump function of all Logging Engines 2. save its status to disk for recovery. The *data_publish* event is triggered by the Torrent Processors and used to notify the Logging Engines to publish their data.

Preliminary tests proved that the system is not overloaded on executing data dumps and, in order to provide fast recovery in the case of a failure, a 1 minute interval for data_dumps was set. In order to provide a close-to-real-time view of the peer-to-peer system, the monitor tracker is run with a 1 minute interval between data_publish events.

3.1 Log Parsing Component

The tracker uses the underlying system libraries for writing information to the log files. This results in log messages being cached to library buffers until these buffers are flushed to disk. Special care is needed when handling the last messages in the log file, as they are most of the times incomplete due to the behaviour of library buffers. To avoid this problem, each read of the log file is being checked and the position in the file is saved if a successful read occurs. If an incomplete entry is detected, the position in the log file is reverted to the last known good position. The position in the log file is saved to allow recovery in case of errors.

The data_publish event is not triggered as long as the end of the log file is not reached. This prevents early publishing on recovery/restart - when a recovery/restart occurs, the part of the log that was not processed is sent to the torrent processors. The data_dump events are being generated as log entries are being processed at whenever a time span of 1 minute is detected between the last dump time and the current log entry timestamp. The data_publish event is suppressed until the end of the log file is reached. When the end of the log file is reached, data_publish events are generated just as the data_dump events.

3.2 Torrent Processor Component

The tracker monitor creates a torrent processor for each torrent that is being monitored to collect and determine torrent specific statistics, both for the entire torrent lifetime and for the last dump interval (1 minute). Each torrent processor starts its own Logging Engines instances, as each torrent can be logged differently. Torrent processors save their state on each data_dump for easy recovery.

The torrent *lifetime statistics* include number of completed downloads, completed downloads by city (in Romania), total bytes downloaded/uploaded in the swarm while *instantaneous per torrent statistics* include: number of peers, entire swarm upload/download speed and others.

The peer information is updated as the BitTorrent announce messages are being processed from the log file. Given that the torrents are configured with a 1 minute announce period, each client peer should send announce messages every minute. Because of system and network load, the peers usually can't respect the 1 minute interval and announcements arrive on the tracker either late or early.

3.3 Logging Engine Component

Logging engines are started by the Torrent Processors and are responsible for streaming data out of the torrent processor. They rely on the data_dump and the data_publish events to do their work.

Using the data_dump event, the logging engines can track the evolution of the torrent status using the statistics provided by the torrent processor. The data can be dumped to disk, saved to a database or saved as part of the logging engine state. Each logging engine can implement its own actions to be executed on receiving the data_dump event. The data_publish event is triggered when the logging engine is expected to publish data. Publishing usually consists of converting the data received through the data_dump event to another format making it accessible to the user or to other publishing services.

4 Data Analysis Using Logging Engines

The data that can be gathered from the Torrent Processors allows multiple types of analysis to be performed. The experiment started with 2 types of analysis being done almost in real-time (1 minute interval): *numeric variations in time* and *geographic numeric distribution*. Each type of analysis requires a separate logging engine.

Numeric Variations in Time are recorded and displayed using RRDtool (round-robin database tool) [5] on the experiment frontend. As its name says, it uses a round-robin database in order to store the data to be plotted as a graph. The frontend displays variations of several parameters such as peers number and swarm speed.

RRDtool is especially useful in the context of tracker log analysis in a non-real-time context, because all logged data must be accompanied by the unix timestamp of when the data was recorded. On tracker monitor recovery (as mentioned 3.1), this can help saving data for events which have already happened.

Figure 1 shows an example for the output of the RRDtool logging engine.

The logging engine for **geographic numeric distribution** of statistic data relies on the peer information structure. This structure is filled with the geographic location of the peer when the peer is first discovered by the torrent processor in the tracker log. In this experiment, the geographic distribution is only rendered for Romania, the country where the experiment was mostly advertised.

On each data_dump event the data passed to the logging engine is stored internally and used during the publish event to generate the map data. The Google Maps API [3] is used for rendering the map data in the frontend, drawing circles on a map with their radius depending on the number of downloads. The map data is published by the logging engine in the JSON format [1].

The **MonALISA logging engine** was later added to the experiment to facilitate browsing and further processing of the logged data. The main issues encoutered with the MonALISA logging engine were: memory utilization and data range limitations. The initial design used multiple instances of the MonALISA agent to communicate with the MonALISA farm. This resulted in a total RAM usage of aproximately 220 MB. In order to reduce the resource utilization, the MonALISA agent code was instantiated as a singleton and used by all MonALISA logging engines. Therefore, the RAM utilization dropped to approximately 80 MB.

Data range limitations problems with them were first detected on this engine. The Python interpreter can automatically handle integers of any size (limited by the amount of memory) while the external logging interfaces used by the logging engines usually limit them (at 32 or 64 bit). The tracker monitor maintains this information internally without any problems. Data possibly causing an interger overflow is scaled in the MonALISA logging engine from bytes to kilobytes.

Time in MonALISA is not bound in the messages sent through the agent. The time of a measurement is considered to be the time when the parameter arrived on the MonALISA farm. To prevent publishing of incorrectly timed events to the synchronous logging engines such as MonALISA, the tracker monitor enables publishing only when the tracker monitor reaches the end of the log file, as mentioned in section 3.1.

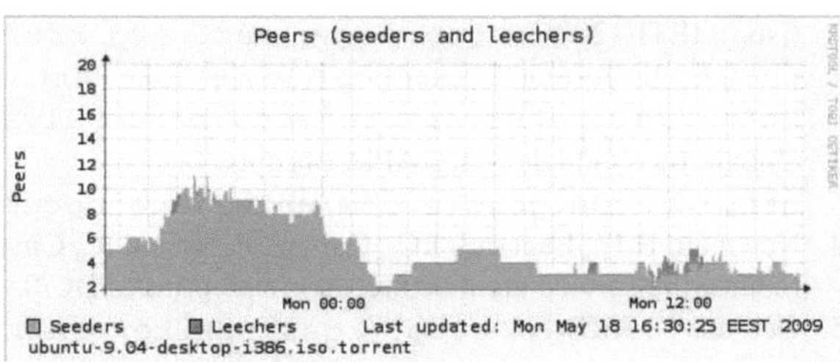

Fig. 1 RRDtool graph showing seeders and leechers

5 Future Development

Preliminary analysis of the data gathered during the experiment showed the need of having a per-tracker view of the activity. The currently implemented granularity level is per-torrent. To solve this problem, virtual torrent processors can be created. The virtual torrent processors would gather all data from all torrents and sum everything up - this would be the only difference compared to a normal torrent processor.

Reparsing of the entire log might be required on certain situations such as creating totally new logging engines or fixing bugs in the tracker monitor. On reparsing the entire log, synchronous logging engines (such as MonALISA) must not receive the updated past events and logging engines that save their state to disk or to a database (such as RRDtool) must be reset/reconfigured as they will be receive all the updated data through data_dump events.

6 Conclusions

Tracker monitoring poses several problems in terms of scalability, robustness and interoperability with external entities. The entire system must be flexible enough to allow extension depending on the experiment needs.

The architectural challenges faced during the initial deployment have been overcome using the solutions presented in this paper. The monitoring solution also provides three types of logging engines, each one posing particular design challenges. They have been described in the paper together with the integration issues they have posed and with the applied solutions.

The tracker monitoring architecture presented in the paper was designed with robustness, scalability and extensibility in mind, making it suitable for large scale BitTorrent tracker monitor analysis. The peer-to-peer network can be analyzed considering both the statistic data variation and the geographical distribution of data, leaving room for other analysis methods to be implemented if needed.

References

1. Crockford, D.: RFC 4627 - The application/json media type for javascript object notation (json), IETF (2006), http://www.ietf.org/rfc/rfc4627.txt
2. Frost, E.: BitTorrent Tracker Log Analyzer, Eike Frost, http://ei.kefro.st/projects/btrackalyzer/
3. Google Inc., Google Maps API Concepts, http://code.google.com/apis/maps/documentation/
4. Newman, H.B., Legrand, I.C., Galvez, P., Voicu, R., Cirstoiu, C.: Monalisa: A distributed monitoring service architecture, Arxiv preprint cs.DC/0306096 (2003)
5. Oetiker, T.: RRDtool, http://oss.oetiker.ch/rrdtool/

A Memory Assistant for the Elderly

Ângelo Costa, Paulo Novais, Ricardo Costa, José Machado, and José Neves

Abstract. In the present-day the ageing population is not receiving the proper attention and care they need, because there aren't enough healthcare providers to everyone. As the elderly relatives have less time to take care of them the healthcare centers are without any doubt insufficient for all, for these reasons there is an extra pressure or demand on the healthcare sector. Focusing on the elderly care, as the human capacity of memorizing events decreases over time, it is intended to promote an active ageing lifestyle for the elderly, where memory assistance tools are vital component. In this paper, it is presented an scheduler which takes charge of the day-to-day tasks and the user agenda.

Keywords: e-Health, Monitoring, Multi-Agent, Scheduling, Memory Assistant.

Ângelo Costa
DI-CCTC, Universidade do Minho, Braga, Portugal
e-mail: angelogoncalocosta@gmail.com

Paulo Novais
DI-CCTC, Universidade do Minho, Braga, Portugal
e-mail: pjon@di.uminho.pt

Ricardo Costa
CIICESI, College of Management and Technology, Polytechnic of Porto, Felgueiras, Portugal
e-mail: rcosta@estgf.ipp.pt

José Machado
DI-CCTC, Universidade do Minho, Braga, Portugal
e-mail: jmac@di.uminho.pt

José Neves
DI-CCTC, Universidade do Minho, Braga, Portugal
e-mail: jneves@di.uminho.pt

G.A. Papadopoulos and C. Badica (Eds.): Intelligent Distributed Computing III, SCI 237, pp. 209–214.
springerlink.com © Springer-Verlag Berlin Heidelberg 2009

1 Introduction

The life expectancy of the world population is increasing and the birth rate of children is decreasing rapidly, as it is acknowledge in a study lead by the United Nations Population Fund (UNPF). The UNPF estimates that the European Population decreased 13% between 2000 and 2050, being the age average being set on the 48 years [6].

The short term memory as referenced by the UNPF reports, states that at the age of 50 the human beings are severely affected by it, being the forgetfulness of events, namely the more recent ones, one of the most occurred symptoms. There is no way of reversing the loss of information so a solution is to use computers that store and, when it is needed, retrieve of all information he/she have saved in it.

Using an agenda or calendar, it can be achieved the help and care intended in this project. Scheduling and storing intelligently the user's activities, communicating with the most important persons to the user, can greatly help the elderly on their daily activities [1, 2].

This project tries to overcome a problem that is present on the society that has little solutions or responses from the academic and corporate platforms. Our goal is to present an innovative solution on the memory assist world.

1.1 Memory Assistant

There is no way of reversing the loss of information so one of the solution is to use systems that can store and, when it is needed, retrieve of all information he/she have saved in it. The urge of developing systems that is able to store important information is a hot topic nowadays. Efforts have been made in introducing software that provides assistance to the user in form of personal computational memory assistance, helps rehabilitating patients suffering from memory impairment after head-injuries and event software that suggests exercises that help the brain by exercising it on the short-term memory area.

1.1.1 Memory Assistant Software

HERMES is aimed to provide cognitive care. It tries to provide an independent living and social participation, by boosting the self-efficacy of the user (patient). The main objective is to develop a system, software and hardware, which will reduce the negative impact of declining cognitive capabilities, particularly memory. An important feature of the this project is the implementation of a system which should build an associations "map" of the different events and, based on that "map", create a pattern that emulates the human memory mechanisms. Despite its ambitious goals, HERMES is still in a very early stage of development. The SenseCam is developed by Microsoft, it is a wearable camera that takes photographs, passively, without the intervention of its user (patient). The objective is to report, in images, the user's day activities and what places he has been. It saves the images into its internal memory, the visioning of them can be as a time lapse video of the user all-day activities. It does not have any kind of intelligence associated and it serves merely users with total memory loss, like Alzheimer's.

2 Agenda Scheduling and Organization

The objective of this work consists in producing a intelligent scheduler that interacts with the user creating a product that will help him remember relevant information or events, i.e. a memory assistant.

For rescheduling an event we see in the agenda what are the least important information and try to move it to other place, considering factors like importance. Therefore the IGenda is more than a intelligent scheduler, it is to be referred to as a memory assistant. It may help specially the ones with loss of memory, by sustaining all the daily events and making known to the user when it is time to put them into effect.

This system will be able to receive information delivered by any platform and to organize it, so that the user will not need to be worried about planning or scheduling specific events and tasks.

This system is (Fig.1) an hierarchy of states and events. It will be built on the base of the Agenda Manager, the Free Time Manager, the Conflicts Manager and the Interface Manager [4].

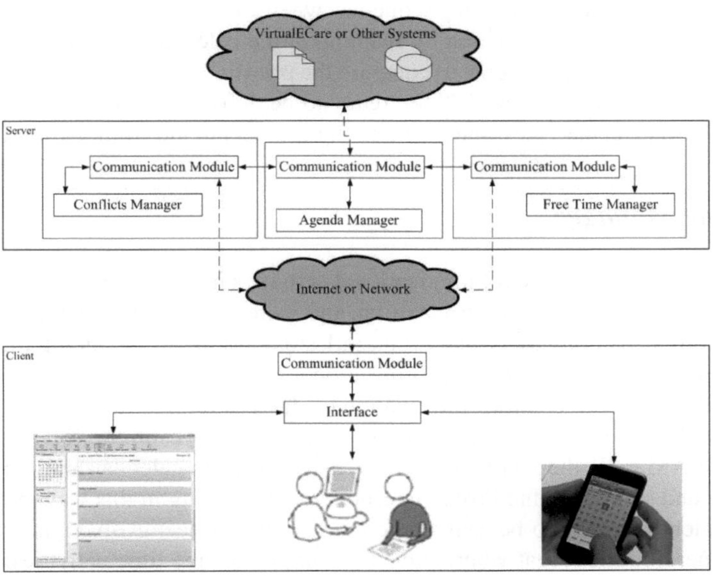

Fig. 1 Modules Scheme of Plan and Scheduling

The communication protocol complies with the FIPA-ACL XML [3] implemented by the JADE Framework. The messages are sent and received through the several modules and clients. As it been implemented in JADE total portability and interoperability is naturally assured.

2.1 Agenda Manager

The Agenda Manager (AM) sets the bridge between the remaining parts of the director system and the scheduling system, using the communication infrastructure available to receive and send the requests. Therefore the AM stands for the starting point of all the work that has to be done. The Agenda Manager consists in a two stage module application. It manages the incoming events to be scheduled and programs the time that triggers the Free Time Manager. It also supports the reception of multiple requests at the same time.

2.2 Conflicts Manager

The Conflicts Manager (CM) will work to make sure that there is any overlap of activities. This module schedules or reorganises the events that are received, verifying that they are in accordance with the rest of the agenda. If there is a collision of different events the collision may be decided by methods of intelligent conflicts management. In case of impossibility of scheduling the event, the error is reported to the sender. The CM works in the following sequence: when an event arrives, the AM sends it to CM. The CM enters in action by reading all the calendar files, parsing the new event and using the CLP engine to compare the priority levels of eventually conflicting events. It is created a new *ICS* file ready to be delivered to the user. At the same time a "new calendar" message is sent to the user and to the administrator.

2.3 Free Time Manager

The Free Time Manager (FTM) will fit recreational activities in the free spaces on the user calendar, in order to create an occupation, and increasing the well being of the user. These activities configure an important milestone for an active ageing on the part of the user, promoting his cultural, educational and conviviality activities. The FTM has a database that contains information of the user's favorite activities, previously checked by the decision support group. The FTM works in the following sequence: the AM calls the FTM, then proceeds reading the calendar and the options available and sends it to the Prolog interpreter, constructing in this way the new activities Calendar, ready to be sent to the user. As the project evolved there where several emerging ideas that where aimed to enrich the free time activities. The two projects that had the most interest where the Time Bank [5] and ePal. The Time Bank consists in trading help and support with the several users in the Time Bank community. It works by having a time deposit, where they set the time they have available to help and support, when support is given by the user the deposit increases, when support is given to the user the deposit the decreased. This is a way of goodwill community, were services are traded, and a simple way of benefit and help, feeling needed at the same time. The extension of Professional Active Life (ePAL) project consists in a group of services available for and to the elderly. The ePAL tries to build a network of active senior professional that are in the retirement and group them following their previous working activities. It creates a work force of highly specialised people, available to contribute to the community with their expertise, with no money income.

2.4 *Interface Manager*

The interface intends to be intuitive and easy to use. The warning method can vary from an active to a selective mode. This means that for some users all of the events will be notified and for other users the warnings show some sensibility. If the user current activity is more important than the event happening, the user will not be notified of the scheduled event, and the event is marked as executed.

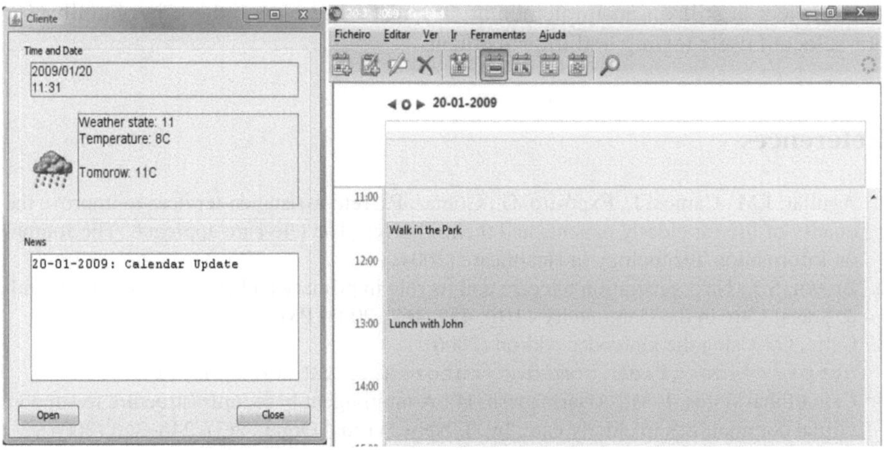

Fig. 2 Screenshot of the Client Interface

3 Social Connections: Objectives

As much centred in the user well-being this project does not forget his relatives. It provides a way to have the user's family aware of his/her health condition, once it gives the relatives a way to add events to the user calendar as common tasks. In a more user focused component, the free time occupation is a vital component of this project, a very large development was and is dedicated to this theme. There are several ways to tackle this problem, our selection was the combination between a database of activities and community service works. In a more intricate way there are two systems that were adopted, the Time Bank and the ePAL. These are the way to enforce socialization and a method to fight loneliness. So, far beyond of being a memory assistance tool, it is also a social enabler, by introducing aspects that go outside the physical limits or range of a simple event scheduling.

4 Conclusions

Although this project had been born to implement one of the branches of the VirtualECare project, it turned independent and auto-sufficient, available to be used in another environments and situations. As the full completion of this project is still a bit far away, some of its major functionalities are already working. This project makes the difference between the other memory assistants because it introduces the component of free time occupation that contains the implementation of social activities that consists in the interaction of the user with other persons to promote a feeling of being still needed as an individual. As the introduction of socialization and social activities are a strong intention it will have to be further exhaustive planning, because it's a complex scheduling system, involving many other people and systems. This project is still on an implementation and tuning phase, there is planning for pre-selected users testing and improvement.

References

1. Aguilar, J.M., Cantos, J., Expósito, G., Gómez, P.: Tele-assistance services to improve the quality of life for elderly patients and their relatives: The tele-care approach. The Journal on Information Technology in Healthcare (2004)
2. Brown, S.J.: Next generation telecare and its role in primary and community care. Health & Social Care in the Community 11(6), 459–462 (2003) PMID: 14629575
3. Caire, G.: Using the xmlcodec add-on (2006), `http://jade.tilab.com/doc/tutorials/XMLCodec.html`
4. Camarinha-Matos, L.M., Afsarmanesh, H.: A multi-agent based infrastructure to support virtual communities in elderly care. Int. J. Netw. Virtual Organ. 2(3), 246–266 (2004)
5. Castolo, O., Ferrada, F., Camarinha-Matos, L.: Telecare time bank: A virtual community for elderly care supported by mobile agents. The Journal on Information Technology in Healthcare, 119–133 (2004)
6. UNFPA: Population ageing and development: Operational challenges in developing countries (2002)

Automatic Ontology Extraction with Text Clustering

Beniamino Di Martino and Pasquale Cantiello

Abstract. This paper presents a technique to automatically derive ontologies which is based on hierarchical clustering of document corpora. The procedure applies to a set of texts forming a domain document corpus and creates a hierarchical structure (tree) where at every node is associated a set of terms derived from the document feature vectors. The labeling of the cluster is made by using a new algorithm presented in this work. The derived terms may represent concepts candidate to build a domain taxonomy from which the hierarchical relationships among the classes of the domain ontology can be extracted. To test the technique shown, has been built a propotype tool named *(OntoClust)*.

1 Introduction and Related Works

In document corpora analysis, and in the semantic web crawling, it is necessary to analyze a lot of documents or web pages, trying to find relations, synonyms, common concepts to create logical taxonomies. Ontology construction made manually by domain experts is obviously expensive.

To automate the ontology building, existing data model, domain ontologies are valuable resources [5]. But most of times there are no glossary or terminology in the interesting domain, so the only source of information is document corpus itself.

An automatic natural language processing technique can be realized to perform the analysis, to classify the information and to extract ontologies. Candidate ontologies can also be matched against reference ontologies for validation.

This paper, after the introduction continues with an analysis of related works in ontology and clustering.

In the section 2 is presented a technique to automatic extract taxonomies with hierarchical clustering from domain corpora. Here a new algorithm is presented to

Beniamino Di Martino
Second University of Naples, Italy
e-mail: beniamino.dimartino@unina.it

Pasquale Cantiello
Second University of Naples, Italy
e-mail: pasquale.cantiello@unina2.it

G.A. Papadopoulos and C. Badica (Eds.): Intelligent Distributed Computing III, SCI 237, pp. 215–220.
springerlink.com © Springer-Verlag Berlin Heidelberg 2009

label the nodes of the cluster and build the taxonomy of concepts, by transferring concepts along a binary tree.

In the section 3, a prototype tool named *OntoClust* that implements the technique is presented. The paper ends at section 4 with conclusions and future work directions.

Related Works

Hotho, Maedche and Staab [12] present a technique to cluster documents in a subjective way. They use ontologies to drive the clustering of the documents, extracting various clusters and letting the user making its choice. The ontology is used with heuristics for feature selection and aggregation. The resulting clusters are not hierarchical, but agglomerative.

Fortuna, Mladenic and Grobelnik [3] reviewed two techniques for topic discovery in collections of text documents. They also presented the *OntoGen* system for semi-automatic data-driven ontology construction. The clustering is not hierarchical, but a K-Means type. Their system is semi-automatic as the user is responsible of choosing the K value.

Cimiano and Staab [1] present an approach for the automatic induction of concept hierarchies from text collections. The introduce the hypernym oracle, which drives the clustering process. The clustering of two documents is made only if there is a common hypernym as suggested by the oracle. The approach is interesting but relies on the sources of information of the oracle (a mix of WordNet, Google and the document itself).

Kashyap, Ramakrishnan, Thomas and Sheth [4] index documents in the cluster by using Latent Semantic Indexing [2]. They also introduce a framework [8] for the Taxonomy and Hierarchy generation. Intra-cluster cohesiveness is defined here as a property of nodes of the cluster, useful to convert binary trees into taxonomies. There are no particular assumptions on how to choose the partition of the range of cohesiveness, so we decide also to make further research on it.

2 The Technique

The technique to analyze document corpora with document clustering and taxonomy extraction is rather standard now in literature, but still actual. We explain our tuning and improvements by describing the steps.

Text Analysis

First of all, a scanning of the document corpora, or (in case of semantic web) a crawling of a web site is done to extract and convert all files to text documents.

Upon each document a lexical analysis is performed to simplify the text and to remove numbers, punctuation and symbols and all the words that are too frequents in the chosen language (so called stop-words), useless to differentiate a document (e.g. articles and conjunctions).

In the *stemming* phase, the documents are normalized by translating all the words into their respective stems: the radixes obtained by cutting any prefix, suffix, plural, etc. These become now the language stems with which is built the vocabulary of the document collection and are used during the indexing phase.

Indexing

In this phase documents are indexed with inverted indexes: for each document is constructed a feature vector, that is a k-length array with real values. Every value is the weight that measures the relevance of the property in the description of the target document.

Given the indexes, for each document is built the relative feature vector. The weights can be defined in different ways, typically by using the local frequency of a word in the document, as proposed by Salton [9]. By using these weights, the occurrence of each word is independent from the length of the document. A value of greater than 0.5 (less than 1) indicates the presence and gives the *importance* of the word in the document. By using these vectors, a *similarity matrix* is constructed, by calculating the distance between documents.

Clustering

The similarity matrix resulting from the previous phases is used to identify groups of similar events (clusters). The clustering algorithm used in this work is the revised Neighbor Joining [10]. The resulting cluster model is a hierarchical binary tree in which the nodes represent the obtained clusters and the leaves the classified events.

The Neighbor Joining, starting from the similarity matrix, recursively computes the distance sums between one event and all the others, and obtains the divergences of the distances among all the possible pairs of events. On the following step, the minimum of the distances identifies the couple of events more similar each other over the entire data-set. The couple is treated as a single event recalculating the distance between this new event and all the others. The algorithm recursively proceeds until the clustering of all the events has been completed.

Taxonomy Extraction

At this point, starting from our binary tree, we should transform it into a taxonomy.

During the clustering process, the *intra-cluster cohesiveness* [4] is calculated as a way to compute the differentiation in meaning between successive levels of the extracted taxonomy. Starting from a set of document vectors $D = d_1, \cdots, d_N$ we have the centroid in eq. 1 and the cohesiveness of the set as in eq. 2.

$$m(D) = \frac{1}{M} \cdot \sum_{i=1}^{M} d_i \tag{1}$$

$$c(D) = \frac{1}{M} \cdot \sum_{i=1}^{M} cos(d_i, m(D)) \tag{2}$$

This cohesiveness measures the degree of attraction in the set and is monotonically increasing from the top of the tree to the leaves. Now, we have a range of values for the cohesiveness, from the root to the leaves. To obtain the taxonomy we can partition this range into N intervals, where N is maximum deepness of the desired taxonomy. By using N threshold levels $\mu_1, \cdots, \mu_N$ we can divide the range into N *iso-cohesiveness* sub-ranges.

For every possible path in the starting tree, $N-1$ nodes are selected: the nodes whose cohesiveness is more close to the thresholds. The nodes within the sub-range are collapsed within their path in their ancestors, according the algorithm presented in [4], giving the requested taxonomy.

Node Labeling

In this phase, starting from the obtained hierarchy, terms should be assigned to new nodes. To properly label the cluster nodes, has been designed a new algorithm that operates on a tree whose nodes are the related feature vectors.

The idea in this algorithm is that every term that is attached to two sons, since it is a common feature, can be transferred to the father, and removed from the sons. By applying this term transfer to the entire tree, we assure that common feature are properly attached to parent concepts. The algorithm works as follows:

The terms with a value in the feature vector below a given threshold are eliminated. This is done to minimize the *noise*. In this phase the values are converted into bitSet, assigning 1 only to the terms with a value greater than the threshold, and 0 otherwise.

A depth-first visit of the tree is done and using a bottom-up approach (starting from the leaves), every node is compared with its brothers; a logical AND between the brothers bitSet is made. Calling *ANDb* the result of this operation, a logical OR between *ANDb* and the bitSet of father of these brothers is done. From now on, the result is associated to the father. The result of a logical XOR between each bitSet and *ANDb* is associated to the brothers.

By using this algorithm, the common words between two brothers node are transferred to their father, leaving only the distinct words to them. Let's suppose that our feature set is composed of five terms (in this order): *house*, *school*, *hand*, *pen* and *sheet*, so each bitSet is composed of five bits. The hierarchy in our tree has the words distributed on it and a possible representation with the bitSet could be that of figure 1. Near each node is shown the relative bitSet.

After applying the algorithm, the result of the AND and OR operation modify the tree as shown in the figure 2. Now the common terms between two brothers are transferred to their father.

Validation and Exporting

At this point we have obtained a taxonomy of related concepts. Now we can use a reference ontology to validate and enrich the taxonomy. In particular, a semantic match is made between the taxonomy and the reference ontology. So its possible to populate the taxonomy by adding the corresponding concepts of the reference

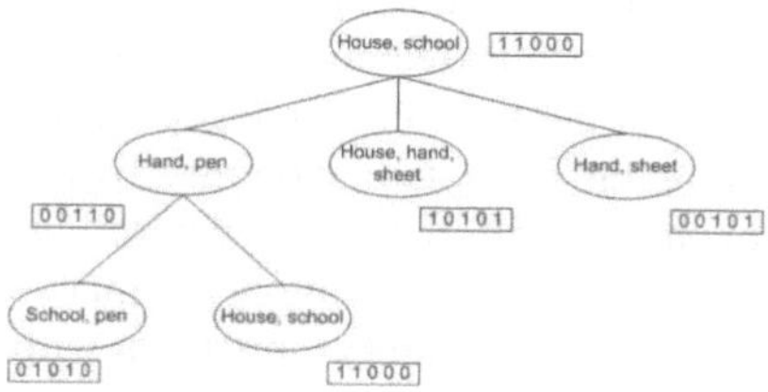

Fig. 1 The tree at starting

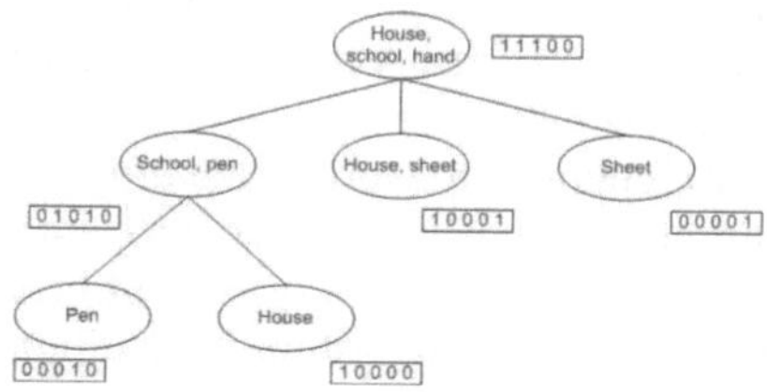

Fig. 2 The tree at the end

ontology. A user should validate the matching. In the end we are ready to export an ontology with the concepts and relations that are now in the taxonomy created and populated during the previous phases.

3 The OntoClust Tool

To test the described technique and the algorithm, we have built a prototype tool, named *OntoClust*. This has been developed in Java language, with the *Eclipse* platform. A rich NLP library named *nlp.unina2* has also been built.

We have used two different stemming tools: WordNet [6] for the English language (interfaced via the Dragon Toolkit [11] and SnowBall that can generate stemmer libraries in Java itself for the Italian language. Due to shortness of this paper, only one snapshot is shown in figure 3. We plan to put the tool on the web as soon as it's more stable.

4 Conclusions and Future Work Directions

In this work we have seen a technique to derive an ontology from a document corpora by using hierarchical clustering. A new algorithm has been developed to show how

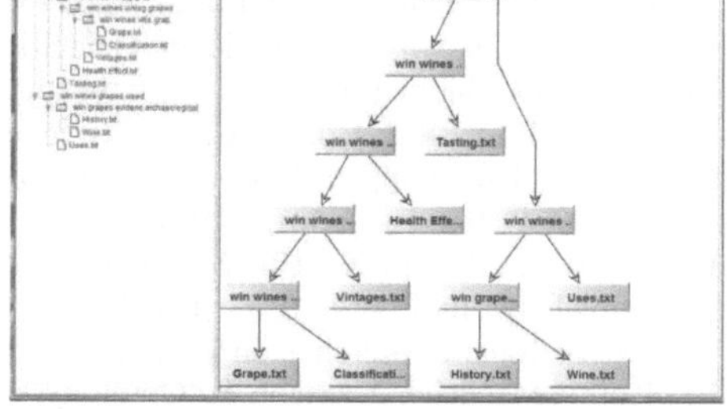

Fig. 3 Snapshot of the clustering phase

to assign terms to the nodes and propagate them from children to parent to extract common concepts. A prototype tool has been built to test the technique. The binary three is converted into a taxonomy through the use of the iso-cohesiveness concept.

We continue to work in the direction of refining the extracted taxonomy, by exploring other types of conversion from tree to taxonomy. Different types of partitioning in the cohesiveness range will be tested. Also we are conducting performance measures on the algorithm. In addition we plan to make test and comparison with `http://clusty.com` search engine.

Acknowledgements. This work has been supported by LC^3 - LABORATORIO PUBBLICO-PRIVATO DI RICERCA SUL TEMA DELLA COMUNICAZIONE DELLE CONOSCENZE CULTURALI - National Project of Ministry of Research MIUR DM17917.

We would thank to Carlo Baia and Angelo Martone for their help in the development of the OntoClust tool.

References

1. Cimiano, P., Staab, S.: Learning Concept Hierarchies from Text with a Guided Hierarchical Clustering Algorithm. In: Proceedings of the Workshop on Learning and Extending Lexical Ontologies with Machine Learning Methods (2005)
2. Deerwester, S., Dumais, S.T., Furnas, G.W., Landauer, T.K., Harshman, R.: Indexing by Latent Semantic Analysis. Journal of the Society for Information Science 41(6), 391–407 (1990)
3. Fortuna, B., Mladenic, D., Grobelnik, M.: Semi-Automatic Construction of Topic Ontology. In: Ackermann, M., Berendt, B., Grobelnik, M., Hotho, A., Mladenič, D., Semeraro, G., Spiliopoulou, M., Stumme, G., Svátek, V., van Someren, M. (eds.) EWMF 2005 and KDO 2005. LNCS (LNAI), vol. 4289, pp. 121–131. Springer, Heidelberg (2006)
4. Kashyap, V., Ramakrishnan, C., Thomas, C., Sheth, A.: TaxaMiner: An Experimental Framework for Automated Taxonomy Bootstrapping. International Journal of Web and Grid Services 1(2), 240–266 (2005)
5. Maedche, A., Staab, S.: Ontology Learning for the Semantic Web. Intelligent Systems, IEEE 16(2), 72–79 (2001)
6. Miller, G.A.: WordNet, a lexical database for the English language, Cognitive Science Laboratory, Princeton University, `http://wordnet.princeton.edu/`
7. Porter, M., Boulton, R., Macfarlane, A.: Snowball, a small string processing language designed for creating stemming algorithms for use in Information Retrieval
8. Ramakrishnan, C., Thomas, C., Kashyap, V., Sheth, A.: TaxaMiner: Improving Taxonomy Label Quality. In: McIlraith, S.A., Plexousakis, D., van Harmelen, F. (eds.) ISWC 2004. LNCS, vol. 3298. Springer, Heidelberg (2004)
9. Salton, G., Buckley, C.: Term weighting approaches in automatic text retrieval. Technical Report TR 87-881, Cornell University (November 1987)
10. Studier, J.A., Keppler, K.J.: A note on the Neighbor Joining Algorithm of Saitou and Nei. Mol. Biol. Evol. 5(6), 729–731 (1988)
11. Zhou, X., Zhang, X., Hu, X.: Dragon Toolkit: Incorporating Auto-learned Semantic Knowledge into Large-Scale Text Retrieval and Mining. In: Proceedings of 19th IEEE International Conference on Tools with Artificial Intelligence, ICTAI 2007 (2007)
12. Hotho, Maedche, A., Stabb, S.: Ontology-based Text Document Clustering. In: KI (2002)

Distributed Information Sharing in Mobile Environments

Joe Gorman and Ulrik Johansen

Abstract. Developing a *distributed information sharing solution (DISS)* for mobile environments is a challenging task if requirements on *functionality, availability, consistency, scalability, robustness, performance and adaptability* are to be taken into account. A combination of different concepts and mechanisms must be applied, and functionality representing high implementation costs must be avoided. In the MIDAS project a conceptual basis for such a solution has been developed and implemented. The paper describes this work, and how its functional and quality properties have been verified in a simulation environment.

1 Introduction

1.1 Background

MIDAS is a research project in the FP6 IST Programme of the European Commission. Its aim is to simplify the development of mobile applications for multiple, co-operating users who make use of a mixture of infrastructure-based communications and ad-hoc communications. An example scenario would be to support emergency workers dealing with a major incident at a location with limited communication facilities.

MIDAS assumes that communications infrastructure may not be available at all locations, necessitating the use of MANETs (mobile ad-hoc networks) by some users. Middleware developed in MIDAS integrates MANETs with infrastructure-based networks. One of the key components is the MDS (MIDAS data space), with origins in the FieldCare project (see [3]). This provides a mechanism to share data between nodes by providing functions which ensure that data entered on any node

Joe Gorman · Ulrik Johansen
Software Engineering, Safety and Security, SINTEF ICT Trondheim, Norway
e-mail: `{Joe.Gorman,Ulrik.Johansen}@sintef.no`

G.A. Papadopoulos and C. Badica (Eds.): Intelligent Distributed Computing III, SCI 237, pp. 221–226.
springerlink.com © Springer-Verlag Berlin Heidelberg 2009

is made available on all other nodes. This is achieved by storing replicas of data on multiple nodes. The MDS manages allocation of replicas, and arranges that requests to access/update data are routed to the "nearest" node containing a replica.

One of the core MDS functionalities is to ensure that all replicas are synchronized with each other. This involves propagating updates to other nodes holding the replica, and dealing with lost or corrupted replicas / updates. A working implementation of these features was produced in the project, and successfully used in two proof-of-concept applications (see [6]). However, this implementation had some limitations concerning scalability, performance and adaptability. The DISS solution described in this paper is an alternative solution to the part of MDS responsible for distribution and sharing of information between nodes. It aims to provide a solution that addresses quality related requirements more thoroughly.

1.2 Requirements to DISS and Main Challenges

The overall goals for the DISS solution broadly correspond to the goals defined for the Bayou Architecture ([1]), where the type of solution should be "update anywhere-anytime-anyway". All goals except those concerning conflict resolution are considered crucial, while conflict resolution is taken care of otherwise in MIDAS (see [7] and [4]).

The main *functional requirements* for the DISS are:

- To provide users with an information sharing solution which provides functionality similar to a distributed *Relational Data Base Management System (RDBMS)*.
- *Multi-master solution*: Applications at different nodes are allowed to make updates to the same information simultaneously.
- *Replication control*: The same information may physically be stored at different nodes (i.e. replicas), and it shall be possible to control at which nodes replicas exist.
- *Dynamic node topology*: Nodes may attach to or detach from the DISS at any time.

Quality related requirements to be explicitly considered in the DISS design and implementation are:

- *Consistency*: It is sufficient to achieve eventual consistency between replicas of the same replicated object, i.e. different replicas eventually reach the same value if update submissions are stopped, and if there is connectivity between all nodes holding replicas of the object.
- *Availability*: Availability shall be controllable by making it explicitly controllable at which nodes replicas shall exist.
- *Scalability*: A large number of replicas can exist without this leading to a significant decrease in performance.
- *Robustness*: The solution must be able to withstand network instability and disruption, and varying and high workload. It must be possible to cope with network partitioning.

- *Adaptability*: The solution must be capable of autonomous adaptation to varying network infrastructure and connectivity, and to the processing and storage resources of the nodes, in order to achieve good overall performance.

An important aspect of MIDAS is that only eventual consistency is required. This means that there may be periods (perhaps even lengthy periods) where inconsistencies may exist between replicas, meaning that full ACID properties will not be fulfilled. This is because:

- Achieving full consistency would necessitate use of the two-phase commit protocol, and this is not feasible in the envisaged operational environment. The number of nodes may be very large, and some nodes may be isolated for long periods. Completion of a commit could therefore take a very long time, during which the update would not be available at any node.
- In the type of applications for which MIDAS is intended, data *availability* is considered to be more important than full consistency (e.g. it is better for emergency workers to have slightly out-of-date information than none at all because the system is in an extended wait for a commit operation).

Our approach to meeting the challenge of addressing all of these requirements, and experimental results from using it, are described below.

2 The Proposed Distributed Information Sharing Solution

We have implemented a solution combining several concepts and mechanisms:

Optimistic Replication

An update made at one node is propagated asynchronously to other nodes containing replicas of the table. If some of the other nodes are temporarily unreachable, updates are nevertheless carried out at the nodes that are reachable, but using protocols which ensure that the update gets propagated to the unreachable nodes when they later become available. This is similar to the lazy group replication concept defined by [5]. It can lead to inconsistencies between nodes - but this is in line with the requirement on only eventual consistency.

Overlay two tier topology

This is an established technique (see [5]) in which nodes are classified as either root nodes or leaf nodes, with each leaf node being assigned to a specific root node. Root nodes synchronize with all other root nodes, and with their own leaf nodes; each leaf node synchronizes only with its root node. The motivation for this is to limit connectivity between nodes for synchronization, in order to achieve a scalable solution.

In the approach described in [5] pessimistic replication is used between root nodes, which are assumed to be always connected. We cannot make that assumption in the operational environment envisaged, so in our approach we use optimistic replication for both leaf and root nodes.

The overlay two tier topology is dynamic in our approach, maintained by a role determination algorithm using available information about network characteristics.

Delta Update and Recover Update

We assume that - most of the time - attempts to propagate updates from node to node will be successful A delta update protocol implements this with minimum use of resources, and without explicit acknowledgment messages that would generate more traffic. Use of sequence numbers enables receiving nodes to detect lost updates.

When a node detects an inconsistency, a recover update protocol is initiated. This is bi-directional between a pair of nodes, involving detailed data exchanges between them. It defines three different levels (each involving different levels of resource usage), with mechanisms to determine which level is appropriate.

Automated Data Allocator

The overall effectiveness of the solution depends critically on the allocation of replicas to nodes. A balance must be sought between *availability* (increased by having many replicas), and *performance* (decreased by having many replicas, as network traffic for synchronization increases). This is a complex optimization problem, and depends on how the stored data is actually used by applications: which applications on which nodes access which data and how often.

Our approach includes a separate Data Allocator model which makes decisions on allocation of replicas to nodes based on dynamic monitoring of usage patterns and overall policies defined statically by application designers using knowledge about the application.

3 Experimental Results

Full scale testing of our solution with real users and large numbers of mobile nodes was not possible within the resource limitations in the project. A simulation environment was used, involving multiple instances of the DISS middleware implementation (one per simulated node), and simulator modules for the network (simulating capacity and connectivity) and traffic load (insert/delete/update operations at different nodes).

The simulation environment was used to investigate the scalability and performance of the DISS under varying conditions concerning number of nodes, system load and allocation of leaf nodes/root nodes.

Figure 1 shows one key set of results. The three different lines show differing total numbers of insert/delete/update operations. Part (a) shows how the monitored total system load increases linearly with the number of nodes (i.e. acceptable scalability). Part (b) shows the normalized system load, and how this decreases with

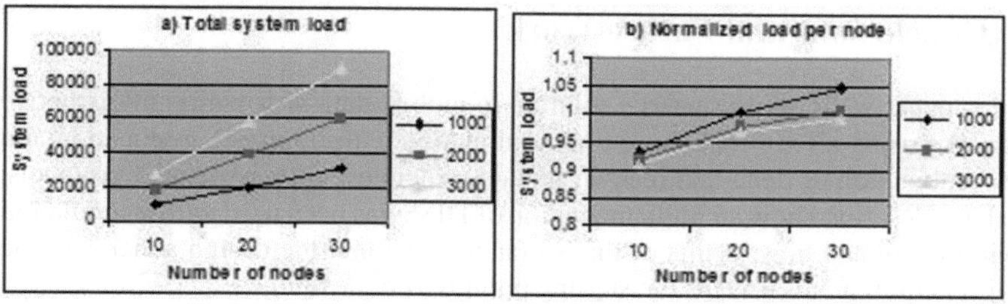

Fig. 1 System load as a function of varying number of nodes and workload

increasing workload (because the administrative overheads related to establishing network topology are less significant at higher workloads.)

We also investigated how total system load varies according to the number of leaf nodes per root node, and concluded that the effect is minimal. More detailed experiments showed how load distribution between lead and root nodes is affected by allocation of node roles; root nodes should limit the number of associated leaf nodes.

The overall conclusion from our experiments is that scalability and performance are acceptable.

4 Related Work

[1] specifies a set of goals for supporting data sharing between mobile users. These are functional goals, independent of specific implementations in a system solution. The goals cover support for portable computers with limited resources, high availability, reaching eventual consistency, detection and resolution of update conflicts, permitting disconnected clients, and giving users ultimate control over the placement and use of databases. All of these are issues that have been addressed in DISS.

[2] gives a comprehensive survey about optimistic replication: what it is, where it has been used, and what are the challenges of using it. Issues discussed include what shall be the replicated objects, propagation of updates to replicas, detection and resolution of conflicts, concurrency control, eventual consistency, communication topology, and controlling replica divergence. All of these have served as main contributions to the DISS design.

[5] defines eager and lazy replication and introduces the two-tier replication concept having a root tier of connected "base nodes" which are always connected and use pessimistic replication, and a set of "mobile nodes" which are disconnected much of the time and which use optimistic replication with the base nodes. In DISS this principle of two tier replication has been adopted, but in a context where both types of nodes have high demand for being available and keeping their replicas up-to-date, and where it is not possible to assume that the base nodes can always be connected. Our solution therefore uses optimistic replication for all nodes.

5 Conclusions and Future Work

In this paper we have presented a solution to mobile data distribution in disruptive networks using a two tier solution in combination with optimistic replication and the combination of delta and recover update principles, with associated protocols. We have described how an implementation of DISS has been used with a simulation environment, and how results of this so far indicate that the design adheres to the quality requirements used as the starting point for our design.

Some remaining work is necessary to confirm that the solution completely adheres to the requirements set, and that the solution can be applied in practical situations:

- Simulate varying network conditions and dynamic leaf/root topologies.
- Investigate scalability properties more thoroughly by carrying out experiments with much larger numbers of nodes.
- Implement an advanced topology management algorithm as an extension of the currently used simple algorithm, and to do simulations of alternative topologies and with different capabilities for node network connections.
- Define a protocol that can reliably detect lost delta updates, even in cases where it is the last update that it lost (since using sequence numbers does not suffice). This solution must not imply a significant performance overhead.
- Use the solution in real field testing, with real devices and users. This is feasible as the solution has already been prepared for integration with the rest of the MIDAS middleware.

References

1. Demers, A., Petersen, K., Spreitzer, M., Terry, D., Theimer, M., Welch, B.: The Bayou Architecture. Support for Data Sharing among Mobile Users. In: Proceedings of the workshop on Mobile Computing Systems and Applications, Santa Cruz, California (December 1994)
2. Saito, Y., Shapiro, M.: Optimistic Replication. ACM Computing Surveys V(3) (2005)
3. Gorman, J., Walderhaug, S., Kvålen, H.: Reliable Data Replication in a Wireless Medical Emergency Network. In: Anderson, S., Felici, M., Littlewood, B. (eds.) SAFECOMP 2003. LNCS, vol. 2788, pp. 207–220. Springer, Heidelberg (2003)
4. Gorman, J., Wienhofen, L.: Common Notion of Time in the MIDAS MANET: Arrogant Clocks. In: Norsk Informatikkkonferanse 2008, Kristiansand, Norway, Proceedings, pp. 129–140 (2008)
5. Gray, J., Helland, P., O'Neil, P., Shasha, D.: The Dangers of Replication and a Solution. In: SIGMOD 1996 6/96 Montreal, Canada, pp. 173–182 (1996)
6. Plagemann, T., Munthe-Kaas, E., Skjelsvik, K., Puzar, M., Goebel, V., Johansen, U., Gorman, J., Marin, S.: A Data Sharing Facility for Mobile Ad-Hoc Emergency and Rescue Applications. In: 27th International Conference on Distributed Computing Systems Workshops, ICDCSW 2007 (2007)
7. Plagemann, T., Munthe-Kaas, E., and Goebel, V.: Reconsidering Consistency Management in Shared Data Spaces for Emergency and Rescue Applications. In: BTW-MDM 2007, Model Management and Metadaten-Verwaltung, workshop under GI-Fachtagung fr Datenbanksysteme in Business, Technologie und Web, Achen, Germany (2007)

Cost of Cooperation for Scheduling Meetings

Alon Grubshtein and Amnon Meisels

Abstract. Scheduling meetings among agents can be represented as a game - the Meetings Scheduling Game (MSG). In its simplest form, the two-person MSG is shown to have a price of anarchy (PoA) which is bounded by 0.5. The paper defines the *Cost of Cooperation* (CoC) for meetings scheduling games, with respect to different global objective functions. For an "egalitarian" objective, that maximizes the minimal gain among all participating agents, the CoC is non positive for all agents. This makes the MSG a *cooperation game*. The concepts are defined and examples are given within the context of the MSG. A game may be revised by adding a mediator (or with a slight change of its mechanism) so that it behaves as a cooperation game. Thus, rational participants can cooperate (by taking part in a distributed optimization protocol) and receive a payoff which will be at least as high as the worst gain expected by a game theoretic equilibrium point.

1 Introduction

Scheduling meetings between two or more people is a difficult task. Despite the advances offered by modern world electronic calendars these are often limited and serve as passive information repositories. As a result, a dedicated person is usually hired to handle this task. Previous attempts to automate the meetings scheduling problem (MSP) employ one of two extreme approaches: the cooperative approach and the competitive one. Cooperative methods for solving MSPs perform a distributed search for an optimum of a global objective [5]. The competitive approach investigates game theoretic equilibria of suitable games and designs strategies for the competitive agents [1]. One previous attempt to combine the two approaches

Alon Grubshtein · Amnon Meisels
Dept. of Computer Science
Ben Gurion University of the Negev
Beer-Sheva, 84105, Israel
e-mail: {alongrub, am}@cs.bgu.ac.il

G.A. Papadopoulos and C. Badica (Eds.): Intelligent Distributed Computing III, SCI 237, pp. 227–236.
springerlink.com © Springer-Verlag Berlin Heidelberg 2009

was introduced in [3], which empirically attempted to introduce selfishness into a cooperative protocol for the Meeting Scheduling Problem (MSP). Instead of searching for a solution which is socially optimal and potentially harmful for some agents, alternative global objectives are examined. Such solutions maintain an important goal - they provide acceptable gains to each one of the agents.

The present paper attempts to introduce cooperation into a self interested scenario by using a simplified meetings scheduling game (MSG). The use of game theory for studying self interested interactions provides a sound mathematical basis for the analysis of strategic situations under a set of (commonly) acceptable axioms (cf. [8]). A fundamental result of game theory is the existence of equilibrium points which define an action (or strategy) profile that is stable (i.e., no participant can gain by deviating from this profile). There are many definitions of various types of stable points, and unless otherwise specified we will assume the common "pure strategy Nash Equilibrium" (NE). A NE is simply a set of actions (assignments) for each agent in which no agent can gain by unilaterally changing its value [8].

Investigating the simple MSG, one shows that the NE of the MSG may be different than its optimal points. This is similar to routing games (cf. [11]). Consequently, the widely accepted efficiency measure, known as the *Price of Anarchy (Stability)* [4, 9, 12, 8, 11], for the simple MSG is different than unity. For scheduling meetings and its underlying game, we present user oriented criteria, and formulate a game property based upon it - the Cost of Cooperation (CoC). This property can motivate selfish users to cooperate by presenting *each one* of the players a guaranteed cooperative gain which is higher than its worst equilibrium value.

2 Meetings Scheduling by Agents

The Meeting Scheduling Problem (MSP) involves a (usually large) set of users with personal schedules and a set of meetings connecting these users. It is an interesting and difficult problem from many aspects. In the real world, agents have a dual role. They are expected to be both self interested (i.e. seek the best possible personal schedule) but also cooperative (otherwise meetings will not be scheduled).

MSP users have personal calendars which are connected by constraints that arise from the fact that multiple groups of the overall set of users need to participate in meetings. Meetings involve groups of users and each meeting includes at least two participants. Meetings can take place in different places and carry a different significance for their participants. All participants have their personal schedules and need to coordinate their meetings. Each user is represented by an agent and the set of agents coordinate and schedule all meetings of all users. The final result is a set of updated personal schedules for all agents/users (a global calendar). One uses utilities to represent the significance that users attach to meetings and to certain features of the resulting schedule, as it appears in their personal calendars.

Two very different approaches were studied for solving MSPs. The first employs Distributed Constraints Satisfaction (DCSPs) and Optimization (DCOPs) [5, 6, 14]. In these studies, agents are expected to be *fully cooperative* and follow the protocol regardless of their private gains. The outcome of such protocols is a solution whose **global utility** is optimal. The global utility does not necessarily account for the quality of the personal schedules of the participants.

An alternative approach is offered by researchers in the field of *Game theory*. Here, agents are rational, *self-interested* entities which have different (and often conflicting) utility functions (cf. [8]). A large share of the game theoretic research related to MSPs emphasizes the underlying mechanism of the interaction [2, 1, 10]. The basic assumption of these studies is that the gain of agents from their scheduled meetings can be represented by some universal currency. Moreover, the assumption accepts a uniform exchange rate for some monetary means and *unscheduled* meetings. These fundamental assumptions seem very unrealistic for scheduling meetings among people. However, if one is ready to accept this monetary model many game theoretic results and mechanisms can be applied.

The approach of the present study examines simple game theoretic mechanisms to restrict or predict agents behaviors.

3 The Meetings Scheduling Game (MSG)

Imagine a pair of self-interested users, each with her own preferences about her personal schedule. Many agreements need to be reached in order to generate a viable and consistent schedule. One way to enforce agreements is to pose the problem as a game with rules that enforce schedules that are then accepted by the participants. In a way, this is analogous to routing by a player that accepts the delay as a given result of the route it selected. One can think of this as the mechanism of the game.

We begin by defining a simple MSG. The game includes two players coordinating a meeting. The meeting can be scheduled in the morning or in the evening. Each player places her bid for scheduling by specifying her preferences (or expected gains) from each schedule. This list of preferences takes the form of a tuple $b_i = \langle x,y \rangle \in \{0,1,...B\} \times \{0,1,...B\}$ for the two time slots, where the first value (x) corresponds to the evening time slot and the second number (y) corresponds to the morning time slot. A higher number indicates a higher gain from having the meeting at the corresponding time slot and therefore a stronger desire to meet at that time slot. Once both bids are presented, a time slot is selected by the game's mechanism according to its decision rule, and the participants are informed of the decision. There are many possible decision rules, but for now we use the "utilitarian" approach [8] - the time slot selected is the one with the highest sum of bids on it (or when sums are equal, outcomes have equal probabilities). In other words, the selected time slot has the highest total gain for the players. The payoff of each participant is defined by the player's utility function (preference) for a time-slot.

We limit ourselves to the case where players are never indifferent to the results, i.e., they will always prefer some time slot over the other. Any bid of the form $\langle x,x \rangle$ is thus excluded (it results in a scheduling based on the opponent's preference).

The first approximation of the MSG assumes that when the meeting is held at a time that is not preferred by a player, her gain from it will equal zero. We will later remove this assumption. We distinguish between the payoffs of players when the meeting is held at the player's most desired time, and say that player 1's payoff in such a case is m, while player 2's payoff for an analogous case is k. This enables us to treat users that are essentially different.

Figure 1 depicts the simplest possible MSG for two players. The descriptive power of a strategy is limited to two values of preference, either 0 or 1 (preferred). In this example player 1 prefers to have the meeting in the morning. When player 2 also prefers the morning time-slot both players have the same dominant strategy - $\langle 0,1\rangle$. That is, bidding $\langle 0,1\rangle$ will *always* yield a higher payoff for both players.

When player 2 prefers the evening time-slot her dominant strategy becomes $\langle 1,0\rangle$ (player 1's dominant strategy remains $\langle 0,1\rangle$). This can immediately be translated into an equilibrium point resulting from playing the dominant strategy. When the desires of both players are the same, the equilibrium payoff is m,k, and when these desires conflict the equilibrium payoff is $\frac{m}{2},\frac{k}{2}$ (i.e., expected values).

Fig. 1 Payoff matrices of a simple MSG. Player 1 prefers a morning meeting, and player 2 preferences can be either the same (a), or the opposite (b)

	$\langle 0,1\rangle$	$\langle 1,0\rangle$
$\langle 0,1\rangle$	m,k	$\frac{m}{2},\frac{k}{2}$
$\langle 1,0\rangle$	$\frac{m}{2},\frac{k}{2}$	$0,0$

(a)

	$\langle 0,1\rangle$	$\langle 1,0\rangle$
$\langle 0,1\rangle$	$m,0$	$\frac{m}{2},\frac{k}{2}$
$\langle 1,0\rangle$	$\frac{m}{2},\frac{k}{2}$	$0,k$

(b)

If a NE is the expected outcome of an interaction, a valid question to ask would be "How globally efficient or inefficient is this solution?". This question is the focus of a large body of work on "the Price of Anarchy" (PoA) [8, 9, 4, 12, 11]. The PoA is a measure of the inefficiency of an equilibrium point, and is defined as the ratio between the cost of the *globally worst* NE, to the cost of the *globally optimal* solution. Studies of the price of anarchy originate with routing games (cf. [11]).

It is interesting to see that the same question can be formulated for MSGs. What is the PoA for MSGs ? In reality, a player does not know her opponent's private preferences and this uncertainty should be reflected in our analysis. However, we will consider the above simple example as two different games for the sake of simplicity. In the first game the desires of both opponents align and in the unique equilibrium point the payoffs are m,k. The global utility in this case is $m+k$. Clearly this is also the optimal solution when considering a cooperative mechanism. Hence, $PoA = 1$.

When the desires of the opponents conflict, the global gain in the unique equilibrium point is $\frac{m+k}{2}$. This is not necessarily the optimal solution: when $m < k$ the optimal solution's global utility is k. In this case the PoA moves further away from unity as k increases, [1] and we have: $PoA = \frac{m}{2k} + \frac{1}{2}$. That is, *the Price of Anarchy for the simplest MSG* is bounded by $\frac{1}{2}$.

[1] Unlike the routing minimization problem [11, 12], MSP is a maximization problem, hence the PoA is always smaller than 1.

	$\langle 0,1\rangle$	$\langle 0,2\rangle$	$\langle 1,0\rangle$	$\langle 1,2\rangle$	$\langle 2,0\rangle$	$\langle 2,1\rangle$
$\langle 0,1\rangle$	m,y	m,y	$\frac{m+x}{2},\frac{k+y}{2}$	m,y	x,k	$\frac{m+x}{2},\frac{k+y}{2}$
$\langle 0,2\rangle$	m,y	m,y	m,y	m,y	$\frac{m+x}{2},\frac{k+y}{2}$	m,y
$\langle 1,0\rangle$	$\frac{m+x}{2},\frac{k+y}{2}$	m,y	x,k	$\frac{m+x}{2},\frac{k+y}{2}$	x,k	x,k
$\langle 1,2\rangle$	m,y	m,y	$\frac{m+x}{2},\frac{k+y}{2}$	m,y	x,k	$\frac{m+x}{2},\frac{k+y}{2}$
$\langle 2,0\rangle$	x,k	$\frac{m+x}{2},\frac{k+y}{2}$	x,k	x,k	x,k	x,k
$\langle 2,1\rangle$	$\frac{m+x}{2},\frac{k+y}{2}$	m,y	x,k	$\frac{m+x}{2},\frac{k+y}{2}$	x,k	x,k

(a)

	$\langle 0,1\rangle$	$\langle 0,2\rangle$	$\langle 1,0\rangle$	$\langle 1,2\rangle$	$\langle 2,0\rangle$	$\langle 2,1\rangle$
$\langle 0,1\rangle$	m,y	m,y	$\frac{m+x}{2},\frac{k+y}{2}$	m,y	$\frac{m+x}{2},\frac{k+y}{2}$	m,y
$\langle 0,2\rangle$	m,y	m,y	$\frac{m+x}{2},\frac{k+y}{2}$	m,y	$\frac{m+x}{2},\frac{k+y}{2}$	m,y
$\langle 1,0\rangle$	$\frac{m+x}{2},\frac{k+y}{2}$	$\frac{m+x}{2},\frac{k+y}{2}$	x,k	x,k	x,k	x,k
$\langle 1,2\rangle$	m,y	m,y	x,k	m,y	x,k	$\frac{m+x}{2},\frac{k+y}{2}$
$\langle 2,0\rangle$	$\frac{m+x}{2},\frac{k+y}{2}$	$\frac{m+x}{2},\frac{k+y}{2}$	x,k	x,k	x,k	x,k
$\langle 2,1\rangle$	m,y	m,y	x,k	$\frac{m+x}{2},\frac{k+y}{2}$	x,k	x,k

(b)

Fig. 2. Payoff matrices for a more general MSG with different mechanisms - "utilitarian" (left) and "egalitarian" (right). Also note that B=2, and that player 1 prefers a morning meeting, and player 2 prefers an evening meeting.

This behavior is also expected in more realistic form of the two-players two time-slots MSG in which the players may express their strategies in terms of multiple preference values. In such a case, the limit on the maximal preference of each player - the value B - is greater than 1. In such cases the MSG has at least one equilibrium point and one can find the worst possible value that it can have. Next, comes the formulation of these results in the form of lemmas and the outlines of their proofs.

Lemma 1. *Every participant in the MSG described above has one (possibly weak) dominating strategy, or bid, which is composed of the value B for the preferred time slot, and 0 for the remaining time slot.*

The proof is simple and requires showing that if an agent assigns B to the preferred time slot and 0 otherwise (i.e., each player's action would be either $\langle B,0\rangle$ or $\langle 0,B\rangle$), her opponent can only force a draw, which will result in a fair toss of coin. If both players assign this action, no unilateral deviation will result in a higher payoff.

Lemma 2. *There are only two possible equilibrium outcome values to the above MSG, (m,k) and $(\frac{m}{2},\frac{k}{2})$*

By examining all five possible outcome values, one can easily rule out assignments which lead to $(0,0)$, $(m,0)$ or $(0,k)$. Thus we are left with (m,k) and $(\frac{m}{2},\frac{k}{2})$.

From these two simple lemmas, it is clear that when preferences coincide, the worst equilibrium payoff is (m,k). In this case the price of anarchy (PoA) is 1 [11]. In fact, this is the only equilibrium value due to the dominant strategies. When the preferences for time slot of the two players are in conflict, the worst equilibrium payoff of both players is $(\frac{m}{2},\frac{k}{2})$. If (without loss of generality) $m < k$, then the PoA decreases to a value lower than 1 as k increases (and is bounded below by $\frac{1}{2}$).

We now proceed to generalize this game, and allow players to assign any non-negative gain when their less preferred time slot is selected by the game mechanism. More specifically, we define for player one the payoff for a non optimal time slot as $0 \leq x \leq m$, and for player two $0 \leq y \leq k$, as demonstrated in the example in figure 2a.

The lemmas hold in this case (the tie's payoff is slightly revised). This can easily be understood by noticing that the best response to an opponent's strategy basically remains the same. Just as before, ties which result in a random selection by the mechanism (coin flip) and a value of $\frac{m+x}{2}$ or $\frac{k+y}{2}$, are always preferred over losing (x or y respectively).

As a result, when the preferences of the two agents are the same, the worst NE (also the best) has a value of $m+k$ which is also optimal - leading to a PoA of 1. When preferences contradict, the value of the stable point is $\frac{m+x}{2}+\frac{k+y}{2}$. This is not necessarily optimal. For example, if $\frac{m-x}{2}<\frac{k-y}{2}$ than the optimal result can be $x+k$, resulting in

$$PoA = \frac{m+y}{2(x+k)} + \frac{1}{2}$$

Combining these results we reach several intermediate conclusions. The first is that when players have contradicting preferences, the PoA is bounded below by $\frac{1}{2}$, and depends on the relationship between the players payoffs. The second is that when the two players have the same preferences, the PoA is 1. Our analysis also indicate that the PoA in the MSG depends on several factors:

- The agents' private payoffs (i.e. m,k,x,y). This motivates the use of a mechanism which maps payoffs to a uniform or universal scale which can further bound this value. Such a scale can be the quality of a schedule as described in [3]. Indeed, when the individual payoffs are equal, the PoA becomes unity.
- The *mechanism* of the MSG. Figure 2 depicts two possible mechanisms: a "utilitarian" mechanism (Figure 2a), and an "egalitarian" one (Figure 2b) which selects a time slot that maximizes the minimal bid on it. This leads to a different game, with a different NE.
- The PoA may also be affected by the designer's perception of optimality. For example, given the MSG of figure 2a, solutions yielding the gain $\langle x,k\rangle$ are optimal when considering the "utilitarian" approach (maximal combined payoffs). However, if one's perception of optimality is that it maximizes the lowest gain (e.g.,"egalitarian"), then the optimal solution gain for this MSG (Figure 2a) is $\langle \frac{m+x}{2},\frac{k+y}{2}\rangle$. Optimizing different objective functions can produce substantially "better" results [3].

The MSG toy example can become more realistic along several dimensions:

1. The maximal preference bid (B) - while this value increases the assignment space of each player, we have shown that it does not affect the behavior of the two-players, two-time-slots game.
2. The number of participants. Although not presented here, adding more players does not change the fact that these have a (possibly weak) dominant strategy.
3. The number of time slots. Adding more time slots requires several modifications to the game described above (for example, in case of a tie, will the mechanism randomly pick any of the possible time slots? just those that were tied? Can two time slots receive the same bid value? etc). Even after adding the required adjustment, finding the PoA will still rely on the utility that each agent associates with a time slot. In the case of three time slots, if each agent's utility from a meeting will be x for the least preferred time slot, z for the most preferred time

slot, and y for the remaining time slot, one can state that if $x+z \geq 2y$ then a (weak) dominant strategy exists for the players (under some set of rules).
4. The number of meetings. Note that the natural scenario of incremental addition of meetings defines a different game which is an iterated game.

4 The Cost of Cooperation

Following this detailed tour of the Price of Anarchy for MSGs, an interesting question arises: what exactly is the meaning of the PoA for MSGs? In other words, what is the meaning of "anarchy" in the context of MSG. After all, "anarchy" for an MSG is the natural situation because agents want their best possible personal calendar. The "cooperative" method of optimizing some global goal is not very natural. Why would self interested users wish to optimize some global objective ? There is no sense of "public savings" or "minimization of overall cost", as in the case of routing games. Players in a MSG are only interested in cooperation as a form of mediation between them. This implies that examining efficiency in terms of some global value does not fit the agent's point of view.

The present paper proposes a different measure. Instead of addressing the "social" cost, the focus here is to quantify the cost (or gain) of a single agent, from participating in any cooperative protocol.

Definition 1. An agent's *Cost of Cooperation*(CoC) with respect to $f(x)$ is defined as the difference between the lowest gain the agent can have in any of the NEs of the underlying game (if any, otherwise zero) and the lowest gain an agent receives from a (cooperative) protocol's solution x that maximizes $f(x)$.

The intuition behind this definition is quite simple. A user about to interact with other users can *expect that the combined spontaneous action taken by all participants will result in a NE*. Since there can be more than one such stable points the user anticipates the worst case for herself, i.e., the personal lowest gaining NE. For her to join a cooperative protocol (and use a multi user distributed optimizing protocol), she uses knowledge about her possible payoffs. This information is provided by the CoC. When the CoC value is negative, the user knows that she *stands to gain* from joining the cooperative protocol.

As a simple example, consider the famous Prisoners' Dilemma game as depicted in Figure 3. In this problem, a single stable action profile exists - $\langle D,D\rangle$. When both participants select this action, each has a gain of 1. In the optimal solution that maximizes the overall gain (e.g., the *sum of all personal gains*) - $\langle C,C\rangle$ - the gain for each agent is 4. This corresponds to a CoC of -3 for each participant, which implies that both agents gain from following this cooperative solution.

The $f(x)$-CoC for a set of users is defined as a tuple, in which every element CoC_i corresponds to the CoC of agent A_i. Defining the CoC for an entire system can be used to motivate the participation of users in the cooperative protocol. A negative CoC (i.e., gain), provides a self-interested user with a strong incentive to join the cooperative protocol. One may even venture to call this *Individual Rationality* [8].

Fig. 3 The Prisoners' Dilemma game matrix

	Cooperate	Defect
Cooperate	4, 4	0, 6
Defect	6, 0	1, 1

4.1 Cooperation Games

Let us define special strategic situations in which the CoC vector of all participants has non positive values.

Definition 2. A game is an $f(x)$-*Cooperation game*[2] if there exists a solution to the game in which the CoC (with respect to $f(x)$) of all agents is non positive.

For example, the simple MSG of Figure 2a, is a cooperation game with respect to a Max-Min objective function (i.e., maximizing the minimal gain).

Given a general MSG, one may want to change it into a cooperation game (if it is not one already). This can be achieved by defining the cooperative goal (e.g., the objective function) in a specific manner, by changing the mechanism of the interaction itself, or by adding an interested party to the interaction. Some work in the direction of the latter, was recently reported in [13, 7]. *Mediators* are introduced as parties wishing to influence the choice of action (e.g., strategy) of participants which are not under their control. Mediators cannot enforce payments by the agents or prohibit strategies, but can influence the outcome of an interaction by offering to play on behalf of some (or all) of the participants. By doing so a mediator commits to a pre-specified behavior [13].

An interesting property of Routing mediators [13] is that they are capable of possessing information about the actions taken by agents. In a two player strategic situation, one may use the following routing mediator to generate a stable solution: the revised game includes mediated actions and each agent may either play its game as before, or let the mediator play for it. When the mediator plays for an agent, it will always choose to assign the bid which will result in the minimal payoff to the agent's *opponent*. If both agents use the mediator, the mediator assigns the interaction which results in the lowest value that is at least as high as any value that results from the play of the agent's opponent.

Two important attributes of the new mediated interaction are the existence of at lease one, new, pure strategy NE (the assignment $\langle Med, Med \rangle$), and the existence of an action profile improving the gains of all agents (i.e. it is now a cooperation game). For example, consider the two agents interaction depicted in figure 4a. In this interaction, two agents must choose between playing U or D, L or R. The payoffs values are not specified, but their order is: $A_1 > A_2 > ... > A_8$.

Our former restriction to pure strategies now results in a scenario which has no stable points. This means that in every solution one player stands to gain from deviating. The underlined values in figure 4a represent the best response that each agent has, in view of her opponents strategy. For example, if the first agent selects

[2] Not to be confused with cooperative games.

	L	R
U	$\underline{A_6},A_7$	$A_3,\underline{A_5}$
D	$A_8,\underline{A_2}$	$\underline{A_1},A_4$

(a)

	L	R	Med
U	$\underline{A_6},A_7$	$A_3,\underline{A_5}$	A_6,A_7
D	$A_8,\underline{A_2}$	$\underline{A_1},A_4$	$A_8,\underline{A_2}$
Med	$\underline{A_6},A_7$	$A_3,\underline{A_5}$	$\underline{A_3},\underline{A_5}$

(b)

Fig. 4. A two agent strategic situation with ordinal payoffs - with and without a mediator

the U strategy when her opponent plays L, the second agent will change its assignment to R (resulting in D by the first agent and L again by player two). Going back to our former line of reasoning, this implies that an agent does not know what to expect from such an interaction, and any gain is plausible. However, by adding our previously described mediator, this situation is improved, as depicted in figure 4b.

When playing for the columns agent, the mediator "threatens" the row player - it always picks the move L, with the worst outcome for her (A_6,A_8 instead of the possible A_3,A_1). The opposite is true for the row player. Here if the mediator is selected it plays U so that it guarantees the worst outcome for the columns player (A_7,A_5). However, when both players use the mediator, it selects the action profile $\langle U,R\rangle$. This assignment is used because it is a valid assignment for which the column's payoff is A_5 (which is not worse than A_7 and A_5), and the row's payoff is A_3 (better than either A_6 or A_8). The end result is a transformation of the original game (by introducing a mediator) into a cooperation game. That is, if agents choose to participate in a utilitarian optimization protocol (cooperate) the end result will be $\langle D,R\rangle$, with a payoff of A_1 to agent one and A_4 to agent two. This improves the expected gain of *each one of the agents* from playing the game in Figure 4b.

5 Discussion

The paper discusses two opposite extreme approaches that are inherent to many multi agent scenarios - cooperation and competition. In order to investigate these a a simple scheduling problem was formulated in the form of a simple game and analyzed. The maximal preference bid and general payoffs are incrementally added and their impact on the stable points and the PoA of the interactions are examined. Our analysis leads naturally to a new measure that quantifies the cost/gain to agents from participating in a cooperative protocol. The Cost of Cooperation (CoC) as defined in the paper further defines a game property that is dubbed "Cooperation game". Participants in a cooperation game may be better off cooperating than playing (selfishly) out the game.

Acknowledgements. The research was supported by the Lynn and William Frankel Center for Computer Science,and by the Paul Ivanier Center for Robotics.

References

1. Crawford, E., Veloso, M.: Mechanism design for multi-agent meeting scheduling. Web Intelligence and Agent Systems 4(2), 209–220 (2006)
2. Ephrati, E., Zlotkin, G., Rosenschein, J.S.: A non manipulable meeting scheduling system. In: Proc. Intern. Workshop on Distributed Artificial Intelligence, Seatle, WA (1994)
3. Gershman, A., Grubshtein, A., Meisels, A., Rokach, L., Zivan, R.: Scheduling meetings by agents. In: Proc. 7th Intern. Conf. on Pract. & Theo. Automated Timetabling (PATAT 2008), Montreal (August 2008)
4. Koutsoupias, E., Papadimitriou, C.: Worst-case equilibria. In: Proc. of the 16th Annual Symposium on Theoretical Aspects of Computer Science, pp. 404–413 (1999)
5. Maheswaran, R.T., Tambe, M., Bowring, E., Pearce, J.P., Varakantham, P.: Taking dcop to the real world: Efficient complete solutions for distributed multi-event scheduling. In: Proc. 3rd Intern. Joint Conf. on Autonomous Agents & Multi-Agent Systems (AAMAS 2004), NY, New York, pp. 310–317 (2004)
6. Modi, J., Veloso, M.: Multiagent meeting scheduling with rescheduling. In: Proc. 5th workshop on distributed constraints reasoning DCR 2004, Toronto (September 2004)
7. Monderer, D., Tennenholtz, M.: Strong mediated equilibrium. Artificial Intelligence 173(1), 180–195 (2009)
8. Nisan, N., Roughgarden, T., Tardos, E., Vazirani, V.V.: Algorithmic Game Theory. Cambridge University Press, Cambridge (2007)
9. Papadimitriou, C.: Algorithms, games, and the internet. In: STOC 2001: Proc. of the 33rd annual ACM symposium on Theory of computing, Hersonissos, Greece, pp. 749–753 (2001)
10. Petcu, A., Faltings, B., Parkes, D.: M-DPOP: Faithful distributed implementation of efficient social choice problems. Journal of AI Research (JAIR) 32, 705–755 (2008)
11. Roughgarden, T.: Selfish Routing and the Price of Anarchy. MIT Press, Cambridge (2005)
12. Roughgarden, T., Tardos, É.: How bad is selfish routing? J. ACM 49(2), 236–259 (2002)
13. Rozenfeld, O., Tennenholtz, M.: Routing mediators. In: IJCAI, Hyderabad, India, pp. 1488–1493 (2007)
14. Wallace, R.J., Freuder, E.: Constraint-based multi-agent meeting scheduling: effects of agent heterogeneity on performance and privacy loss. In: Proc. 3rd workshop on distributed constrait reasoning, DCR 2002, Bologna, pp. 176–182 (2002)

Actor-Agent Communities: Design Approaches

S.M. Iacob, C.H.M. Nieuwenhuis, N.J.E. Wijngaards,
G. Pavlin, and J.B. van Veelen

Abstract. We describe several principles for designing Actor-Agent Communities (AAC) as collectives of autonomous problem solving entities (software agents and human experts) that self-organize and collaborate at solving complex problems. One of the main distinctive aspects of the AAC is their ability to integrate in a meaningful way the expertise and reasoning of humans with different information processing algorithms performed by software agents, without requiring a unique and complete description of the problem and solution spaces.

1 Preliminaries

Many architectures for multi-agent systems have been proposed, each excelling in one or more functional or non-functional parameters, such as distributed processing, negotiation, response time, self-organization, etc. (see e.g. [2]). Nevertheless, if one abstracts from specific functionality and implementation platform, agents can be regarded as individual software programs which get activated by certain triggers in their environment (be it user commands, or some sensor values). A multi-agent system (MAS) can thus be regarded as a software program with different asynchronous threads (the agents), which can be coordinated either globally or locally. A distributed MAS is in principle only different form a MAS in that the individual agents may run on different physical platforms. At this abstraction level the different agent systems can be described by a single reference architecture, as proposed by FIPA [8]. From an application perspective, all agent-based systems have a request-response interaction model, where agents respond to requests from users (actors) or

S.M. Iacob · C.H.M. Nieuwenhuis · N.J.E. Wijngaards · G. Pavlin · J.B. van Veelen
Thales Research and Technology Netherlands, D-CIS Lab,
Postbus 90, 2600 AB Delft, The Netherlands
e-mail: {sorin.iacob,kees.nieuwenhuis}@icis.decis.nl,
{niek.wijngaards,gregor.pavlin}@icis.decis.nl,
bernard.vanveelen@icis.decis.nl

G.A. Papadopoulos and C. Badica (Eds.): Intelligent Distributed Computing III, SCI 237, pp. 237–242.
springerlink.com © Springer-Verlag Berlin Heidelberg 2009

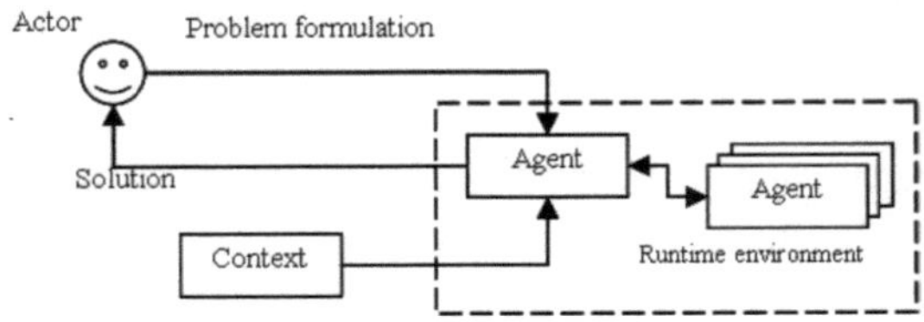

Fig. 1 Generic interaction model in MAS

other agents, and possibly react to the environmental context (see Fig. 1). In general this interaction can be replicated for a multitude of actors and agents. The autonomy of agents refers to their ability of performing some of the problem-solving steps without requiring explicit input from a human operator. We address the use of MAS for solving complex problems, where complexity refers to the solution process, and not to the computational complexity of the algorithms. Complex problems cannot be decomposed into a finite number of simpler problems that each be solved by an algorithm, either because of lacking adequate models, or because it contains uncomputable elements. The common approach to solving such problems relies on approximations through heuristics (or more generally, on computing in the limit [3]). In data-driven approaches these heuristics can in principle be found through training and learning algorithms, provided that enough training data is available [5]. However, the utility of these solutions is mostly limited to classification problems.

In this work we propose a new approach to autonomous cooperative problem-solving systems, where agents and humans form a problem-solving community. Complex problems are detected, formulated and solved jointly by humans and autonomous agents.

2 Actor-Agent Communities

Actor-Agent Communities (AAC) are socio-technical information systems that deliver a solution for otherwise intractable information processing problems. At the highest abstraction level an AAC can be regarded as a collection of autonomous problem-solving entities, with specific problem-solving capabilities resulting from their individual knowledge (or world models), interaction modalities, and (limited) capabilities and resources for information processing. A coherent behaviour can be induced on an amorphous collection of such autonomous entities by assigning them a common goal. This may be regarded as a set of quantities that describe a particular state of the environment (possibly including states of the entities themselves). Depending on the richness and complexity of their respective world models, each entity will maintain a particular (partial) representation of these goals. Problems can now be defined as mismatches between a target state of the environment and an observation thereof. The entity that discovers a problem may not be able to solve it all by itself, in which case other entities will be asked to contribute. Teams can thus be formed whose purpose is solving that particular problem. In order for two or more entities to be able to team up, their world model need to partially overlap (achieve a "common ground"). The initiation and completion of such self-organization is

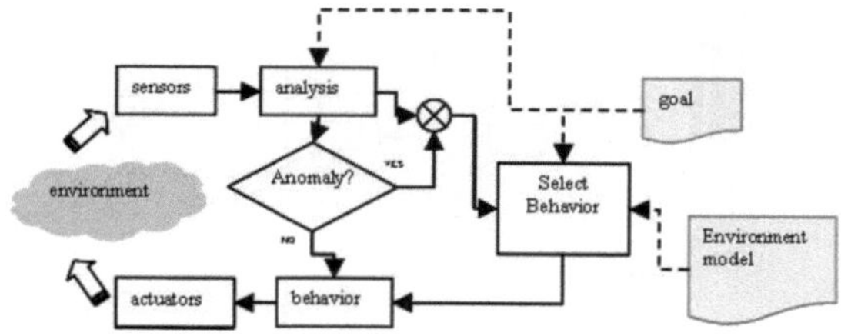

Fig. 2 A simplified functional model for autonomous problem-solving entities

only determined by the problem detected in the environment, the overlap in world models, and capabilities of the entities.

3 A Functional Model for Autonomous Problem-Solving Entities

An autonomous problem-solving entity is defined as a set of inputs (sensors) and outputs (actuators) and a set of states. A set of state transitions is defined as a function of a system goal, a system-specific world model, and the current inputs. The goals are regarded as static parameters that define a target state of the system (i.e. the environment and the entity itself). A state is characterized by a number of (heterogeneous) parameters, which the entity can measure or estimate, and whose meaning is described semantically in the entity's world model. A heterogeneous distance function is defined that allows the entities to detect and react to anomalies. The entity tries to perform a sequence of predefined state transitions in an attempt to bring the system to one of the desired states. A problem solving entity makes use of an explicit world model that consists of a set of labels for the sensed data, possible data patterns, possible actions, etc., and the relationships between all these. In principle, such a world model could be regarded as an ontology fragment (in fact it also contains quantitative knowledge and complex functions). Both the goal of the entity and the state of the environment can be represented by arbitrary semantic constructs of tuples (concept, value) connected with any of the logical operators $\vee$, $\wedge$, or . Values are specified using relational operators (=, $<$, $>$, etc.).

Example: The input data for an agent is a video stream from a camera. The agent can measure the average luminance of the observed scene, can detect multiple instances of cars and people, and can map the apparent coordinates of the detected objects on absolute geographical coordinates. This agent's ontology fragment can be described as in Fig. 3 by a minimal set of primary concepts (i.e. those that label the physical quantities or objects which the agent can detect) and some useful generalizations and abstractions. Some concepts are further generalized in some other ontology, which may remain unspecified (the "ext" connectors). The link between the primary concepts and the physical interfaces of the entity is performed through predefined algorithms (e.g. feature extraction, pattern recognition), which are

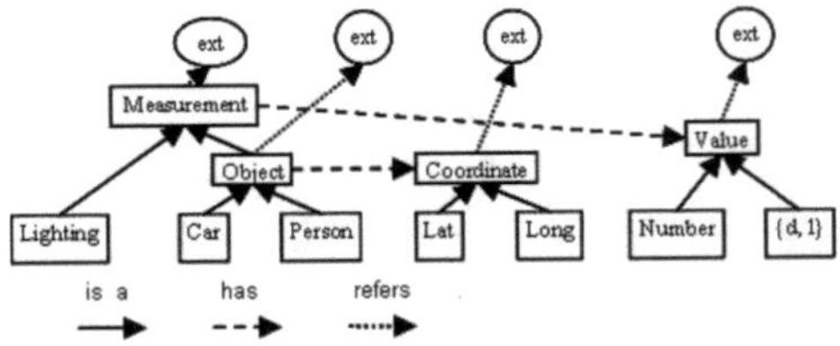

Fig. 3 Example of a local ontology fragment

regarded as implicit (or procedural) knowledge of the entity. A goal can be represented in this case as a target state of the environment:

((Lighting, dark) $\wedge$ ((Car, <1) $\vee$ (Person, <1))) $\vee$ ((Lighting, light) $\wedge$ (Car, <10) $\wedge$ (Person, <20)).

Once an anomaly is detected the entity selects and executes one of the predefined algorithms for restoring the state of the system. From this perspective, the entity behaves in a purely reactive manner. However, for complex problems it may not be efficient, or even possible to define all the algorithms required for coping with all the anomalies that the entity can detect. Recall from above that an environment model describes not only relationships between the measured environment variables, but also how they can be acted upon. This latter part of the world model (i.e. the causal ontology, [7]) forms the basis on which the entity can select an appropriate behavior given a certain state of the system. Ideally, such a causal ontology would establish relationships between the available actuators of an entity and all its primary perceptual concepts, meaning that the entity is capable of influencing all the environmental parameters that it can measure. This, of course, is not always possible, so some actions may need to be specified in a causal ontology fragment that links to external ontologies.

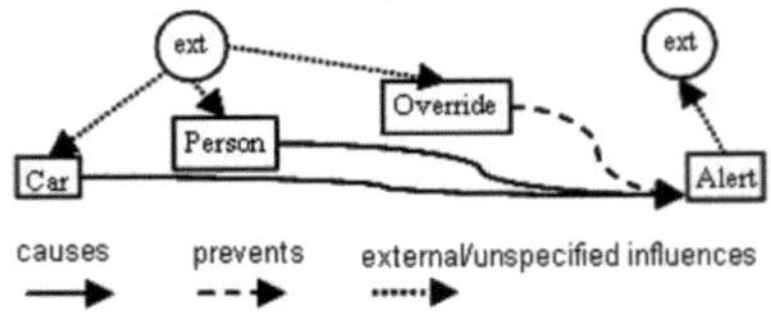

Fig. 4 A causal ontology fragment; external influences are only exerted on primary concepts (they interface with the environment). The internal causal relations are inherently nave given the limited ontologies. "Override" allows an external event to alter randomly the behavior of this entity.

4 AAC Self-organization

As explained earlier, a goal can be formulated as a set of target states of a system (i.e. environment and AAC entities) expressed as (heterogeneous) vectors. If an entity has $m \leq n$ primary concepts (see section 3) that describe the same state parameters as m of the goal vector's components, then the entity will acquire a subgoal which is a projection of the original set of target states on the m – dimensional subspace defined by the set of overlapping state parameters. However, the problem is more complex that just defining a linear map. Indeed, under the assumption that each entity has an incomplete and independently generated world model, a direct mapping between primary concepts is not a trivial task, as indicated by the ongoing research in the field of semantic matching and ontology alignment (see e.g. [1], [4]). Given the space limitations we do not extend the present discussion in this direction. Assume that suitable methods exist for estimating similarities or inclusion relations (e.g. subsumptions, intensions) between concepts. Obviously, finding these relations is only possible when the local ontologies of the entities in a given community partially overlap. For this reason this should be regarded as a hard requirement for the design of an AAC. Assume therefore that the community goal, expressed as a set of vector values indicating the desired states of the system, can be measured by two entities (see Fig. 5). The state vectors of these entities may contain additional parameters, and some other parameters may need to be provided by other entities (indicated by the grayed boxes). Nevertheless, a community goal can in principle be fulfilled when all state parameters can be measured, which means that all leaf nodes in the goal splitting multi-tree represent primary concepts for those entities. In such a case the community is perceptually complete. However, this does not mean that the goal can be effectively fulfilled. In order for this to happen, it is necessary that the participating entities possess the effectual capabilities required for bringing each of the state parameters to a desired value. The analysis of the requirements for effectual completeness is similar to that for perceptual completeness. The problem that remains is how to cope with the absence of a unique ontology for the whole community. Although we have assumed earlier that bilateral semantic similarities can be evaluated with existing techniques, it is still not obvious how a large number of entities can meaningfully work together at solving a complex problem. To explain this, we start by recalling that the goal vector is just a set of values corresponding to a semantic construct within the ontology of a certain entity. When a different entity tries to interpret this goal vector it actually parses the ontology of the originating entity, and tries to match some of the concepts in the source ontology with concepts in its own ontology. This second entity can derive a sub goal that it can fulfil either by its own, or with the help of some other entities. In the latter case it generates an additional goal containing those elements of its local goal which it cannot directly measure (i.e. are not primary concepts) and posts it as a new community goal. Eventually, if the community is perceptually complete, an entity will be able to fulfil a sub-goal all by itself. Then the entity which formulated the higher level goal can also fulfil a higher-level goal, and so on, up to the level of the original goal.

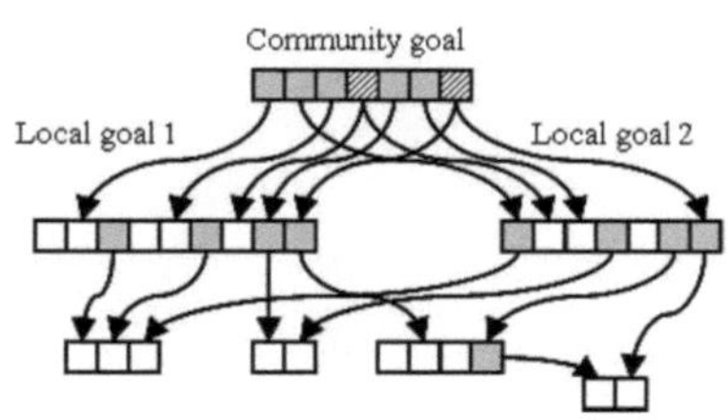

Fig. 5 Possible mappings between state vectors

5 Discussion

The idea of integrating human users and software agents into a team is not completely new. In [6] Sycara and Lewis proposed a solution based on a set of specialized agents and a coordination framework where agents coordinate the communication between humans, but also contribute to the problem solving process. This is achieved through a common knowledge model shared by all agents. Problem solving capabilities and resources, task decomposition and behaviours are all defined at design time. An advertisement mechanism allows agents to find each other at runtime and respond to collaboration requests. The system is very efficient in coordinating multiple tasks and in coping with resource limitations, but requires a full definition of a common knowledge model, tasks sets, and behaviours.

The AAC approach proposed here provides a truly decentralized solution for a meaningful integration of human reasoning and software algorithms. The self-organization is based on goal decomposition and partial overlaps of the world models of the AAC entities.

References

[1] Doan, A., Halevy, A.: Semantic Integration Research in the Database Community: A Brief Survey. AI Magazine 26(1), 83–94 (2005)
[2] Ferber, J.: Multi-Agent Systems, an introduction to distributed artificial intelligence. Addison-Wesley, Reading (1999)
[3] Gold, M.E.: Limiting Recursion. Journal of Symbolic Logic 30(1), 28–48 (1965)
[4] Kalfoglou, Y., Schorlemmer, M.: Ontology Mapping: The State of the Art. The Knowledge Engineering Review Journal 18(1), 1–31 (2003)
[5] Lathrop, R.H.: On the Learnability of the Uncomputable. In: Proc. 13th Intl. Conf. on Machine Learning, pp. 302–309 (1996)
[6] Sycara, K., Lewis, M.: Integrating Agents into Human Teams. In: Human Factors and Ergonomics Society Annual Meeting Proceedings, Cognitive Engineering and Decision Making, pp. 413–417 (2007)
[7] Terenziani, P.: Towards a Causal Ontology Coping with the Temporal Constraints between Causes and Effects. Int. J. Human-Computer Studies 43, 847–863 (1995)
[8] Zhou, B.-H., Li, C.-C., Zhao, X.: FIPA agent-based control system design for FMS. Int. J. Adv. Manuf. Technol. 31, 969–997 (2007)

A Trusted Defeasible Reasoning Service for Brokering Agents in the Semantic Web

Kalliopi Kravari, Efstratios Kontopoulos, and Nick Bassiliades

Abstract. Based on the plethora of proposals and standards for logic- and rule-based reasoning for the Semantic Web (SW), a key factor for SW agents is reasoning task interoperability. This paper reports on a framework for interoperable reasoning among agents in the SW that deploys third-party trusted reasoning services. This way, agents can exchange arguments, without conforming to a common rule or logic paradigm; via an external reasoning service, the receiving agent can grasp the semantics of the received rule set. The paper presents how a multi-agent system was extended with a third-party trusted defeasible reasoning service, which offers agents the ability of reasoning with incomplete and inconsistent information. In addition, a brokering trade scenario is presented that illustrates the usability of the approach.

1 Introduction

The *Semantic Web* (*SW*) [5] is a rapidly evolving extension of the WWW, in which the semantics of information and services is well-defined, making it possible for people and machines to understand Web content and satisfy their requests. SW research is currently focusing on logic, reasoning and proof. *Intelligent agents* (*IAs*) can be favored by SW technologies [8], because of the interoperability SW offers. The integration of *multi-agent systems* (*MAS*) with SW technology will significantly affect the use of the Web; its next generation will feature groups of intercommunicating agents traversing it and performing complex actions on behalf of their users.

A core setback in agent interoperation is the variety in representation and reasoning. Despite KIF's efforts [7], there is still no globally agreed knowledge representation and reasoning formalism for agents. For SW agents, on the other hand, we can safely assume that *OWL* could be the global knowledge exchange language.

Kalliopi Kravari · Efstratios Kontopoulos · Nick Bassiliades
Dept. of Informatics, Aristotle University of Thessaloniki, GR-54124 Thessaloniki, Greece
e-mail: {kkravari,skontopo,nbassili}@csd.auth.gr

G.A. Papadopoulos and C. Badica (Eds.): Intelligent Distributed Computing III, SCI 237, pp. 243–248.
springerlink.com © Springer-Verlag Berlin Heidelberg 2009

As for rule-based reasoning, there is a variety of proposals [6], [9], [14]. Thus, a key factor for the success of SW systems and agents in particular is reasoning task interoperability among multiple, heterogeneous web entities exchanging rule bases to justify their positions.

Reasoning interoperability among agents is often achieved by translating the received rule set into the receiving agent's formalism. This, however, can only be accomplished, when the two agents use the same rule formalism with different syntax, or when one formalism can be semantically translated into the other (e.g. translation between defeasible logic and Datalog rules [3]).

On the other hand, we propose a simpler approach that does not rely on semantic interoperability, but on exchanging the rule base model, instead. This way, agents can exchange their arguments, without conforming to a common rule paradigm or logic. The receiving agent can use an external reasoning service to grasp the semantics of the result set of the received rule base. A critical assumption, of course, is that reasoning services are trusted and are hosted by authoritative organizations, like *W3C* or *RuleML.org*. More specifically, this paper presents how a JADE MAS was extended with *defeasible reasoning* (*DR*) [11], i.e., the ability to reason with incomplete and inconsistent information. A DR service with a reputation mechanism was implemented as a JADE agent. The approach is generic, since any reasoner can be deployed as an agent-service in the system. Moreover, the paper presents a use case brokering trade scenario that illustrates the usability of these technologies.

In the rest of the paper, the implemented MAS is presented, focusing on the reasoning service that is based on DR-DEVICE, the core reasoning engine deployed in this paper. Section 3 presents the use case scenario and the paper is concluded with final remarks and directions for future work.

2 Defeasible Reasoning Service Agents

Defeasible reasoning constitutes a simple rule-based approach for efficient reasoning with incomplete and inconsistent information [11]. It offers enhanced representational capabilities and low computational complexity.

The DR-DEVICE defeasible logic reasoner [3] employs an OO RDF data model that treats properties as encapsulated attributes of resource objects, providing more compact representation and increased query performance. DR-DEVICE supports a rule language that extends RuleML with defeasible logic elements. The URL of a defeasible logic rule base is accepted as input. The facts for the rule program are contained in RDF documents, whose addresses are declared in the rule base. When the rule base is submitted, the designated facts are downloaded and the inference process commences. Rule conclusions are materialized inside DR-DEVICE and are then exported as an RDF document, including corresponding RDF/S definitions.

In our approach, reasoning services are wrapped by an agent interface, the *Reasoner*, allowing other IAs to contact them via ACL messages. The Reasoner launches associated reasoning engines that perform inference. Here, DR-DEVICE was chosen, but our plans involve integrating a range of reasoning engines that use various logics. In essence, the Reasoner is merely a service and not an autonomous agent, used fir integrating it into JADE.

The Reasoner constantly stands by for new requests (ACL messages with a "*REQUEST*" communication act). As soon as it gets a valid request, it launches DR-DEVICE that processes the input data (i.e. rule base) and returns an RDF document containing the results. Finally, the Reasoner returns the above result through an "*INFORM*" ACL message.

2.1 Trust Metric

There are various trust metrics [10]; our approach adopts a combination of *Sporas* (the most widely used) and *CR* (one of the most recently proposed), while using a decentralized reputation mechanism. Each agent keeps the references given by other agents and calculates the reputation value, according to the formula:

$$R_{i+1} = \frac{1}{\theta}\sum_{1}^{t}\Phi(R_i)R_{i+1}^{other}(W_{i+1} - E(W_{i+1})), \tag{1}$$

where $\Phi(R) = 1 - \frac{1}{1+e^{\frac{-(R-D)}{\sigma}}}$ and $E(W_{i+1}) = \frac{R_t}{D}$. W_i represents user i's rating, t is the number of ratings the user has received, θ is a constant integer > 1, R^{other} is the reputation of the user giving the rating, D is the reputation value range and σ is the acceleration factor of damping function Φ. W_i is based on four coefficients: $Correctness(Corr_i)$, $Response\ time(Resp_i)$ and $Flexibility(Flex_i)$; the latter refers to the Reasoner's flexibility in input parameters. The evaluation of the coefficients is based on user standards and their ratings vary from 1 to 10. The final rating value is the weighted sum of the coefficients (Equation 2), where a_{i1}, a_{i2}, and a_{i3} are the respective weights and $nCorr_i$, $nResp_i$ and $nFlex_i$ are the normalized values for correctness, response time and flexibility, accordingly:

$$W_i = a_{i1}nCorr_i + a_{i2}nResp_i + a_{i3}nFlex_i. \tag{2}$$

New users start with $R_i = 0, R_i \in [0,3000]$, while $W_i \in [0.1,1]$. As soon as the interaction ends, the Reasoner requests a rating. The other agent responds with a new message containing both its rating and its personal reputation and the Reasoner updates its reputation, according to Equation 1.

2.2 Agent Communication Protocols

The protocol describes the allowed performatives in agent communication: the two main performatives in our approach are "*REQUEST*" that refers to the action of submitting a request to perform a certain action, given certain preconditions, and "*INFORM*" that deals with returning the result of a previously submitted request to perform an action. A third, "*NOT-UNDERSTOOD*" performative is used if a non-appropriate message is received. The protocol is represented as a finite state machine with discrete states and transitions.

3 Use Case: A Brokering Scenario

DR is useful in various applications, like brokering and bargaining [4], which are also extensively influenced by agent-based technology [13]. Here, a DR brokering scenario [2] is described that demonstrates the functionality of our approach.

3.1 Scenario Overview

A MAS is formed by three independent parties, represented by IAs: (a) the *customer* (called Carlo) is a potential renter who wishes to rent an apartment based on his requirements (e.g. location, floor) and preferences, (b) the *broker* possesses a list of available apartments, along with their specifications (stored as an RDF DB). His role is to match Carlo's requirements with the apartment features and propose suitable flats, (c) the *reasoner* is an independent, trusted third-party agent-based service, that uses DR-DEVICE in order to infer conclusions from a defeasible logic program and a set of facts and produces the results as an RDF file.

The broker does not wish to reveal his DB to customers, as it's one of his most valuable assets. However, the schema (RDF/S file containing properties and data types) *must* be exposed, so that customers can formulate their requirements. After inspecting the schema, Carlo expresses his requirements in defeasible logic (in the DR-DEVICE RuleML-like syntax) and his agent sends them to the broker, in order to retrieve all available apartments with the proper specifications.

The broker cannot directly process Carlo's defeasible logic requirements, as it internally uses a different logic, so a trusted third-party reasoning service is requested and is retrieved from the framework directory service or, alternatively, recommended by the customer. The broker sends Carlo's requirements to the Reasoner, along with the URI of the RDF DB containing all available flats, and stands by for the list of proper apartments. Alternatively, the broker could send the file itself instead of the URI. Also, not necessarily *all* the DB flats are available.

The Reasoner launches DR-DEVICE, which processes the above data and returns an RDF document, containing the apartments that fulfill all requirements. The result is sent back to the broker's agent and the latter, consecutively, sends it to Carlo's agent. Meanwhile, the broker can process the results in order to filter out some, using his own negotiating strategy. Eventually, the broker sends a rating to the Reasoner, updating its reputation accordingly.

Eventually, Carlo receives the list of appropriate flats and has to decide which one he prefers. However, he does not want to send his preferences to the broker, because the latter might not present him with the optimum choices. Thus, Carlo's agent sends the list of acceptable apartments and his preferences (again as a defeasible logic rule base) to the Reasoner. The latter calls DR-DEVICE, gets the single most appropriate flat and proposes the best transaction. Carlo's agent could have an internal engine (e.g. [1]), but here we emphasize its ability to use any suitable external reasoning engine. The procedure ends, when Carlo sends his rating to the Reasoner and can now safely choose, based on his requirements and preferences.

3.2 Brokering Protocol

Although FIPA provides standardized protocols, none supports 1-1 automated brokering. Thus, a brokering protocol was implemented that encodes the allowed sequences of actions for the automation of the brokering process. The protocol (Fig. 1) is based on specific performatives that conform to the FIPA ACL specification. S_0 to S_6 represent states of a brokering trade and E is the final state. Predicates *Send* and *Receive* represent interactions that cause state transitions. The sequence of transitions for the customer is $S_1 \rightarrow S_2 \rightarrow S_3 \rightarrow S_4 \rightarrow E$, while, for the broker it is $S_0 \rightarrow S_1 \rightarrow S_2 \rightarrow S_3 \rightarrow E$. In case an agent receives a wrong performative, it sends back a "*NOT-UNDERSTOOD*" message and the interaction is repeated.

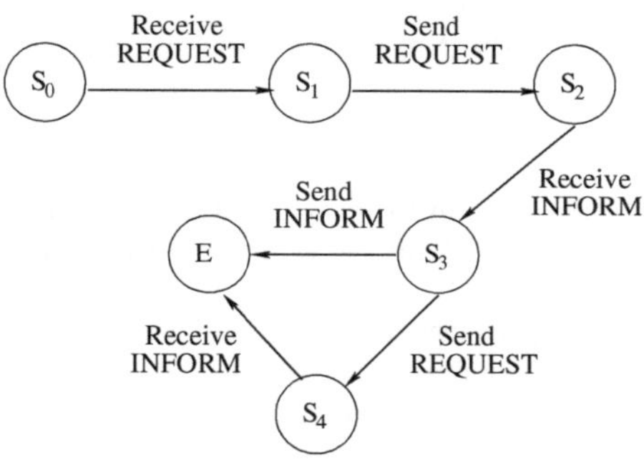

Fig. 1 Agent brokering communication protocol.

4 Conclusions and Future Work

This paper argued that reasoning interoperability among SW agents can be achieved via trusted, third-party reasoning services to provide reasoning capabilities in a variety of logic and rule-based formalisms. Towards this direction, a JADE MAS was presented, whose main component is a DR service implemented as an agent. Also, trust metrics have been studied and a reputation mechanism for the reasoning service has been embedded. Finally, the paper presented a use case brokering trade scenario that illustrates the usability of the proposed approach.

A related effort is the *Rule Responder* (*RR*) [12] project that builds a service-oriented methodology and a rule-based middleware for interchanging rules in virtual organizations. RR exhibits the interoperation of distributed platform-specific rule execution environments via Reaction RuleML as a platform-independent rule interchange format. The main differences with our approach are: (a) RR assumes a unique rule interchange language, while in our approach this is not necessary,

(b) in RR rule engines are incorporated inside agents, whereas we deploy them as in-dependent service-agents, and (c) our framework is totally FIPA-compliant, whereas RR introduces a new RuleML-based agent communication interface.

Concerning future directions for our approach, we would like to verify our architecture's capability to adapt to a variety of additional scenarios, like negotiation. Another goal is to integrate a variety of reasoning engines, in order to test and evolve the reasoning interoperating capabilities of our framework.

References

1. Antoniou, G., Skylogiannis, T., Bikakis, A., Bassiliades, N.: Dr-brokering - a defeasible logic-based system for semantic brokering. In: Proc. IEEE International Conference on E-Technology, E-Commerce and E-Service, pp. 414–417. IEEE, Los Alamitos (2005)
2. Antoniou, G., van Harmelen, F.: A Semantic Web Primer. MIT Press, Cambridge (2004)
3. Bassiliades, N., Antoniou, G., Vlahavas, I.: A defeasible logic reasoner for the semantic web. IJSWIS 2(1), 1–41 (2006)
4. Benjamins, R., Wielinga, B., Wielemaker, J., Fensel, D.: An intelligent agent for brokering problem-solving knowledge. In: Mira, J. (ed.) IWANN 1999. LNCS, vol. 1607, pp. 693–705. Springer, Heidelberg (1999)
5. Berners-Lee, T., Hendler, J., Lassila, O.: The semantic web. Scientific American (May 2001)
6. Boley, H.: The ruleml family of web rule languages. In: Alferes, J.J., Bailey, J., May, W., Schwertel, U. (eds.) PPSWR 2006. LNCS, vol. 4187, pp. 1–17. Springer, Heidelberg (2006)
7. Genesereth, M., Fikes, R.: Knowledge interchange format version 3.0 reference manual. Technical report, Logic Group, Comp. Sc. Dept., Stanford Univ. (1992)
8. Hendler, J.: Agents and the semantic web. IEEE Intelligent Systems 2(16), 30–37 (2001)
9. Kifer, M.: Rule interchange format: The framework. In: Calvanese, D., Lausen, G. (eds.) RR 2008. LNCS, vol. 5341, pp. 1–11. Springer, Heidelberg (2008)
10. Macarthur, K.: Trust and reputation in multi-agent systems. In: Proc. AAMAS 2008, May 12-16 (2007)
11. Nute, D.: Defeasible reasoning. In: Proc. 20th International Conference on Systems Science, pp. 470–477. IEEE Press, Los Alamitos (1987)
12. Paschke, A., Boley, H., Kozlenkov, A., Craig, B.: Rule responder: Ruleml-based agents for distributed collaboration on the pragmatic web. In: Proc. 2nd International Conference on Pragmatic Web, pp. 17–28. ACM, New York (2007)
13. Skylogiannis, T., Antoniou, G., Bassiliades, N., Governatori, G., Bikakis, A.: Dr-negotiate - a system for automated agent negotiation with defeasible logic-based strategies. DKE 63(2), 362–380 (2007)
14. Wagner, G., Giurca, A., Lukichev, S.: R2ml: A general approach for marking up rules. In: Marchiori, M., Fages, F., Ohlbach, H. (eds.) Principles and Practices of Semantic Web Reasoning, Dagstuhl Seminar Proceedings 05371 (2005)

Self-Built Grid

Eusebiu Marcu

Abstract. This paper introduces a new type of grid and gives an example using this type of grid for an authentication mechanism using the biometric holographic signature. We start by stating the definition of this new type of grid and some of its properties; after that, we will define a new biometric authentication architecture that uses this type of grid using Windows Communication Foundation Application as the main management node for client's requests for authentication and finally, we shortly describe the biometric signature authentication process.

1 Introduction

In this paper we propose a different biometric authentication architecture (defined in [1]) and a new grid computing architecture - we will name it *self-built grid* - that will be defined, analyzed and used for a biometric authentication mechanism. Current architectures are [1]:

- Store on Server, Match on Server - the authentication process runs on server side
- Store on Client, Match on Client - the authentication process runs on client side
- Store on Device, Match on Device - the authentication process runs on a device
- Store on Token, Match on Server - data is kept on a token but the authentication process runs on server side
- Store on Token, Match on Device - data is kept on a token but the authentication process runs on a device
- Store on Token, Match on Token - the authentication process runs on a token.

In the end, we will discuss about some methods to create this architecture more secure and show the experimental results.

Eusebiu Marcu
Research and Development Department, SOFTWIN, No. 1/VII Pipera-Tunari, Nord City Tower, Voluntari - Ilfov, Romania
e-mail: marcueusebiu@gmail.com

G.A. Papadopoulos and C. Badica (Eds.): Intelligent Distributed Computing III, SCI 237, pp. 249–255.
springerlink.com © Springer-Verlag Berlin Heidelberg 2009

2 Grid Computing

"Grid is a type of parallel and distributed system that enables the sharing, selection, and aggregation of geographically distributed "autonomous" resources dynamically at runtime depending on their availability, capability, performance, cost, and users' quality-of-service requirements".(definition in [2]).

The biometric authentication mechanism involves specific algorithms for stating the identity of the person that uses this type of authentication. Some algorithms depend on biometric technology. In any case, these algorithms are time and resource (CPU and memory) consuming.

2.1 *Self-Built Grid*

A **self-built grid** is a grid where client machines become nodes inside the grid and take part in solving the grid's task. In our case, for an application that uses e-authentication, the task that must be solved by the grid is authenticating, in an acceptable time, all clients' authentication requests. Figure 1 shows a self-built grid at a time moment.

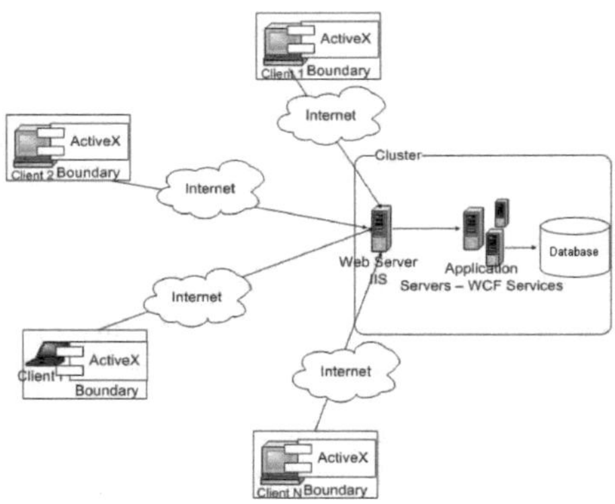

Fig. 1 Self-built grid at a time moment

As shown in the figure, each client is connected to a web server that sends the request to some application servers, grid's main nodes and each client connected to the web server is a common node inside this grid.

The self-built grid architecture differs from a classic architecture because the common nodes are not always in the same location or local network.

2.2 *Self-Built Grid Properties*

There are a number of properties for this type of grid:

- grid's size (number of nodes) varies in time; each client that requests an authentication increases the grid's size;

- a new client becomes a node inside the grid - becomes a common node and increases the grid load;
- a new client node (common node) takes over the task of loading the grid;
- a client who is to meet the demands leaves the grid automatically - grid unloading;
- if a client node leaves the grid, neither the grid nor the main job are influenced (the remaining tasks are still computed).

3 A New Biometric Authentication Architecture

Each architecture has it's advantages and disadvantages. For match on server - store on server, there'll be a problem when the number of simultaneous requests made by clients is very large. For match on client - store on client, exists some security issues.

Having in mind self-built grid architecture model, we define a new biometric authentication mechanism.

3.1 Client-Server-Client

This architecture model means that a template (under BioAPI 2.0 Standard definition [5]), of a number of samples, is stored inside a database in some format and the matching is made on client's machine.

Knowing that the client nodes inside a self-built grid can solve a task (in this case a matching between a template and a biometric sample to match), we define the new authentication architecture model in the following manner: a client solves the task for authenticate other client (authentication in **three** points) - see Figure 2.

The work flow is the following:

- two or more clients want to authenticate themselves against the system at the same time or in a very short interval
- clients requests are taken by the server, the server associates an ID to that request, a time stamp and adds this newly created structure into a queue
- the queue is verified to see if there's only one request and in this case the server will process the request and will authenticate the client and the answer will be sent to the client
- the thread that inserts the structure (Client's A Structure) in the queue removes a waiting structure that contains all necessary data for authentication, like clients template and sample to authenticate and set the *IsAtClient* property to true), request which belongs to other client (Client's B Structure)
- the authentication is made inside client's A installed ActiveX and creates a new request for the server to send the answer
- the newly created thread sets the result of the structure in a waiting queue according to the request ID - sets *IsOver* to true and signals the clients' A waiting thread to complete its second request

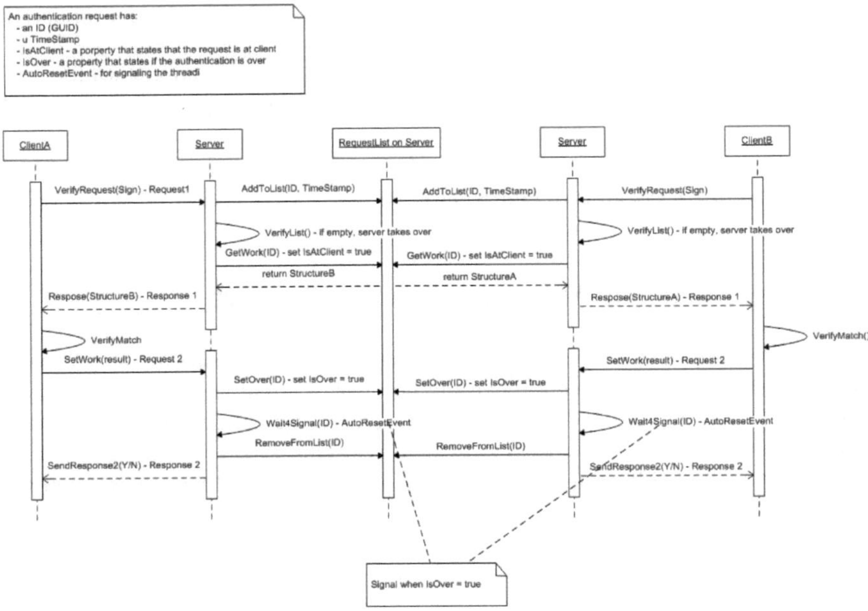

Fig. 2 Authentication in three points)

- final step, the answer is sent to the client A.

It is clear that is this model the server becomes only a manager of client's requests and it will intervene when an exception is thrown or the number of resends is achieved.

3.2 Grid's Performance

Let us assume that there is no way that a "biometric template" [5] can be transformed again into the same "biometric sample" that was captured by the biometric device. For enrollment and authentication operations, a biometric sample must be processed. To increase the system performance we can "move" the biometric sample processing on client side and consequently, decreasing server computation work. Adopting this model, the system works only with processed samples and increase data security.

The client opens the acquisition application (which hosts the ActiveX object), the application loads a BioAPI Engine, an user-interface appears, the subject gives a number of samples that are processed by the engine that creates a biometric template.

3.3 Advantages

The proposed architecture has the following advantages:

- the server will be unloaded by all clients' authentication tasks. For an authentication, to maximum time will be:

$$t_{total_{max}} = t_{connectionA} + t_{connectionB} + n_{resends} \cdot (t_{timeout} + t_{connectionC}) + t_{server} \quad (1)$$

  where

$$t_{connectionC} \quad (2)$$

  is the connection time of new client C that gets client's A task. The minimum time is given by:

$$t_{total_{max}} = t_{connectionA} + t_{connectionB} + t_{effective} \quad (3)$$

- the server will be unloaded by all clients' enrollment/process operation
- usage of "processed template" format increases the data security - the original data cannot be rebuilt using the sample features
- when a client computes other ones task, there is no information what so ever about the client that wants the task solved.

3.4 Disadvantages

The proposed architecture has the following disadvantages:

- the presence of the authentication engine on the client machine makes possible the observation of operations made by the engine. A possible countermeasure is the online downloading of the engine.
- there is a vulnerable point before the template is created - there is a possibility that an attacker can break into the engine and save the biometric samples.

3.5 Notes about Security

Even if this type of grid is used as a classic grid inside an Intranet network or as a cloud application, security is important. There are many types of attacks that can

Table 1 Types of attacks

Attack	Countermeasure
man in the middle	encrypting the network traffic
spying on application process	encrypting data on client, decrypting on server
spoofing	using SSL certificates for authentication

be launched against a system that is using this architecture. The following table contains some types of attacks.

4 Experimental Results

In order to test a system that uses this architecture, two types of applications were created: a WCF application as the management core and a console application simulating the client. The console application was installed on a number of 20 computers and, at a given time, an authentication request was sent. Before this system, we had a classic "Store on Server, Match on Server" architecture. The grid method and the classic method are compared below (on X-axis is the number of nodes, on Y-axis is the time, in seconds).

The test was made using the same signature for all requests. Therefore we get an almost "perfect" linear variation (Figure 3).

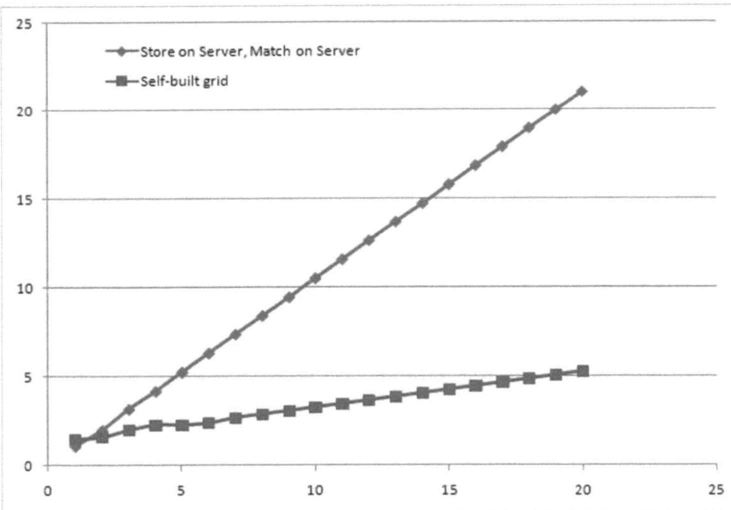

Fig. 3 Self-built grid vs. Classic method

5 Conclusions

This paper's new definition of a grid tries to create the base point of a new type of distributed computing. Of course, this type of grid isn't widely applicable. The new authentication model is derived from match on server - store on server model, the only difference is that the second client solves the task (as it was a part of the server). So, from client's point of view, the server is a multi-machine system.

References

1. International Committee for Information Technology Standards, 2007, Study Report on Biometrics in E-Authentication, INCITS M1/07-0185, March 30 (2007)
2. Grid Computing, Grid Computing Info. Centre, `http://www.gridcomputing.com/gridfaq.html`
3. Foster, I., Kesselman, C.: The grid: Blueprint of a New Computing Architecture. Morgan Kauffman Publishers, San Francisco (1999)
4. Windows Communication Foundation, Microsoft Co. (2007), `http://msdn.microsoft.com/en-us/netframework/aa663324.aspx`
5. BioAPI Consortium, Information technology - Biometric application interface - BioAPI Specification, ISO/IEC 19784-1 (2005)

Obtaining Knowledge Using Educational Data Mining

Cristian Mihaescu, Dumitru Dan Burdescu, Mihai Mocanu, and Costel Ionaşcu

Abstract. Obtaining knowledge is one of the most important tasks in currently developed systems. Regarding this open problem, educational data mining is one of the areas that gather many efforts. This paper presents a custom methodology of obtaining knowledge about learners. The obtained knowledge is based on activity performed within e-Learning environment. The logged activity regards data about how the student answered to test questions. The main task of the procedure regards clustering students such that different pedagogical approach will be used for each cluster. K-means algorithm has been used as clustering method. The final goal is to create a model of analysis which may conclude whether or not an e-Learning platform is capable of classifying students depending on accumulated knowledge.

1 Introduction

This paper presents advances made within an e-Learning platform called Tesys[1]. This platform has initially been designed and implemented only with core functionalities that allowed involved people (learners, course managers, secretaries) to collaborate in good conditions. The platform has built in capability of monitoring and recording user's activity, especially data regarding testing activities. The activity represents valuable data since it is the raw data for our machine learning and modeling process. User's sequence of sessions makes up his activity. A session starts when the student logs in and finishes when the student logs out. Under these

Cristian Mihaescu · Dumitru Dan Burdescu · Mihai Mocanu
University of Craiova, Software Engineering Department, Bvd.Decebal 107, Craiova, RO-200440, Romania
e-mail: {mihaescu,burdescu,mocanu}@software.ucv.ro

Costel Ionaşcu
University of Craiova, Analysis, Statistics and Mathematics Department, A.I.Cuza 13, Craiova, RO-200529, Romania
e-mail: icostelm@yahoo.com

G.A. Papadopoulos and C. Badica (Eds.): Intelligent Distributed Computing III, SCI 237, pp. 257–262.
springerlink.com © Springer-Verlag Berlin Heidelberg 2009

circumstances, a sequence of actions makes up a session. There is a wide range of machine learning data representations that may be used but from all of them we choose clustering. To cluster objects, the common approach is to select some distance measure and then apply a distance-based clustering algorithm e.g. k-medoid methods like CLARANS [4, 5] or a density-based algorithm like DBSCAN [2]. In [3], a method for producing a good initial mixture is presented which is based on multiple sampling. It is empirically shown that using this method, the EM algorithm achieves accurate clustering results. The final goal is clustering students. Student's clustering may have a predictive value in the sense that from the activities a student has made there may be pulled conclusions about his learning proficiency.

2 Tesys Application Platform

We have developed an e-Learning platform that represents is a collaborative environment for students, professors, secretaries and administrators. Secretary users manage sections, professors, disciplines and students. The secretaries have also the task to set up the structure of study years for all sections. The main task of a professor is to manage the assigned disciplines. The professor sets up chapters for each assigned discipline by specifying the name and the course document, and manages test and exam questions for each chapter. The platform offers students the possibility to download course materials, take tests and exams and communicate with other involved parties like professors and secretaries. All tests and exams are taken under time constraints. The professor sets up a number of seconds necessary to answer questions from a chapter. When a test or exam is taken all the seconds are summed thus obtaining a maximal interval of time in which the student has to finish the test. The elapsed and remaining time are managed on server side and presented to the student after each request. A history of sustained tests is kept for the entire user. A student or a professor may view a sustained test. In this view the formula which computes the grade is presented. For each question it is presented what the student has checked, which was the correct answer, which was the maximum points that could be obtained from that question and which was the number of obtained points. At the end it is presented the final formula and the grade. Besides these core functions for on-line testing other ones are implemented or currently under development. A message board is implemented for professors, secretaries and students to ensure peer-to-peer communication. This facility is implemented within the platform such that no other service (e.g. email server) is needed. All these facilities for students and professors give the platform a good position within involved parties.

3 The Data Source

In this part we shall focus on describing the monitoring capabilities of the platform when running. The activity of users is represented by a series of working sessions. A session starts after a user has logged in and ends when he logs out. Whatever actions a student performs between these two moments are recorded for later analysis. The platform implements two ways of monitoring activity. Since business logic is implemented in Java, the log4j utility package was chosen to be used in order to log specific events. The next lines present how the utility was set up.

```
log4j.appender.R.File=D:/Tomcat/idd.log
log4j.appender.R.MaxFileSize=1000KB
log4j.appender.R.MaxBackupIndex=5
```

These lines state that all the logging process will be done in idd.log file and will have a maximum file size of 100KB in maximum five files. This utility package is also used in debugging process and the logs may be very useful in finding security breaches like unsuccessful attempts to log in or run actions that are not allowed. The main disadvantage of this technique is the semi structured way in which information is stored. This makes the information retrieval and analysis to be not so easy. The second method of monitoring user activity within the platform is through a database table called activity. In this table a record is added each time a user performs an action. In the next table it is presented the structure of activity table.

Table 1 Structure of activity table

Field	Description
id	primary key
userid	identifies the user who performed the action
date	stores the date when the action was performed
action	stores a tag that identifies the action
details	stores details about the action performed
level	specifies the importance of the action

In the table the action is represented by a tag. The detailed explanation of what the tag means is set in a properties file. In the next table there are the accompanied by their description.

Table 2 Tags and their description

Tag	Description
STUDENT_LOGGED	student logged in
STUDENT_FINISH_TEST	student finished test
STUDENT_LOGGED_OUT	student logged out

For each language a separate properties file is created, each file containing the same tags but with description in a different language. The details field stores specific information regarding the action that was executed. For example, if a secretary modifies the profile of a student in the details field there will be stored information about what fields were updated. The level field specifies the importance of the executed action. There are defined three level of importance: 0,1 and 2 where level

0 specifies the critical actions. So far, in activity table there are close to 40000 recorded actions in almost four month of running the platform. At the end of the cycle there are expected almost 100,000 recorded actions.

4 Building Clusters from Data Using Partitioning Methods - The EM Algorithm

There are many different ways for representing patterns that can be discovered by machine learning. From all of them we choose clustering, which is the process of grouping a set of physical or abstract objects into classes of similar objects. Basically, for our platform we create clusters of users based on their activity. There are many clustering methods in the literature: partitioning methods, hierarchical methods, density-based methods, grid-based methods or model-based methods. From all of these we chose to have a closer look on partitioning methods.

Given a database of n objects and k, the number of clusters to form, a partitioning algorithm organizes the objects into k partitions ($k < n$), where each partition represents a cluster. The clusters are formed to optimize an objective partitioning criterion, often called similarity function, such as distance, so that objects within a cluster are "similar", whereas the objects of different clusters are "dissimilar" in terms of database attributes. So, the first step is to define a list of attributes that may be representative for modeling and characterizing student's activity. The database of the platform contains twenty tables. For each student there are computed the values of the needed attributes that will represent the input data for the algorithm. Because we are dealing with numeric attributes, iterative-based clustering is taken into consideration from partitioning methods. The classic k-means algorithm is a very simple method of creating clusters. Firstly, it is specified how many clusters are being thought: this is the parameter k. Then k points are chosen at random as cluster centers. Instances are assigned to their closest cluster center according to he ordinary Euclidean function. Next the centroid, or the mean, of all instances in each cluster is calculated - this is the "means" part. These centroids are taken to be the new center values for their respective clusters. Finally, the whole process is repeated with the new cluster centers. Iteration continues until the same points are assigned to each cluster in consecutive rounds, at each point the cluster centers have stabilized and will remain the same thereafter.

The EM algorithm is implemented in Weka package [6] and needs the input data to be in a custom format called $arff$. Under these circumstances we have developed an offline Java application that queries the platform's database and crates the input data file called *activity.*$arff$. This process is automated and is driven by a properties file in which there is specified what data will lay in *activity.*$arff$ file. The most important step in this procedure is the attribute selection and the granularity of their nominal values. The number of attributes and their meaning has a crucial importance for the whole process since irrelevant attributes may degrade classification performance in sense of relevance. On the other hand, the more attributes we have the more time the algorithm will take to produce a result. Domain knowledge and of course common sense are crucial assets for obtaining relevant results. For a student in our platform we may have a very large number of attributes. Still, in our procedure we used only six: *positive_count* - the number of correctly answered questions; *correct_percent* - the percentage of correctly answered questions from the total number of questions; *total_tries* - the total number of tries; *avg_tries* - medium number of tries per question; *avg_question_time* - on average, how long it takes for a student to answer a question; *total_time* - total time spent on testing.

For each attribute there is obtained one of the five possible nominal values: VF - very few; F - few; A - average; M - many; VM - very many. Here is how the arff file looks like:

```
@relation activity
@attribute positive_count {VF, F, A, M, VM}
@attribute correct_percent {VF, F, A, M, VM}
@attribute total_tries {VF, F, A, M, VM}
@attribute avg_tries {VF, F, A, M, VM}
@attribute avg_question_time {VF, F, A, M, VM}
@attribute total_time {VF, F, A, M, VM}
@data
VF, F, A, A, F, A,
F, A, M, VM, VF, A,
A, M, VM, A, V, F,
VM, VM, A, VM, M, VF, F,
```

As it can be seen from the definition of the attributes each of them has a set of five nominal values from which only one may be assigned. The values of the attributes are computed for each of the 375 students and are set in the @data section of the file. For example, the first line says that the student:

- answered correctly to very few questions;
- the percentage of correctly answered questions is few;
- the student had an average total number of tries;
- the student had an average number of tries per question;
- the student spent few time for answering a question;
- the total time spent on testing is average;

The granularity for the nominal values of the attributes can be also increased. In our study we considered only five possible values but we can consider testing the algorithm with more possible values. This should have great impact on the number of obtained clusters. The time taken by the algorithm to produce results should also increase. In order to obtain relevant results we pruned noisy data. We considered that students for which the number of taken tests or the time spent for testing is close to zero are not interesting for our study and degrade performance and that is why all such records were deleted. After this step there remained only 268 instances.

Running the EM algorithm created four clusters. The procedure clustered 40 instances (15%) in cluster 0, 83 instances (31%) in cluster 1, 112 instances (42%) in cluster 2 and 33 instances (12%) in cluster 3. The final step is to check how well the model fits the data by computing the likelihood of a set of test data given the model. Weka measures goodness-of-fit by the logarithm of the likelihood, or log-likelihood: and the larger this quantity, the better the model fits the data. Instead of using a single test set, it is also possible to compute a cross validation estimate of the log-likelihood. For our instances the value of the log-likelihood is -3.75 which represents a promising result in the sense that instances (in our case students) may be classified in three disjoint clusters based on their activity.

5 Concluding Remarks

We have created a procedure of data analysis which may provide interesting conclusions regarding the classification of students from an e-learning platform. The platform was developed, is currently running and has built in capabilities of monitoring students testing activities. An off-line application was developed for creating the input data files that are analyzed. Data analysis is done using EM clustering algorithm implemented by Weka system. The main goal is clustering of students. Student's clustering may have a predictive value in the sense that from the performed testing activities a student has made there may be pulled conclusions about his learning proficiency. On the other hand, platform's characterization may have as result an estimation of the capability of an e-learning system to grade and order students according to their accumulated knowledge. This analysis is critical for having as conclusion that a system can support generalized tests. In order to achieve this goal many other analysis techniques may be used. If a clustering method is used to label the instances of the training set with cluster numbers, that labeled set could then be used to train a rule or a decision tree learner. The resulting rules or tree would form explicit description of the tree classes. A probabilistic clustering scheme could be used for the same purpose, except that each in-stance would have multiple weighted labels and the rule or decision tree learner would have to be able to cope with weighted instance.

References

1. Burdescu, D., Mihaescu, C.: Tesys: e-Learning Application Built on a Web Platform. In: Proceedings of International Joint Conference on e-Business and Tele-communications, Setubal, Portugal, pp. 315–318 (2006)
2. Ester, M., Kriegel, H.P., Sander, J., Xu, X.: A Density-Based Algorithm for Discovering Clusters, in Large Spatial Databases with Noise. In: Proc. Int. Conf. on Knowledge Discovery and Data Mining (KDD), pp. 291–316 (1996)
3. Fayyad, U., Reina, C., Bradley, P.: Initialization of Iterative Refinement Clustering Algorithms. In: Proc. Int. Conf. on Knowledge Discovery in Databases, KDD (1998)
4. Han, J., Kamber, M.: Data Mining Concepts and Techniques. Morgan Kaufmann Publishers, San Francisco (2001)
5. Ng, R., Han, J.: Efficient and Effective Clustering Methods for Spatial Data Mining. In: Proc. Int. Conf. on Very Large Databases (VLDB), pp. 144–155 (1994)
6. Weka, `http://www.cs.waikato.ac.nz/ml/weka`

TN4PM: A Textual Notation for Process Modelling

Andrei-Horia Mogos and Andreea Urzica

Abstract. In this paper we compare three visual notations for modelling processes, and we propose a textual notation for modelling these processes. Our textual notation can be used just as a modelling notation, but it can also be used to translate the process models from one visual notation to another.

1 Introduction

Process modelling is an essential technique for the design and analysis of processes. Most of the process modelling notations are visual notations; a visual notation is easy to understand and easy to use for people from the business area, but more difficult to use for computer science specialists that need a formalized description in order to translate it into an executable workflow or formal specifications.

Several efforts have been done to translate process models represented in visual notations into formal semantic notations, see for example RAD into process algebra [2]. Other researchers investigate the creation of a textual notations related to a certain visual notation; see for example a textual notation for RAD [8] or a textual notation for BPMN [10].

This paper compares three visual process modelling notations: Business Process Modeling Notation (BPMN), Unified Modeling Language Activity Diagram (UML AD), and Role Activity Diagram (RAD), and propose a textual notation for process modelling, called Textual Notation for Process Modelling (TN4PM) that covers the

Andrei-Horia Mogos · Andreea Urzica
University Politehnica of Bucharest, Splaiul Independentei 313, 060042 Bucharest, Romania
e-mail: andrei.mogos@cs.pub.ro, andreea.urzica@cs.pub.ro

G.A. Papadopoulos and C. Badica (Eds.): Intelligent Distributed Computing III, SCI 237, pp. 263–268.
springerlink.com © Springer-Verlag Berlin Heidelberg 2009

most important visual elements of these three visual notations. TN4PM can be seen as a common process modelling notation for business and computer science.

Our textual notation uses some ideas from GPSS/H. GPSS/H is a simulation language that uses the transaction-flow modelling paradigm [1]. This modelling paradigm uses the following idea: active objects compete for passive resources while traveling through a diagram representing a system [1]. For our textual notation, we used active objects that can communicate with each other using messages. Each active object, called in our notation entity, executes an activity flow; the activity flows can be seen as executed in parallel. In addition, we used from GPSS/H the jump-label technique for modelling the activity flows.

The paper is organized as follows. Section 2 presents the basic concepts related to three visual modelling notations: BPMN, UML AD, RAD. Section 3 introduces our proposed textual notation for process modelling. Section 4 provides a case study. Finally, Section 5 contains some conclusions and future work.

2 Three Visual Notations

In this section we are going to briefly present and compare three process modelling visual notations: BPMN, UML AD, and RAD.

BPMN is supported by Object Management Group / Business Process Management Initiative. BPMN permits easy development of business diagrams, being a notation intended for the use of business analysts without any technical computer science background requirements. This paper discusses BPMN 1.1 [5], BPMN 2.0 being, at the moment, a Request for Proposals. A business process diagram is, essentially, a flowchart composed by: flow objects, connection objects and, optionally, artefacts.

Concerning UML Activity Diagrams, they are intended to model both computational and organisational processes (i.e. workflows) [3]. While BPMN is a notation for high-level modeling, UML AD is oriented more towards execution. The UML specification, version 1.4.2, was developed by Object Management Group [9], being an accepted ISO specification starting with January 2005.

The third notation discussed in this paper is RAD, for which the first concepts were proposed in 1983 by Holt et al [4], and later enriched by Ould [7], [6]. A business process is seen as a set of distinct, concurrent activities corresponding to a set of roles. "A role involves a set of activities which, taken together, carry out a particular responsibility or set of responsibilities" [7].

Table 1 offers a comparison of UML AD, BPMN, RAD, and TN4PM. TN4PM is our textual notation and it will be introduced in the next section.

3 Introducing a Textual Notation

In this section we present a textual notation for business and computational processes. Entity is our term for UML AD - swimlane, BPMN - swimlane, and

Table 1 UML AD, BPMN, RAD, and TN4PM Comparison

UML AD	BPMN	RAD	TN4PM
Swimlane	Swimlane	Role	Entity
Initial state	Start Event	-	Entity <EntityName>
Final state	End Event	-	End
Activity state	Task / Subprocess	Activity	User Defined Action
State	Intermediate Event / Task	State line	Comment line
Wait state	Timer Event	-	Wait
XOR split	Exclusive Gateway	Case Refinement	*XOR_Split*
XOR join	(Exclusive) Gateway	-	*XOR_Join*
AND split	Parallel Gateway	Part Refinement	Split
AND join	(Parallel) Gateway	-	Join
Control Flow	Sequence Flow	State line	If Then Else / Goto, Labels
Signal send / receipt	Message Flow	Interaction	Send / Receive results
Note	Annotation	Label	Comment

RAD - role. The notation is based on entity blocks. Each entity block contains an activity flow. The activity flows related to the entity blocks work in parallel. Entities can communicate with each other using messages. Next, we give the description of an entity block:

< entity_block >::= Entity < entity_name >< block_line >* End
< entity_name >::=< letter >+< digit >*
< block_line >::=< comment_line > | < action_line >
< comment_line > " : " < comment >
< action_line >::=< action > |(< label >< action >)|
|(< action > " : " < comment >)|(< label >< action > " : " < comment >)
< label >::=< letter >+< digit >*
< action >::=< predefined_action > | < user_defined_action >
< predefined_action >::=< sequence_flow_action > |
| < message_action > | < split_join_action > | < wait >
< sequence_flow_action >::=< if_then_else > | < goto >
< if_then_else >::= If"(" < condition > ")"Then < label > Else < label >
< goto >::= Goto < label >
< message_action >::=< send > | < receive >
< send >::= Send"(" < entity_name > "," < message_name > ")"
< receive >::= Receive"(" < entity_name > "," < message_name > ")"
< message_name >::=< letter >+< digit >*
< split_join_action >::=< xor_split > | < xor_join > | < split > | < join >
< xor_split >::= XOR_Split"(" < label >+ ")"
< xor_join >::= XOR_Join"(" < digit > ")"
< split >::= "(" < label > ("," < label >)*")"" = "Split"("")"

$< join >::= Join"(" < digit > ")"$
$< wait >::= (Wait"(" < digit > ")")|(Wait"(" < digit >< unit > ")")$
$< unit >::=< letter >^{+}$
$< user\_defined\_action >::=< action\_name > "(" < arg > ("," < arg >)^{*}")"$
$< action\_name >::=< capital\_letter >< letter >^{*}< digit >^{*}$
$< arg >::=< letter >^{+}< digit >^{*}$

A process representation consists of several entity blocks. An entity block shows the actions flow related to that entity and contains one or more block lines. A block line can be a comment line or an action line. Comment lines are used especially for describing states, between two actions. An action line must contain an action. It can also contain a label and / or a comment.

For a user defined action, the user can specify the name and the parameters of the action. The user can also explain the user defined action using the optionally comment for an action line. As predefined actions we have: sequence flow actions, message actions, split - join actions, and "Wait" action.

The sequence flow actions are "If - Then - Else", and "Goto" and are responsible for conditioned and unconditioned jumps to a specified label. A label uniquely determines a line of an entity block.

We have two message actions: "Send" and "Receive". "Send" is used to send messages to other entities. "Receive" blocks the entity action flow until the entity receives the message from the specified source entity.

Our notation has 4 split-join actions. "*XOR_Split*" splits the action flow of an entity in several action flows, but only one of these new action flows will be executed by the entity. "*XOR_Join*" unifies several action flows, but only one of these action flows will be the result of this action. "Split" splits the action flow of an entity in several action flows, and all these new action flows will be executed by the entity. "Join" unifies several action flows of the entity; as the result of this action, all these action flows will become a single action flow.

We have another predefined action, called "Wait". This action shows that the entity must wait a given amount of time, before continue its action flow execution. In addition, the events can be modelled using entities and messages. We can add to our model several entities that send only messages that simulate the events.

4 Case Study

In order to illustrate the usage of TN4PM and the way it facilitates the translation from one visual notation to another, we provide a simple scenario:

A student wants his paper to be reviewed by the advisor. He writes the document and sends it to the professor in order to receive a critical opinion. After reading the document, the professor may decide to add some remarks, thus modifying the document, or, he may decide the work is ok. In the end, the professor sends the document (modified or not) back to the student.

The model using UML AD is presented in Figure 1.

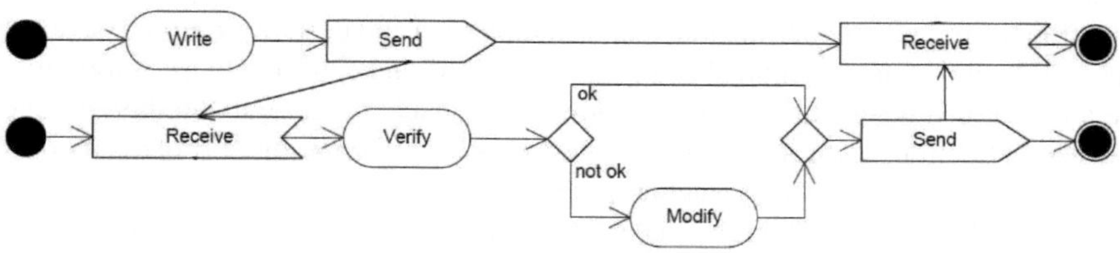

Fig. 1 The UML AD model

Using TN4PM we have the following model:

```
Entity Student
Write(Document)
Send(Document, Advisor)
Receive(Document, Advisor) : waits for the document
End

Entity Advisor
Receive(Document, Student)
Verify(Document) : verifies the document
If (ok_Document) Then Label1 Else Label2
Label1 Send(Document, Student)
       Goto Label3
Label2 Modify(Document) : modifies the document
       Send(Document, Student)
       Goto Label3
Label3 End
```

The model using BPMN is presented in Figure 2.

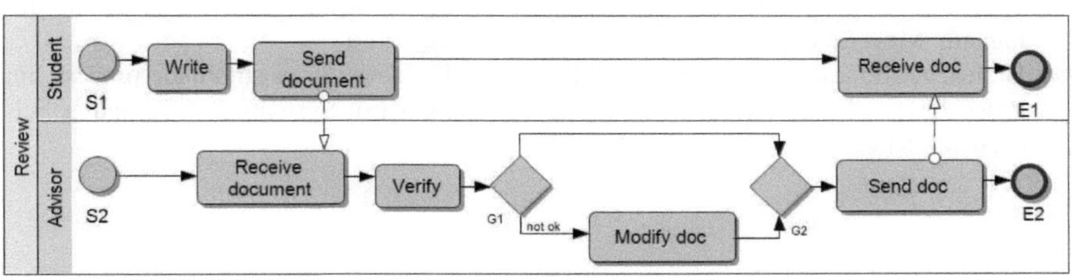

Fig. 2 The BPMN model

5 Conclusions

This paper compares three visual process modelling notations and proposes a textual notation for process modelling that can be understood and used by computer science specialists and by non-technical users. Our notation can be used only as a modelling notation, or can be used for the translation of one visual notation to another. As one can see in Section 4, our notation make easier the transformation of process models between the three discussed visual notations.

A possible future work direction can be to try to make a dynamic visual interface to help users to understand the activity flows and the messages sent and received from the entities. Another future work can be to develop a software system, based on our textual notation, which can automatically translate one visual notation to another. For this goal a first step should be a more exact specification for the user defined actions. Another step should be to add more semantics to the textual notation, possibly using Petri Nets and / or ontologies.

References

1. Wolverine Software, `http://www.wolverinesoftware.com` (last accessed, 2009)
2. Badica, C., Badica, A., Litoiu, V.: Role activity diagrams as finite state processes. In: Proceedings of Second International Symposium on Parallel and Distributed Computing, pp. 15–22 (2003)
3. Dumas, M., Ter Hofstede, A.: UML activity diagrams as a workflow specification language. LNCS, pp. 76–90. Springer, Heidelberg (2001)
4. Holt, A., Ramsey, H., Grimes, J.: Coordination system technology as the basis for a programming environment. Electrical Communication 57(4), 307–314 (1983)
5. Notation, B.: 1.1 Specification. Tech. rep., Technical report, OMG, 2008 (2008)
6. Odeh, M., Kamm, R.: Bridging the gap between business models and system models. Information and Software Technology 45(15), 1053–1060 (2003)
7. Ould, M.: Business Processes: Modelling and Analysis for Re-engineering and Improvement. John Wiley and Sons, West Sussex (1995)
8. Phalp, K., Henderson, P., Walters, R., Abeysinghe, G.: RolEnact: role-based enactable models of business processes. Information and Software Technology 40(3), 123–133 (1998)
9. Specification, O.: Version 1.4, September 2001. Object Management Group, Inc., Framingham, Mass (2001), Internet: `http://www.omg.org`
10. Urzica, A., Tanase, C.: Mapping BPMN to AUML: Towards an automatic process. In: Proceedings of 17th International Conference on Control Systems and Computer Science, pp. 539–547 (2009)

Group-Based Interactions for Multiuser Applications

Carmen Morgado, José C. Cunha, Nuno Correia, and Jorge Custódio

Abstract. We propose a group-based model for the development of interactive applications where multiple users communicate and collaborate with each other and share information. The model defines the organization of a system in terms of entities and groups, each one with a well-defined interface, that mediates all the interactions. Entities can dynamically enter and leave groups, and distinct forms of communication among the group members are allowed, namely based on messages, events and a shared space. This model provides support for explicit groups and a novel concept of implicit groups, which allows the system to automatically adapt the group membership according to the dynamic changes in user profiles. We discuss applications of the model and illustrate its use to support a tourism scenario in a local community.

1 Introduction

The recent increase in the development of applications exploring user mobility, interaction and information sharing, motivates the need for more expressive abstractions, models and mechanisms to enable a transparent and flexible specification of group-based interactions between users. The relevance of group models [3] has been recognized in multiple domains like collaborative work systems, multi agents and more recently interactive Web applications for social networking. We describe a model that supports group abstractions to dynamically manage and organize small and medium location based communities of users. The MAGO model - Modeling Applications with a Group Oriented approach [10, 11], and its computing platform facilitates the organization, access and sharing of contents by multiple users, and enables their dynamic interactions. By considering groups as confined spaces for

Carmen Morgado · José C. Cunha · Nuno Correia · Jorge Custódio
CITI, Departamento de Informática, Faculdade de Ciências e Tecnologia,
Universidade Nova de Lisboa
e-mail: {cpm,jcc,nmc,jfc}@di.fct.unl.pt

G.A. Papadopoulos and C. Badica (Eds.): Intelligent Distributed Computing III, SCI 237, pp. 269–275.
springerlink.com © Springer-Verlag Berlin Heidelberg 2009

user interactions, the model allows to exploit geographic proximity of users, their associated locality issues, privacy and access control issues. We identify two main contributions of this work: (1) to dynamically adjust and control group membership, as well as to define policies to guide the implicit and dynamic group formation according to user profiles; (2) to support a combination of forms of interactions that are required by real life applications, including peer-to-peer, multicast, asynchronous event notification, and access to a shared space, internal to each group.

In section 2 we identify possible requirements of some interactive applications. In section 3 we present the MAGO model, main concepts and functionalities. In section 4 we describe an application for a small place-based community. Finally, conclusions and future directions are presented.

2 Motivation and Requirements

We identified a diversity of scenarios where groups can be used to ease the modeling and development of applications where users form groups, to establish interactions and to share information. Examples arise in working places, travels, tourism spots, shopping centers or any places where "social interactions" may exist. Places like schools campus, museums, shopping centers, restaurants and coffee shops, or airports, are examples where the users may tend to use groups as a way to interact, share information or to achieve a common goal. In most of these places people stay there for a limited period of time and are open to interact with each other. Another requirement is the need to structure the users space, so the services offered can be personalized based on the users preferences.

Similar requirements are also found in applications that manage user collaboration, connections and context awareness [9]. Examples of such applications are on the social software domain Facebook, MySpace or Hi5, and applications promoting the direct communication among users like ICQ, Skype, MSN Messager or Google Talk. On the information sharing domain we have general purpose applications like Flicker, Ringo or YouTube, or domain specific applications like canal*ACCESSIBLE to help the navigation of people with locomotion disabilities using photos captured with mobile devices, or CONFECTIONARY, to create narratives based on user content, such as photos and videos. Also many applications exploit user mobility, allowing users to upload and access contents using handleware devices. Context awareness is a key issue in such applications [9]. In tourism area we can find many examples of interactive applications, like Tech4Tourism [2] where the tourists can navigate through the maps gathering contextualized information (audio, video, photos or explanation texts) based on their actual position and personal interests; InStory [4] allows the users to have more interaction with the system and with each other, sharing information and grouping themselves based on different activities; or Marble [12] where the museum visitors can have access to information concerning the art objects nearby. None of the above however, has the possibility of dynamic creation of groups or offer an integrated platform for communications and data sharing. One goal of our proposed model is to offer, in the same platform, the following functionalities: (i) dynamic management of user profiles; (ii) flexible and dynamic creation of groups; (iii) transparent access to personal and group information; (iv) manipulation of different kinds of media; (v) user interactions and task coordination.

3 The MAGO Model and Characteristics

The MAGO model [10, 5] has the following main characteristics:

- Users are represented as elementary entities and are modeled as Java objects;
- Groups are units for application composition with a well-defined interface, an Universal Group includes all entities and groups defined in the system and where each entity can belong to multiple (explicit ou implicit) groups;
- Groups are dynamic, allowing multiple entities entering and leaving during a group life cycle;
- A group allows distinct forms of communication, namely, direct point-to-point, multicast (messages and asynchronous events), and shared memory;

The model relies on the transparent management of the consistency of views that are observed by the group members, concerning the events related to messages, events, and access to shared state, and event causality. The MAGO model primitives fall into three main classes:

Entities and groups: this includes primitives for managing entities and groups, and dynamic group membership. When creating a new group its defining characteristics are specified: name; admission policy; maximum number of members admitted; type of group; activation and deactivation deadlines. Disconnected operation or unavailability of group members is modeled through offline/online states.

Communications: the model combines three forms of communication within a group:

- Direct: a point-to-point communication, based on the call to an interface method;
- Events: based on a publish-subscribe mechanism [7] and used to multicast messages inside a group. Group members can subscribe to any event type that has been advertised on the group, and receive a notification on event publication. An event can be associated to messages or to changes on internal structures.
- Shared space: based on the concept of a Linda [8] tuple space with corresponding primitives to insert (*update*), remove (*get*) and consult (*consult*) tuples. An additionally primitive (*find*) allows the search of multiple tuples. Each tuple has a name, owner information, the type and application specific information. The model uses events to notify changes on the shared space.

Implicit groups: MAGO provides a higher level of group concept that enables the definition and automatic management of groups based on user properties and attributes (*implicit groups*). The implicit group membership is based on the users currently on the system, their characteristics, and rules previously defined by the application at group creation time. Such rules are simple conjunctions of user attributes and their satisfaction is evaluated by a search engine, always looking for candidate members that match the specified rules. The action of joining an implicit group occurs in three situations: (i) when a new implicit group is created and its characteristics are defined; (ii) when a new entity enters in the system, and specifies its attributes; (iii) when an entity changes its attributes (this last action can also lead to an implicit action of leaving an implicit group).

A layered architecture supports the proposed model as shown in figure 1. The MAGO architecture relies on a low-level Java API - JGroupSpace [6], which is an extension to JGroups [1], and offers primitives for group management, message and event multicast, and an implementation of the Linda tuple space. The MAGO model offers a higher level semantics for interactive applications concerning messages, event publication and subscription, access to the shared tuple space, and also provides the concept of implicit groups and an interface to an information system.

Fig. 1 System layers and modules - MAGO components are included in the gray area in the figure. The bottom layer offers basic communication and system operation support, responsible for connections and management between devices. Middleware layer is responsible for group communications, events and shared space. Information system gives access and manages persistent data manipulated by the system.

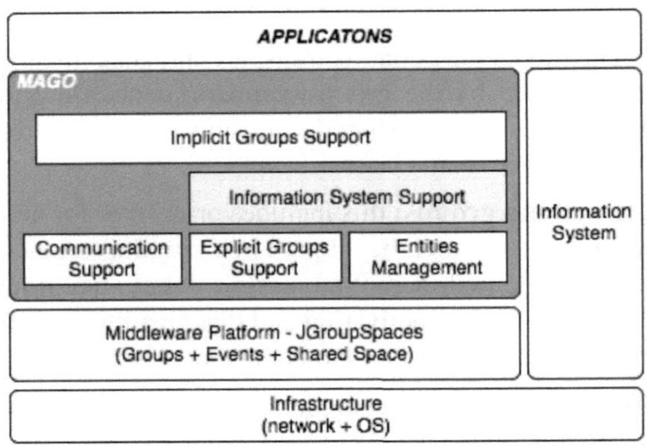

An external information system can store the contents produced and shared by the users, and also information related to user and group attributes, which is used to manage implicit groups. The current prototype supports an interface to an ORACLE Database system. A working prototype of the MAGO architecture and a user interface giving access to the model primitives were implemented in Java, and used to support the experimentation with real applications.

4 A Case Study Application

We describe how MAGO supports a tourism scenario in a local community where groups are used to facilitate user activities and interactions in a cultural heritage setting. This work extends a real world application previously developed in the

InStory project [4]. Tourism applications usually deal with interaction between multiple users, information dissemination and content creation and sharing. In this scenario many users have mobile devices, for example PDAs, cell phones or small laptops and are likely to interact with others to establish groups, and share information and content like photos or videos. Based on the existing InStory architecture we integrated our model as a way to simplify the support for existing functionalities and to develop new ones, where the users (visitors and managers) were modeled as elementary entities.

Visitor groups: are used to distribute information and share photos (or other content) taken during a tour. Assuming that all the visitors have already registered into the system, by invoking the *register* primitive, a typical sequence of actions is:

- a visitor guide creates a new explicit group with name "tour14" using the primitive create (*g1=create(..., "tour14", EXPLICIT, ...)*);
- a visitor (*u1*) can join the group using the primitive *join(u1, g1, ...)*);
- the visitor can access group functionalities: (i) sharing a photo and notify all group members: first update a file on the information system, then create a data object with information specified by the application (for example with the reference to the data file on the information system and a location reference), and finally update data on the "tour14" group shared space (*update(u1, g1, PUBLIC, NOTIFY, data_obj)*); (ii) other members can now access the photo, based on notification information, by first create a data object with the notification information, then consult the data on "tour14" group shared space (*consult(u1, g1, , , NO_BLOCK, data_obj)*) and extract the information system reference to the photo, from the data object, finally get the file from the information system and visualize it on the local device; (iii) disseminate information to all group members: spread a message to the group, by invoking the send primitive (*send(u1, g1, SPREAD, "tour start in 5m")*).

Implicit groups: the implicit group concept in the MAGO model proved adequate to manage the dynamic changes in user profiles and system configuration, as illustrated in the following scenarios.

First a manager (tour guide) creates an implicit group defining the value attributes that all the group members must match. Then by defining the attributes list (*lst=[nationality, "italian"]*) and creating the new group (*create(id_usr,"itV", IMPLICIT, , , lst)*), and finally update the group information on the information system. After group creation, all the users that match the specified attributes are automatically joined to the group. Now the new group can be used, for example, to send personalized messages. If a new visitor has attributes that match the implicit group attributes, he/she automatically becomes a member of the group.

As another example, a manager creates an implicit group, this time with attributes *lst=[local, "Templar Well"]*. Then, each time a visitor arrives at the specified location, the location attribute changes to "Templar Well", and the visitor automatically becomes a member of the group, allowing access to the group resources. When at that location, the visitors can interact with other members of the implicit group, using the group communication mechanisms offered by the model, and can also update and access the contents of the group shared space. When a visitor leaves that location (by changing the location attribute) he/she will be automatically removed from the group "Templar Well".

5 Conclusions and Future Work

Emerging interactive applications increasingly exhibit collaboration and information sharing in multiuser dynamic environments. In order to model those applications, we are exploiting group based abstractions, according to the MAGO approach. The model supports a framework for group management, allowing the structuring of an application in terms of a dynamic collection of entities. Furthermore, groups provide confined interactions spaces, offering different forms of communication (messages, events and shared state), which were found suitable to model the typical application interaction patterns. In order to ease the modeling process, MAGO supports a new concept of implicit groups, allowing an automated management of groups, based on dynamic identification of user profiles. As part of ongoing work, we are completing an implementation of the referred case study. We intend to use MAGO to design and implement other real real world applications with more complex services, namely in the tourism and cultural heritage domains, and in learning environments.

References

1. Ban, B.: JavaGroups - group communication patterns in java. Technical report, Cornell University (July 1998)
2. Chittaro, L., Corvaglia, D., De Marco, L., Gabrielli, S., Senerchia, A.: Tech4tourism (2006), `http://hcilab.uniud.it/t4t/index_en.html`
3. Chockler, G.V., Keidar, I., Vitenberg, R.: Group communication specifications: a comprehensive study. ACM Comput. Surv. 33(4), 427–469 (2001)
4. Correia, N., Alves, L., Correia, H., Romero, L., Morgado, C., Soares, L., Cunha, J.C., Romão, T., Eduardo Dias, A., Jorge, J.A.: Instory: a system for mobile information access, storytelling and gaming activities in physical spaces. In: ACE 2005: Proceedings of the 2005 ACM SIGCHI International Conference on Advances in computer entertainment technology, pp. 102–109. ACM Press, New York (2005)
5. Cunha, J.C., Morgado, C.P., Custódio, J.F.: Group abstractions for organizing dynamic distributed systems. In: Euro-Par 2008 Workshops, pp. 450–459 (2009)
6. Custódio, J.: JGroupSpace- Support for group oriented distributed programming (in portuguese). Master's thesis, Faculdade de Ciências e Tecnologia, Universidade Nova de Lisboa (2008)
7. Eugster, P.T., Felber, P.A., Guerraoui, R., Kermarrec, A.-M.: The many faces of publish/subscribe. ACM Comput. Surv. 35(2), 114–131 (2003)
8. Gelernter, D.: Generative communication in linda. ACM Trans. Program. Lang. Syst. 7(1), 80–112 (1985)

9. Kernchen, R., Bonnefoy, D., Battestini, A., Mrohs, B., Wagner, M., Klemettinen, M.: Context-awareness in mobilife. In: Proceedings of the 15th IST Mobile and Wireless Communication Summit. Joint Workshop - Capturing Context and Context Aware Systems and Platform (2006)
10. Morgado, C.: A group-based model for interactive distributed applications (in portuguese). PhD thesis, Faculdade de Ciências e Tecnologia, Universidade Nova de Lisboa, Lisboa (2009)
11. Morgado, C., Correia, N., Cunha, J.C.: A group-based approach for modeling interactive mobile applications [poster]. In: International ACM Conference on Supporting Group Work - Group 2007 (November 2007)
12. Santoro, C., Paterno, F., Ricci, G., Leporini, B.: A multimodal mobile museum guide for all. In: Mobile Interaction with the Real World (MIRW 2007) -Workshop @ MobileHCI 2007 (September 2007)

Blueprint: A Continuation-Based Web Framework

Alex Muscar and Mirel Cosulschi

Abstract. A relatively recent trend in the web framework world is the use of continuations to model user-interaction. This paper shows the design and implementation of such a framework using a dialect of the functional language Scheme. We will show that Scheme, or any other Lisp-like language, is an adequate language for web development due to its use of *s-expressions* and its functional nature. Our web framework will use *s-expressions* to represent layout information in a manner similar to HTML. The generated web pages are represented as XML and are then transformed to XHTML by means of an XSL transformation.

1 Introduction

Despite all its drawbacks, many web developers continue to use PHP as their main tool in web site development. There are some exceptions though. Recently, languages like *Ruby*[1] and *Python*[2] have been looked upon as web development alternatives. *Ruby on Rails*[3], a web framework that automates most of the mundane tasks of developing a web site, has gained a lot of momentum in the past few years.

Although not so well known as Ruby, *Lisp*[4] is a suitable language for web development. *S-expressions*[5], a textual form of representing structured data akin to XML, and its use of symbolic computation make it especially appropriate for representing web-pages in a declarative way, that is more natural and closer to the HTML language. The web framework we discuss in this paper (to which we refer from now on as *Suncube*) will take advantage of the aforementioned features of the language and will represent web-pages as s-expressions which can be thought of as

Alex Muscar · Mirel Cosulschi
Faculty of Mathematics and Computer Science, University of Craiova, A.I. Cuza 13, Craiova, Romania
e-mail: `muscar@gmail.com`, `mirelc@central.ucv.ro`

[1] Ruby - http://www.ruby-lang.org/en/
[2] Python - http://www.python.org/
[3] Ruby on Rails - http://www.rubyonrails.org/
[4] Lisp - http://www.lisp.org/alu/home
[5] S-expressions - symbolic expressions

G.A. Papadopoulos and C. Badica (Eds.): Intelligent Distributed Computing III, SCI 237, pp. 277–282.
springerlink.com © Springer-Verlag Berlin Heidelberg 2009

lists of symbols. Given Lisp's functional heritage, functions are first class citizens. Suncube makes use of this to implement explicit continuations, which are used to model user-interaction in order to facilitate the implementation of web dialogues [6, 1]. As shown in [7], Scheme can be used for web development related tasks.

In the next section, we will take a closer look at the continuation-based approach as it was implemented in two other projects. Section 3 is dedicated to the design of the *Suncube* framework and an example of its usage. We will conclude our paper in Section 4 and also show possible enhancements to the framework.

2 Previous Works

In the previous years two main directions have evolved in the field of web languages. The first one transforms a web application written in a source language into a mixture of *Javascript* and HTML. Examples include *Links* [4], the framework proposed in [8] and *Volta*[6]. The other approach is to use the source language *per se*, eventually accompanied by a web framework, as a web programming language. It is the case of languages that use continuation-based frameworks.

The most well known continuation-based web framework is *Seaside*[7], developed using the *Smalltalk*[8] programming language. In Seaside, pages are built as a collection of smaller *components* [9], each one holding its own state. Continuations are used to control the flow between the components that make up a page and maintain the state of the web application on the server.

Components are Seaside's answer to the biggest problem of web development: the HTTP protocol does not preserve state between transactions (the web is stateless). State is preserved across page views through components, thus the task of managing the state of the web application is partially handled by the framework. Another aspect of the components is that they encourage code reuse. By splitting a web application's functionality across smaller pieces, the chances are that a certain piece can handle a sufficiently general task so it can be used as a building block in another page.

Seaside uses closures to specify the action to be taken when a link is clicked or a form is submitted. In Seaside, HTML markup is generated programmatically, following the philosophy of separating the content from the presentation.

Another example of continuation-based framework is *Weblocks*[9], written in *Common Lisp*[10]. Like Seaside, Weblocks organizes web pages into smaller pieces called *widgets*. A widget can potentially be anything that is rendered by the browser [2].

Another key concept of the Weblocks framework is *the view*. Views encapsulate functionality to serialize data structures defined by the user to HTML sequences

[6] Volta - http://labs.live.com/volta/
[7] Seaside - http://www.seaside.st/
[8] Smalltalk - http://www.smalltalk.org/main/
[9] Weblocks - http://common-lisp.net/project/cl-weblocks/
[10] Common Lisp - http://common-lisp.net/

such as tables or forms. This approach is taken to separate content from presentation by making the developer writes HTML code as little as possible.

Both frameworks presented above share the following design goals:

- the lack of state of the HTTP protocol is addressed through the use of closures
- pages are made up out of smaller pieces i.e., components or widgets
- the content is separated from the presentation

As can be seen in the section dedicated to Suncube, our framework adheres mainly to the same principles.

3 The Suncube Framework

The Suncube framework is developed in a dialect of Scheme called Blueprint. Blueprint inherits Schemes' simple, yet powerful, semantics. The dialect differs from the Scheme way by introducing Common Lisp-like macros, partial application, direct indexing of strings and hash tables and a shortened, *Arc*[11]-like [5], notation for unary anonymous functions. We will now look at each of these features that take Blueprint apart from Scheme.

Suncube uses continuations to maintain state across web requests in the same manner as Seaside or Weblocks. This has been proved to be a fruitful approach with good results. A proof of this is the growing number of continuation-based frameworks e.g, Seaside, Weblocks, *PLT Scheme Server*[12] [10]. Regarding the structure of a web page, Suncube stays true to the functional concept: it does not introduce a new layer of abstraction as is the case of components or widgets, it uses functions instead. Of course, Blueprint being a Lisp dialect, macros are also an appropriate way of abstraction.

As for the separation of the content from the presentation, Suncube is halfway there. By default, content manipulation is intermingled with the description of the page, because in the author's opinion, a more coherent view of the application is in the advantage of the programmer. The application logic can be separated from the presentation layer, again, through the use of functions.

To illustrate the use of the framework, we will take the Arc challenge[13] posed by Paul Graham: *Write a program that causes the URL said (e.g. http://localhost: port/said) to produce a page with an input field and a submit button. When the submit button is pressed, that should produce a second page with a single link saying "click here." When that is clicked it should lead to a third page that says "you said: ..." where ... is whatever the user typed in the original input field. The third page must only show what the user actually typed. I.e. the value entered in the input field must not be passed in the URL, or it would be possible to change the behavior of the final page by editing the URL.*

---

[11] Arc - http://www.paulgraham.com/arc.html

[12] PLT Scheme Server - http://docs.plt-scheme.org/web-server/index.html

[13] The Arc Challenge - http://www.paulgraham.com/arcchallenge.html

First we will give the answer in Blueprint using the Suncube framework, then we will discuss the features of the framework based on this example.

Listing 1. Answer to the Arc challenge in Blueprint

```
(site arc-challenge
 (define-page said
   (form :method "post" #:action
     (lambda ()
       (link #:target
         (lambda ()
           (text (format "{0}" (user-data 'name))))
       (text "click here")))
     (input :name "name")
     (submit :value "Go"))))
```

3.1 Web Application Structure

Every web application lives in a site, which is characterized by a name. The macro `site` takes a name and the contents of the site — in this example we have only one page, but a site can have as many pages as necessarily; also the site can contain any valid Blueprint expression.

Pages are lists of elements. They are also characterized by a symbol that will be used as their URL in the web application. A page is introduced by the macro `define-page`.

An element of a web page can have two kinds of attributes: *static* and *dynamic*. Static attributes, like normal XML attributes, are textual, meaning they are not evaluated and are introduced by a semicolon. `:method` is a static attribute of the element `form`.

Dynamic attributes are evaluated. If the value of a dynamic attribute is a closure a new continuation will be created and stored in the site's pages list with an unique name. They will expire after their first use and are introduced by a sharp semicolon. `#:action` is an example of such a dynamic attribute. The same goes for the `#:target` attribute of the link in the aforementioned continuation.

Next we will describe the way a Suncube application works. When a request is made by a user, the web server calls the application's handler, which is defined in the implementation of the macro `site` and can be seen in Listing 2. The handler receives as its arguments the requested page's name, the output stream of the server and the data passed by the user through *POST* or *GET*.

First, the handler sets the global variables `user-data` and `out-port` to the data sent by the user, respectively to the output stream of the server, so that they can be accessible to the web application. Next it checks if the requested page exists. In case of a positive result, the handler will output the XML header of the document followed by the document, otherwise an error page will be displayed.

Listing 2. The web application handler

```
(lambda (request port data)
  (if data
    (set! user-data data))
    (set! out-port port)
    (let ((requested-page (pages request)))
      (if requested-page
        (begin
          (xml-header port)
          (write-tag port nil 'document)
          (requested-page port)
          (write-tag port t 'document))
        (port-formatted-write port "Page {0} not found." request))))
```

The way the page is rendered is very simple due to its representation as a list of symbols and can be observed in Listing 1, which is the actual implementation of the macro page in the framework. The list is iterated and every element is expanded — it is serialized — as XML on the output port. The process is recursive so the entire page will be rendered at the end of the process.

The next step is the XSL transformation which is made by a globally defined template, unique for the framework. After the transformation, the page is a valid XHTML that can be displayed by the browser.

4 Conclusions and Future Works

The result of the framework implementation is a succinct, yet easily maintainable and extensible piece of software. Blueprint proved to rise up to the task which shows that functional languages, in general, and Lisp, in particular, are appropriate vehicles for web development. Blueprint is an experimental prototype so it has some obvious shortcomings. Despite its caveats, the experience of implementing the Suncube framework with the help of Blueprint was a promising one.

Feedback received from the web developers that used the prototype was positive. Based on their feedback, one direction for improvement for Suncube is the introduction of *Ruby on Rails*-like *scaffolding*[14]. Coupled with automatic XHTML generation for the structure of the database, such as insert and update forms, this would reduce further the numbers of lines of code necessary to implement a web application and would rise the overall productivity.

Another valuable observation from the testers was the possibility to define custom XSL templates. Since the framework offers support for defining new tags, it makes sense to include this feature in a future release. A possible improvement would be

[14] Scaffolding in Ruby on Rails - http://wiki.rubyonrails.org/rails/show/Scaffold

the integration with a Javascript framework, e.g. Prototype[15] or Mootools[16]. Another future direction that seems very interesting is the one presented in [12, 13], but this implies a whole rethinking of the framework.

Acknowledgements. The work reported was partly funded by the Romanian National Council of Academic Research (CNCSIS) through the grant CNCSIS 375/2008.

References

[1] IBM developerWorks: Use continuations to develop complex Web applications (2004), `http://www.ibm.com/developerworks/library/j-contin.html`
[2] Akhmechet, S.: Weblocks User Manual (2007), `http://trac.common-lisp.net/cl-weblocks/wiki/UserManual`
[3] R6RS Editors Committee: Revised 6 Report on the Algorithmic Language Scheme (2007)
[4] Cooper, E., Lindley, S., Wadler, P., Yallop, J.: Links: Web Programming Without Tiers (2006)
[5] Graham, P.: Arc tutorial (2008), `http://ycombinator.com/arc/tut.txt`
[6] Graunke, P.: Web Interactions. In: European Symposium on Programming, pp. 238–252 (2003)
[7] Graunke, P., Krishnamurthi, S., Van Der Hoeven, S., Felleisen, M.: Programming the Web with High-Level Programming Languages. In: Sands, D. (ed.) ESOP 2001. LNCS, vol. 2028, pp. 121–136. Springer, Heidelberg (2001)
[8] Hanus, M.: Putting Declarative Programming into the Web: Translating Curry to JavaScript. In: Proc. of the 9th International ACM SIGPLAN Conference on Principle and Practice of Declarative Programming, PPDP 2007 (2007)
[9] The Seaside Project: Seaside Framework Tutorial (2008), `http://www.seaside.st/documentation/tutorials`
[10] Krishnamurthi, S., Walton Hopkins, P., McCarthy, J.: Implementation and Use of the PLT Scheme Web Server. Higher-Order and Symbolic Computation 20, 431–467 (2007)
[11] Ducasse, S., Lienhard, A., Renggli, L.: Seaside -A Multiple Control Flow Web Application Framework. In: ACM International Conference Proceeding Series, vol. 178, pp. 163–172 (2006)
[12] Piancastelli, G., Omicini, A.: A Logic Programming Model for Web Resources. In: Proceedings of 4th International Conference on Web Information Systems and Technologies (WEBIST 2008), pp. 158–164 (2008)
[13] Piancastelli, G., Omicini, A.: A Multi-Theory Logic Language for the World Wide Web. In: Garcia de la Banda, M., Pontelli, E. (eds.) ICLP 2008. LNCS, vol. 5366, pp. 769–773. Springer, Heidelberg (2008)

[15] Prototype Javascript Framework - http://www.prototypejs.org/
[16] Mootools - http://mootools.net/

Towards Reliable Trust Management in Insecure P2P Environments

Miroslav Novotny and Filip Zavoral

Abstract. The rapid growth of communication technologies in the past years has enabled development of collaborative peer-to-peer applications. The reliable collaborative application must implement some methods which help to detect malicious entities and avoid collaboration with them. Such systems are based on building trust relationships between entities; their design is difficult due to a decentralized architecture and general distrust among unknown entities. In this paper we analyze requirements for reliable trust management in an insecure environment like Internet. We propose criteria which reliable trust management should meet and we compare several published systems according to these criteria. We also propose guidance on design and evaluation of trust management techniques.

1 Introduction

The reliable collaborative application must implement some methods which help to detect malicious entities and avoid collaboration with them. The design of such system is difficult due to decentralized architecture and general distrust between unknown entities. The only way how to predict future behavior of an entity is based on observation of its past behavior or information about its past behavior obtained from other entities. The application relies on the assumption that the entity, which proves its honesty in the past, will continue in the honest service.

The part of the collaborative application which deals with entities' behavior is called trust management (TM). Most TM are based on reputations. There are also policy-based and social network-based [1] TM but this paper focuses solely on reputation-based TM. The reputation-based TM collects information about transactions between entities, calculates reputations and disseminates the calculated values.

Miroslav Novotny
Charles University in Prague
e-mail: novotny@ksi.ms.mff.cuni.cz

Filip Zavoral
Charles University in Prague
e-mail: zavoral@ksi.ms.mff.cuni.cz

G.A. Papadopoulos and C. Badica (Eds.): Intelligent Distributed Computing III, SCI 237, pp. 283–288.
springerlink.com © Springer-Verlag Berlin Heidelberg 2009

Each entity is able to create trust relationships with other entities and according to this trust make a decision of cooperation. The trust represents confidence in quality and reliability of the entity within a context of a given service.

The reliable TM should be able to precisely predict the future peer behavior. This is difficult to achieve in an environment where any party can attempt to exploit the system for its own benefit. The malicious entities can also try to subvert TM to not lose chance to abuse the system. The robust TM must not only be able to establish correct trust between entities but also protect itself.

This paper focuses on TM in an insecure environment like Internet. We analyze all parts of TM, cooperation between individual parts and their requirements. Our work contributes to understanding how individual components make for overall success of TM task. We establish several simple criteria to each component and discuss how to contribute to reliability of TM. We also compare several published TM according to the proposed criteria.

The rest of the paper is organized as follows. We present layered architecture of TM in section 2. Sections 3, 4, and 5 analyze individual layers and establish the criteria. Finally, section 6 concludes the paper with some discussion on future work.

2 The Architecture of Trust Management

For better understanding of the task of TM we divide its function into three-layered architecture. These layers are inserted between an application logic and P2P overlay: secure P2P layer, information handling layer, and formulation and calculation layer. The secure P2P layer provides secure communication between entities. It ensures integrity of messages and unique identification. The information handling layer gathers all information needed for calculation and alternatively disseminates the calculated values among other entities. The calculation layer is responsible for deriving trust from obtained information. Previously published techniques can be used on each layer. These techniques often differ in requirements which they put on lower layers. We analyze the requirements of each layer with regard to higher layer and reliability of TM.

3 Secure P2P Layer

In this section we describe requirements on secure communication expected by higher layers of TM. We divide them into basic and additional. Three basic requirements: Each entity has a unique identification, each entity can prove its identity, and the system ensures the message authenticity and integrity. The first requirement is met naturally by P2P overlay which use this identification to route messages in the P2P network. The next two requirements can be easily met using asymmetric cryptography and entity's identification derived from a public key. There is no problem with fulfillment of these requirements, but some TM require additional ones. Four additional requirements: The entity cannot choose its own identification, one physical node or user can act as the only one entity, the system ensures the message confidentiality, and the entity identification cannot be connected to a physical node or user.

The first additional requirement relates to TM where the entity identification influences its roles in a calculation and dissemination process. For instance, the identification of the entity A, which performs a special function for the entity B, is derived from identification of the entity B. The malicious entity which has an ability to choose an arbitrary identification can easily impersonate into a special role towards a target entity. The second requirement is an effort to reduce the risk of Sybil attacks. It is difficult to meet this requirement because a reliable solution requires a unique unchanging identification of physical nodes or users. Such identification either does not exist or the entities are not willing to provide it in an untrustworthy network. The solution of this problem is often postponed to higher layers.

The last two requirements ensure privacy and anonymity. Some TM require these properties to function correctly, but they are also desired properties of a majority of the collaboration applications in general. The confidentiality can be easily guaranteed but there are difficulties with anonymity, especially in conjunction with the second requirement which is in opposite. We formulate following criterion about the secure P2P layer.

Criterion 1: The secure P2P layer should meet all basic requirements and these additional requirements which are required by higher layers.

We analyze requirements which are put on this layer in several different TM. The result of this comparison is in table 1. Some TM implement its own secure P2P layer, in such cases the table shows whether it implements this requirement.

4 Information Handling Layer

The information handling layer gathers all information needed for the calculation layer. The amount of this information significantly influences the quality of calculated values. The informed decision must be motivated by a complete set of relevant information. We address the issue of spreading information in the network, discuss what information is relevant to entity decision and whether all this information are available in time of the decision.

The first question is what information is relevant to the decision. The answer is related to the requirements of the calculation layer. The basic unit is a piece of information about one transaction. The transaction is an operation between two entities (source and destination) when both entities can rate the benefit of the transaction for them. This rating is spread in the network as a feedback. The entities which collect feedbacks can compute trust toward some entity and use this value for their own decision or spread it in the network as an opinion. Obtaining of a feedback or an opinion is considered as next transaction which can be also rate.

The calculation layer determines which information is relevant for one decision. We separate this information into three groups. First-hand information: All feedbacks and opinions which include the entity under consideration. Transitive information: First-hand information and all feedbacks and opinions which help to judge the quality of the first-hand information. All information: All feedbacks and opinions created in the network.

The first-hand information includes only direct opinion towards the entity. The calculation layer has only a little information about reliability of this information. The second group, transitive information, includes information about quality of feedbacks or opinions which the sources of first-hand information provided in the past. The calculation layer can use this information for determining the expected reliability of obtained first-hand information. The last group contains all available information, the calculation layer can calculate trust towards all entities at once. The task of the information handler is to provide all relevant information.

Criterion 2: The information handling layer should ensure that the calculation layer has all information relevant to decision in time of decision.

This criterion cannot be evaluated alone. The well-designed calculation can profit more from incomplete information than inferior calculation from complete information, especially if some information is spurious.

Although the issue of identification of misleading information is the task of the calculation layer, we can minimize the misleading information dissemination in the information handling layer. We explore the path of a feedback from its creator to the entity which use this feedback for its own decision. There can be some entities on this path which have the ability to modify the feedback. For instance, the entities which collect several feedbacks, calculate trust and distribute it as their opinion. This distributed computation significantly reduces the overhead connected with feedback dissemination but also increases the risk of an unauthorized modification. One criterion on reliable TM is a number of such intermediaries. We distinguish three situations. Distributed calculation: There are several intermediaries on the path which collect available feedbacks and opinions, calculate trust and pass it as opinion to the next entity on the path. Trust agent: The trust agent collects all feedbacks about one entity and calculates trust towards this entity. All other entities in the network accept trust calculated by the trust agents. Direct path: The feedbacks are delivered to the entity without modification. The final entity is the only one on the path which calculates trust.

The distributed calculation reduces the overhead connected with feedback dissemination, but each of intermediaries can manipulate its part of the calculation. On the other hand, the direct path eliminates the possibility of an unauthorized modification, except on the feedback source, but it is connected with a significant overhead. The trust agent solution can be considered as a trade-off between previous two solutions. The role of the trust agent is pivotal and its function is mostly duplicated on several entities. The duplication can reduce the risk of being deceived by a malicious trust agent, the final entity contact all trust agents and simply use a majority voting. One trust agent for all entities is used in a partially decentralized system, where this trust agent is not a regular peer.

The distributed calculation represents biggest risk of an unauthorized feedback manipulation. On the other hand, the direct path eliminates this risk. So we formulate following criterion:

Criterion 3: All feedbacks are spread either on direct path or by trust agents duplicated on sufficient number of entities.

Table 1 compares several TM according to the criteria defined in this section.

Table 1 Comparison of trust managements

TM	Secure P2P Layer				Information handling layer			Note
	Own ident.	Multiple ident.	Confidentiality	Anonymity	Relevant inf.	Available inf.	Feedback path	
Credence06 [2]	x	I	x	x	T	N	D	
XuHe07 [3]	x	x	x	x	A	A	TA	Partially decentr.
Lee03 [4]	x	R	x	x	T	N	D	
Eigen03 [5]	R	R	x	x	A	AD	DC	
PeerTrust04 [6]	R	R	x	x	A/T	A/AD	TA	
Lee05 [7]	R	R	x	x	F	A	TA	
Gupta03 [8]	x	x	x	x	F	A	TA	Partially decentr.
P2PRep06 [9]	x	x	x	x	F	N	D	
hiREP06 [10]	x	R	I	I	F	A	TA	No calculation
TrustMe03 [11]	x	R	I	I	F	A	TA	No calculation

Secure P2P layer: x = Do not require, R = Require, I = Implement
Relevant information: F = First-hand information, T = Transitive information, A = All info.
Information available: A = All relevant information, AD = All information but some with delay, N = System does not ensure that all relevant information are available.
Feedback path: DC = Distributed calculation, TA = Trust agents, D = Direct path.

5 Formulation and Calculation Layer

The formulation is the mathematical description of information about entity past behavior in the form enabling transformation into a trust value. The calculation describes the process of this transformation, it receives input information and produces a trust value. The formulation determines the theoretical boundaries of the system. The number of described aspects of peer behavior and accuracy of these descriptions determine the range of information available for the calculation. The high-quality formulation enables more precise and robust calculation. The calculation must also deal with false information, because all information originate from unknown entities, it is highly possible that some of them are false or misleading. This false information should influence trust as little as possible.

In our previous work [12] we propose the matrix model which provides several criteria to evaluate the calculation and formulation parts of TM. This model is basically a simulation model which evaluates characteristics of individual TM.

6 Conclusion

In this paper we define three basic criteria to evaluate trust management. It does not include all challenges faced by present collaborative applications but tries to summarize basic requirements on each layer of TM. In this context we understand the calculation layer as a fundamental part of TM and two lower layers as means of gathering reliable information. These three criteria address the requirements on these two layers to provide all necessary information for the calculation layer in a high possible quality. The next criterion should come from the matrix model proposed in our previous work and investigate the calculation layer.

We compare ten TM with respect to these criteria. The calculation layer must be evaluated by a simulation model which is currently under development. The ultimate goal is to provide methods which allows to analyze the reliability of TM.

Acknowledgements. This work was supported in part by the Czech Science Foundation (GACR), grant number 201/09/H057.

References

1. Suryanarayana, G., Taylor, R.N.: A Survey of Trust Management and Resource Discovery Technologies in Peer-to-Peer Applications, Tech. Rep. UCI-ISR-04-6 (2004)
2. Walsh, K., Sirer, E.G.: Experience with an object reputation system for peer-to-peer filesharing. In: Proceedings of the 3rd conference on Symposium on Networked Systems Design & Implementation, p. 1. USENIX Association (2006)
3. Xu, Z., He, Y., Deng, L.: A Multilevel Reputation System for Peer-to-Peer Networks. In: Proc. of the Sixth International Conference on Grid and Cooperative Computing, pp. 67–74 (2007)
4. Lee, S., Sherwood, R., Bhattacharjee, B.: Cooperative peer groups in NICE. In: IEEE Infocom, pp. 523–544 (2006)
5. Kamvar, S.D., Schlosser, M.T., Garcia-Molina, H.: The Eigentrust algorithm for reputation management in P2P networks. In: Proceedings of the 12th international conference on World Wide Web, pp. 640–651 (2003)
6. Xiong, L., Liu, L.: Supporting Reputation-Based Trust for Peer-to-Peer Electronic Communities. IEEE Transaction on Knowledge and Data Engineering 16, 843–857 (2004)
7. Lee, S.Y., Kwon, O., Kim, J., Hong, S.J.: A reputation management system in structured peer-to-peer networks. In: 14th IEEE International Workshops on Enabling Technologies: Infrastructure for Collaborative Enterprise, pp. 362–367 (2005)
8. Gupta, M., Judge, P., Ammar, M.: A reputation system for peer-to-peer networks, doi:10.1145/776322.776346
9. Aringhieri, R., Damiani, E., De Capitani Di Vimercati, S., Paraboschi, S., Samarati, P.: Fuzzy techniques for trust and reputation management in anonymous peer-to-peer systems, doi:10.1002/asi.20392
10. Liu, X., Xiao, L.: hiREP: Hierarchical Reputation Management for Peer-to-Peer Systems. In: International Conference on Parallel Processing (ICPP 2006), p. 289 (2006)
11. Singh, A., Liu, L.: Anonymous management of trust relationships in decentralized P2P systems. In: Proc. of Third International Conference on Peer-to-Peer Computing, pp. 142–149 (2003)
12. Novotny, M., Zavoral, F.: Matrix Model of Trust Management in P2P Networks. In: Proceedings of the IEEE International Conference on Research Challenges in Information Science, pp. 519–528 (2009)

Middleware Support in Unmanned Aerial Vehicles and Wireless Sensor Networks for Surveillance Applications

Edison Pignaton de Freitas, Armando Morado Ferreia, Carlos Eduardo Pereira, and Tony Larsson

Abstract. This paper presents the architecture of a middleware that provides an intelligent interoperability support to allow the integration and cooperation among Wireless Sensor Network (WSN) nodes and small Unmanned Aerial Vehicles (UAVs) implementing a surveillance system. The motivation for this study is that the cooperation among distinct types of sensor nodes to achieve common goals can notably enhance the results obtained in surveillance operations. A discussion around the requirements of such systems is also presented, supporting the design decisions and the choice of the techniques employed to develop the middleware. Finally, preliminary results are presented.

1 Introduction

A surveillance system is an application that has potential to benefit from sensor networks. Surveillance systems make use of data aggregation, fusion and analysis solutions of different kinds. This is not only related to the direct observation of the

Edison Pignaton de Freitas
School of IDE, Halmstad University, Halmstad, Sweden
e-mail: edison.pignaton@hh.se

Armando Morado Ferreia
Electrical Engineering Department, Military Institute of Engineering, Brazil
e-mail: armando@ime.eb.br

Carlos Eduardo Pereira
Institute of Informatics, Federal University of Rio Grande do Sul, Brazil
e-mail: cpereira@ece.ufrgs.br

Tony Larsson
School of IDE, Halmstad University, Halmstad, Sweden
e-mail: tony.larsson@hh.se

G.A. Papadopoulos and C. Badica (Eds.): Intelligent Distributed Computing III, SCI 237, pp. 289–296.
springerlink.com © Springer-Verlag Berlin Heidelberg 2009

phenomena themselves, but also to the awareness that is needed about the surrounding environment conditions that may affect the phenomena or its observation. The setup and adaptation of such a system poses several challenges related to the cooperation among the nodes and handling of changes that may take place in the nodes and in the network to perform the desired measurements.

The proposal presented is based on an agent-oriented approach to setup and manage surveillance missions. This is done by providing adaptation via a middleware that congregate the use of agents in order to reason about the application needs, as well as environment and network conditions, in order to accomplish the user needs.

Section 2 discusses related works. Section 3 presents the components of a surveillance system. Section 4 describes the system and the middleware conception, structure and services, while Section 5 presents preliminary results. Section 6 concludes and gives direction of future works.

2 Related Work

Agilla [1] is one of the precursors in the use of mobile agents in middleware for WSN. This approach uses agents that can move from one node to another, carrying services to be deployed on the destination nodes. It also allows multiple agents to run in the same node. In our approach the use of agents is not restricted to move services around the network but also to help in the network setup, reflection and adaptability.

In [2] an approach that uses Artificial Intelligence techniques to configure an underlying middleware is presented. This approach uses the concepts of missions and goals to plan the allocation of tasks in nodes of the network. Our approach differs because it uses agents to provide different intelligent services inside the middleware.

AWARE [3] is a project that proposes a middleware whose goal is to provide integration of the information gathered by different type of sensors, including WSN and mobile robots. Our proposal aims also at addressing heterogeneous sensors, but also concerns like QoS and QoD. Moreover, the autonomy to the sensor nodes, by using an agent-orient approach, is also a plus of our work in relation to AWARE.

3 Problem Statement

The problem investigated in this study is how to provide intelligent communication and interoperability support to sensor networks composed of low-end ground sensor nodes and small UAV carried sensor nodes applied to surveillance of large areas.

Small UAVs cannot carry the same load as normal airplanes or larger UAVs, such as Predator or Global Hawk. This affects directly the communication capabilities, as they are not able to carry much fuel or large batteries. The communication must thus be used in an efficient way, since it is a significant energy consumer. This does also restrict their operational range, constraining the cooperation possibilities among them.

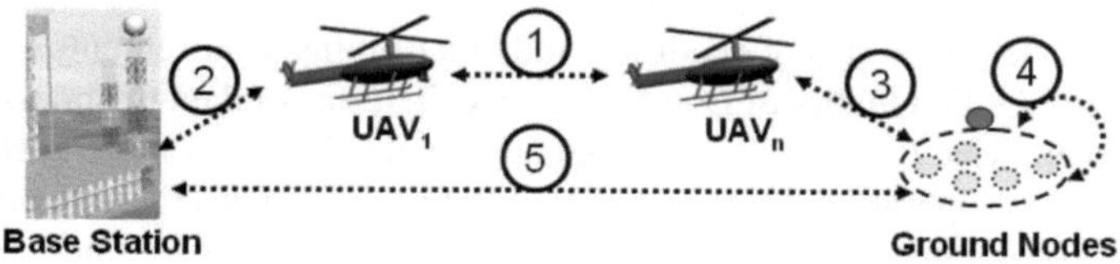

Fig. 1 Communication types among the different elements that compose a surveillance system.

In order to study the communication features required for such systems, an analysis of the types of messages that flow in the network is required. Considering first the UAVs, there are three main classes of messages, according to the type of peer node: 1) *UAV - UAV*: time critical control data, such as formation patterns and tasks assignment negotiation and sensor data exchange (e.g.: sensor fusion); 2) *UAV - Base Station*: commands, mission statements and payload data; 3) *UAV - Ground Nodes*: alarms alerting occurrence of phenomena of interest, request and reply about data produced by the low-end nodes.

Considering the low-end nodes, besides the communication that they have with the UAVs, which was already counted above, there are other two types of communications: 4) *Among Ground Nodes*: which is the classical problem studied in WSN area; and, 5) *Ground Nodes - Base Station (sink)*: which is also a classical problem in WSN research area, but in a surveillance system such as those considered in this work, an additional feature is added, as this communication may be relayed via UAVs. Figure 1 illustrates the types of messages exchanged among the different nodes, in which the numbers labelling the arrows correspond to the presented list.

UAVs fly over the surveillance area following a given pattern, alone or in teams, depending on the type of the mission. They can be sparsely distributed over the area, making possible the occurrence of temporary isolated and disconnected sub networks or nodes. This also holds for the low-end nodes, as nodes may fail, disconnecting group of nodes from the rest of the network. Moreover, new nodes may be deployed in the network. These assumptions state that nodes come and leave the network in an arbitrary way, changing the network topology very often, which requires mechanisms to overcome the problems created by this instability, such as service discovery.

The use of different sensors brings a possibility to enrich the information to be offered by the system as a whole, but it also requires their coordinated cooperative action, in order to fulfill the user requirements. This requires an intelligent behavior of the nodes in order to compose their capabilities and look at common goals, shared by all members of the network.

Real-time properties must be considered in the communication among nodes in this kind of sensor networks. Messages' priorities must be translated into priorities of the tasks that will handle them, according the current operational context, and the type of communication. Different types of messages have different priorities, for example, critical control data has higher priority in relation to sensor data report. Accordingly, this requires a scheme to handle priorities as well as a control of QoS (Quality of Service), in terms of deadlines and other timing requirements on monitoring and control. QoD (Quality of Data) assurance, in terms of data freshness and accuracy, is also an issue that must be handled, as a tradeoff based on the user requirements and the real operational conditions may take place. These issues do also require an intelligent behavior of the supporting platform, in order to tune the communication according to the context. This context awareness requires control of several parameters that crosscut different parts of the platform, from the decision making mechanisms, to the access of hardware devices.

Due to the nodes' resources constraints, memory and energy usage are concerns that must be taken into account. The first requires a minimal system to be installed in each node. However, further changes and adaptations must be possible in order to adapt and customize the system according to current needs. About energy concern, small UAVs platforms such as MBL Trochoid [4] have limited resources that must be carefully used, e.g. having energy autonomy to fly only 18 minutes.

4 System Description

A main idea of the surveillance system proposed is to make it easier for a user to define a mission and setup a VANET and a WSN as to accomplish it. For this purpose a Mission Description Language (MDL) is used to specify missions at a high-level of abstraction; without having to bother about details, such as the choice of sensors that will handle a given mission. For more about MDL, readers are referred to [5]. The directions established in a mission are entered via a user interface (Mission Specification Console), which uses knowledge about the application domain (via a Domain Specific Database) and information about the network (via a Deployed Resource Description) to translate the MDL statements into a formal specification of the mission.

MDL specifications are interpreted within an application framework and act as application programs that run on the network nodes. A mission can be broken down into a set of node-missions (sub-missions that can be assigned to individual nodes or group of them), depending on the mission complexity, and then this set is sent to the network via mobile-agents, called mission-agents that carries and represents a mission or node-mission. When receiving a set of node-missions, the nodes autonomously decide which node-mission each one will perform as to accomplish the mission as a whole. In each node, the node-missions, in effect representing the applications from the nodes' point of view, will run on top of a middleware. Figure 2(a) presents this overall scenario, while part (b) presents the middleware layers that will be discussed further.

4.1 Middleware Architecture

Figure 3 below presents the middleware architecture conception, in which ovals represent services and functions, the rectangle represent the reasoning mechanism,

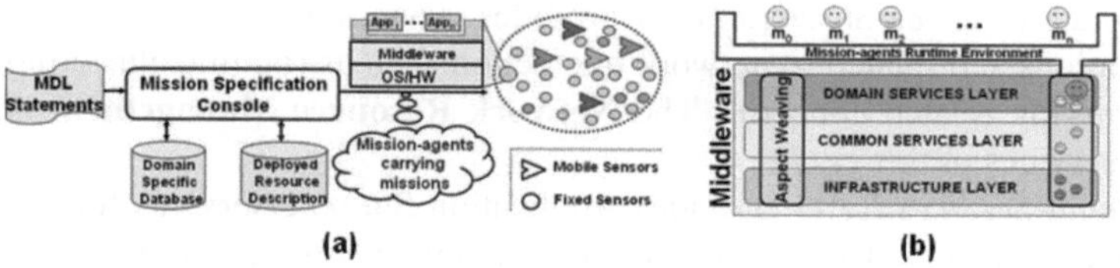

Fig. 2 (a) Overall system architecture. (b) Middleware Layers.

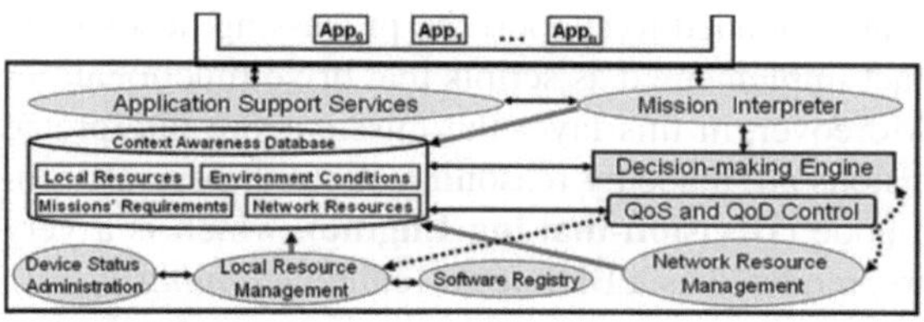

Fig. 3 Middleware conception overview.

the cylinder represent the data base with the pertinent information that is used by other elements of the middleware, and the rounded corners rectangle represent the non-functional requirements that affect the other elements.

Local Resource Management handles node resources in terms of their installed software as well as usage, status and conditions of sensor, energy and communication devices. The **Network Resource Management** handles the network conditions and the use of shared network communication resources. The information provided by these two elements is stored in the **Context Awareness Database**, which receives also information from the **Mission Interpreter** about the requirements of the running applications that represent the missions. The **Decision-making Engine** is responsible for reasoning about the information contained in the database about the context (local and network resource plus the environment conditions) in order to meet mission requirements. The **Application Support Services** offer the necessary means to implement the actions described in the missions. The **QoS and QoD Control** provides monitoring and adaptation of the elements in the middleware in order to comply with the required quality in the service and data expressed by the user needs, which come with the mission requirements.

4.2 Middleware Services

In order to architecturally organize the middleware presented above, it is divided in three layers, according to functional goals in each of them; see Figure 2(b).

The bottom layer is called *Infrastructure Layer*. It is responsible for the interaction with the underlying operating system and hardware devices. This layer encapsulates the **Local Resource Management**, by handling local resources.

The intermediate *Common Services Layer* provides services that are common to different applications and related to network interaction such as routing, clustering and other networking related concerns. The **Network Resource Management** is handled by the services of this layer.

The top *Domain-Services Layer* provides Application Support Services for domain specific needs, such as data fusion and data semantic support, in order to allow production of application-related information from raw data processing. Fuzzy classifiers, special kinds of mathematical filters (e.g. Kalman Filter) and functions that can be reused among different applications in the same domain are found in this layer. These functionalities are intended to support the processing needs from the applications, which are in fact implemented as scripts that drive functionalities provided by the middleware. Moreover, in this layer the missions are interpreted (**Mission Interpreter**) and decisions are made by reasoning about the overall conditions of the network and the node (**Decision-making Engine**), which is a very important part of the middleware, modeled as a BDI agent (called *planning-agent*), which has its beliefs feed by the information contained in the **Context Awareness Database**. For more details about this subject, readers are referred to [5].

One of the system requirements is that the middleware must fit in resource constrained nodes. The use of a component-based development and mobile-agents (service-agents) addresses this need. By using components, a middleware product-line is implemented, from which customized sensor node specific variants of the middleware can be provided. The service-agents provide capability to add middleware services at runtime; more details and an example are found in [6].

Non-functional requirements, related to QoS and QoD, may affect different middleware services hosted in different layers. They thus crosscut the middleware spreading their handling mechanisms. Due to their crosscutting characteristics, they are addressed by means of aspects that concentrate their handling in a modular and scalable way. For more details about the specific real-time middleware handling, readers are referred to [7].

4.3 Middleware Kernel

In each node, a minimal set of middleware services is installed. It provides the minimal intelligence and interoperability that a node needs in order to be integrated into the network. This set is called the *middleware-kernel* (or just kernel) and is composed by components that may be weaved by aspects at static or dynamic time, according to the non-functional crosscutting requirements that affect them.

The kernel presence in the *Domain Services Layer* is minimal (having only the *planning-agent* and the **Mission Interpreter**), as the services that will compose this layer will be tailored according to the node. In the *Common Services Layer*, the kernel presents the following services: (1) *Link Metric*; (2) *Clustering*; (3) *Routing and Messaging*; (4) *Service Discovery*; and (5) *Neighbor List*. In the *Infrastructure Layer* of the middleware, the following components are found: (1) *Clock and Timers*; (2) *Scheduler*; (3) *Memory Manager*; (4) *Service Registry*; (5) *Device Resource Manager*; and (6) *Operation Mode Manager*.

5 Simulation Results

Simulations of the mission dissemination mechanism using the mission-agents mentioned in Section 4 were performed using ShoX [8], which is a Java-based wireless network simulator. The goal of these simulations was to assess how effective the mission dissemination by agents is, using the proposed middleware support. This is possible to be assessed by comparing the actual number of engaged nodes to perform the mission with the optimum value.

For the performed simulations, total of 20 runs, the used setup modeled a network of 8000 nodes randomly distributed on an area with dimensions 5Km x 5Km, in which 2000 are able to perform a mission, which means that these nodes have the sensing capabilities required for that mission. For simplicity, the stated mission was small enough in order to be handled by a single and tiny mission-agent that fits in a communication packet of the IEEE 802.11b, which was the standard used in the performed simulations. Figure 4(a) presents a sample of the nodes' distribution.

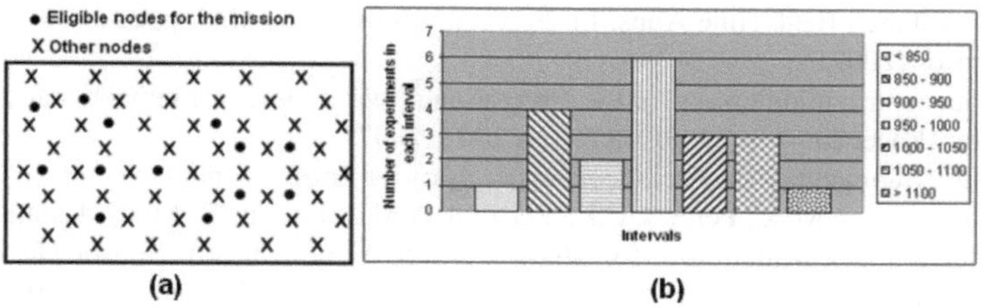

Fig. 4 (a) Sample of the nodes distribution. (b) Results from the experiments divided by intervals.

Figure 4(b) shows the distribution of the simulation runs by the intervals that represent the number of nodes engaged in the mission. It is possible to observe that in the majority of the runs stayed in the intervals close to target optimum value, which is of 1000 nodes. The worst results presented values around 15% from the optimum one, but as it is possible to observe, they were just two or 10% of the 20 runs. Besides the information presented in the figure, the average number of engaged nodes for the presented set of simulations were 970, which is very close to the target value, having a standard deviation of 77.9 nodes, which is a good result.

6 Conclusion and Future Directions

This work presented middleware architecture to support interoperability among heterogeneous small UAV carried sensor nodes and constrained ground sensor nodes which compose an area surveillance system. A detailed discussion around the requirements in terms of cooperation and communication among the nodes were presented. The design rationale of the middleware was presented, being related with the fulfillment of the requirements discussed in Section 3. Finally, the middleware structure and services were outlined and preliminary results presented.

Simulation of the other middleware features are being conducted, as well as the study of how integrate aspect-oriented handling in the simulations in order to provide the evaluation of all features together, before the final implementation.

Acknowledgements. Edison Pignaton de Freitas thanks the Brazilian Army for the grant to follow the PhD program in Embedded Real-time Systems at Halmstad University in Sweden, in cooperation with Federal University of Rio Grande do Sul in Brazil.

References

1. Fok, C.-L., Roman, G.-C., Lu, C.: Rapid development and flexible deployment of adaptive wireless sensor network applications. In: Proceedings of the 24th ICDCS 2005. IEEE, Los Alamitos (2005)
2. Schmidt, D.C., et al.: A Decision-Theoretic Planner with Dynamic Component Reconfiguration for Distributed Real-Time Apps. In: Proceedings of 8th ISADS, pp. 461–472. IEEE, Los Alamitos (2007)
3. Gil, P., et al.: Data centric middleware for the integration of wireless sensor networks and mobile robots. In: Proceedings of 7th ROBOTICA 2007 (2007)
4. MSB Co. web site. Submeter-scale aircraft, `http://spyplanes.com`
5. Freitas, E.P., Wehrmeister, M.A., Pereira, C.E., Ferreira, A.M., Larsson, T.: Multi-Agents Supporting Reflection in a Middleware for Mission-Driven Heterogeneous Sensor Networks. In: Proceedings of 3rd Agent Technology for Sensor Networks Workshop (2009)
6. Freitas, E.P., Wehrmeister, M.A., Pereira, C.E., Larsson, T.: Reflective middleware for heterogeneous sensor networks. In: Proceedings of 7th Workshop on Adaptive and Reflective Middleware (ARM 2008), pp. 49–50. ACM, New York (2008)
7. Freitas, E.P., Wehrmeister, M.A., Pereira, C.E., Larsson, T.: Real-time support in adaptable middleware for heterogeneous sensor networks. In: Proceedings of International Workshop on Real Time Software (RTS 2008), pp. 593–600. IEEE, Los Alamitos (2008)
8. Lessmann, J., Heimfarth, T., Janacik, P.: ShoX: An Easy to Use Simulation Platform for Wireless Networks. In: Proceedings of 10th ICCMS, pp. 410–415. IEEE, Los Alamitos (2008)

SimRad.NBC – Simulation and Information System for Rescue Units at CBRN Disasters

Karin Rainer, Nadine Sturm, Stefan Schönhacker, and Gerhard Chroust

Abstract. The importance of effective response to chemical, biological,radiological and nuclear (CBRN) dangers is growing constantly. These hazards are not directly detectable by human senses and thus no inborn reactions exist. As a consequence, special tools to detect these dangers have to be utilized. Although some "simulation games" try to portray certain levels of management of CBRN incidents, there is a substantial lack of realistic and holistic training instruments for First Responders. The project SimRad creates the basis for a user centered, systematic technology for training and communication in emergency situations using a process view. This includes real life Best Practices of effective management of CBRN crises and First Responder education. It provides the foundation for near-realistic simulations for training of individual subprocesses and communication methods by the use of software intensive simulation technologies such as Mixed Reality.

1 Motivation and Objectives

SimRad.NBC (Simulation- and Information system to manage Rescue units – with focus on NBC threats creates the foundations for satisfying the current user needs for practice oriented simulation and a communication framework for First Responders in CBRN emergency scenarios. At the moment, practically no systems exist which provide a near-reality training for all emergency responders from different

Karin Rainer · Nadine Sturm
Research Institute of the Red Cross, Nottendorfer Gasse 21, 1030 Vienna, Austria
e-mail: {karin.rainer,nadine.sturm}@w.roteskreuz.at

Gerhard Chroust
Institute of Telecooperation, J. Kepler Univ. Linz, Altenbergerstr. 69, 4040 Linz, Austria
e-mail: gc@sea.uni-linz.ac.at

G.A. Papadopoulos and C. Badica (Eds.): Intelligent Distributed Computing III, SCI 237, pp. 297–303.
springerlink.com © Springer-Verlag Berlin Heidelberg 2009

fields and organizations. Though there is a certain amount of "simulation games" available for tactical disaster management (like 'Emergency' or 'Emergo Train') First Responders still revert to cost intensive and thus infrequent real life training exercises. This is necessary to get the required practice in handling tools, setting correct measures and to assess the situation and its potential dangers. This paper extends the ideas and concepts presented in [3].

The first major step in the project was the scientific investigation of pre-existing international studies to generate relevant input for the selection of apt technical approaches. This literature analysis finally led to a matrix of different factors with positive, negative impacts regarding the direction and the loading they have – even considering both effects on the trainee respectively the First Responder. The raw data was transformed to a library of parameterized modules of appropriate simulation tools for the training of emergency responses, allowing validation and flexible use of combinations and interferences. Those outputs led to a set of generally applicable practices and guidelines which were compiled to a set of operational rules.

The valuable input of end users, their suggestions and ratings were implemented via a series of expert interviews. Experienced trainers as well as trainees involved in CBRN response were asked for their assessment of emergency communication, influences on the First Responders and current models of training and practice in respect of simulating "invisible" CBRN dangers. New and astonishing findings (to be published on the SimRad Web page `www.simrad.at` in summer 2009) add on the preliminary literature research and reveal the right path to the aims of the project. In the course of the analysis similarities and differences between the different groups of First Responders are revealed and show the overall needs of all response organizations. To grant this access the consortium of SimRad.NBC consists of the Schwechat Fire Department, the CBRN-Department of the Red Cross Upper Austria and the Research Institute of the Red Cross together with two SME R&D-partners and a university institute specialized in systemic and holistic aspects. As a last preliminary step a holistic and systemic analysis of established tools for simulating common emergency scenarios is conducted.

Based upon these three pillars – the matrix of factors and influences, the interview series and the simulation state of the art – a final collection of the most practicable, useful and effective technologies for an end user centered simulation and information system for rescue units at CBRN disasters is generated.

2 The SimRad Project

Following major work packages are identified within the SimRad project:

- Aggregation: Collecting experiences and Best Practices from end users with core competency in CBRN response, modeling and expressing the results in a uniform way.

- Evaluation: Different Best Practices must be verified and validated on their own merit in relation to each other. Checking and optimizing the compatibility of individual Best Practices is of key importance.
- Education/Training: Once the set of Best Practices – including strategies – is established and simulation models are developed, the appropriate training supported by simulation technology will lead to a higher level of preparedness for CBRN emergencies. Consequently the suitability and effectiveness of the performed actions are meant to rise.
- Use in actual emergencies: Strategic simulation models might also be utilized for management decisions in real CBRN emergencies. Possible outcomes and consequences of decisions and actions can be pre-estimated.

These four work packages interact with and support each other. Continuous re-evaluation will be necessary both in improving training and real experiences and due to emergencies and new threat types or scenarios.

3 Simulation Types

Different types of simulation seem fit for this purpose. They differ in cost, technical complexity, simplicity of application as well as in understanding and effectiveness for training. Some key types are mentioned in [3].

First Responders interventions can be seen as combinations of intertwined processes which can be split into individual subprocesses. As an important option for training, individual subprocesses can be substituted by simulated ones. For such a simulation dynamic models are necessary which can be animated, executed or enacted. There are three software-intensive modeling techniques available:

- Complex Mathematical Simulations: abstracted Virtual Reality, supported by System Dynamics. Models can be used to show long-term effects of certain emergencies and corresponding counter-measures such as distribution of dangerous pollutants, the impact of certain types of contamination or its duration.
- Virtual Reality: environments without correspondence to a real environment [1, 5, 6, 7], e.g. Second Life where all users are represented by virtual figures ("avatars"). The advantage is that it allows virtual interaction and training of many participants.
- Mixed Reality: a combination of reality and simulation, e.g. Augmented Reality (AR) [2, 4, 8]. AR deals with the combination of real-world and computer-generated data, often with the help of translucent glasses to overlay images over a real world scenario (see figure 1). It offers the aspects of a field experiment but allows introducing aspects which do not exist in reality. Thus, it seems to be one of the most promising approaches for training purposes because actions can be taken in a real surrounding, supported by modern technology.

Figure 1 shows an ocular which is able to detect contaminations in a landscape and to give additional information to First Responders. Temperature and data about

Fig. 1 Augmented Reality – adding information in the glasses

radiation are "projected" on the glasses and enrich the real surrounding with additional overlapping information.

Each of the technologies describes some aspects of an emergency situation and allows to explain and to train useful behavior for First Responders. We can observe that the discussed models and related simulations are increasingly enriched with growing similarity to realistic situations and the level of user involvement. Referring to the survey analysis of the expert interviews, the most realistic training environments still seem to be field experiments and Augmented Reality settings. It should also be mentioned that more complex scenarios – and technologies – not necessarily imply a higher learning effect, as users might be distracted by secondary effects.

4 The SimRad Simulation Framework

The Best Practices collected in the course of the project are evaluated practically in training settings. Through the plug-in technology of SimRad the exchange of different simulation tools for specific First Responder groups is enabled. Component-based development – already effectively and usefully adopted in software development – is also used for selecting and combining different technological approaches into a holistic simulation and communication tool. For example simulation of very basic actions such as contamination and decontamination of material and persons will be provided as well as the strategic deployment and management of emergency units on a larger scale. The practical necessities to implement these different levels of disaster management will be taken into account and manifested in different hardware and software components.

The plug-in technology of SimRad will make it possible to exchange modules for different First Responder groups, using the same look-and-feel and operational handling of the simulation framework. Several types and levels will be taken into account in order to answer divergent user needs:

Fig. 2 SimRad plug-ins

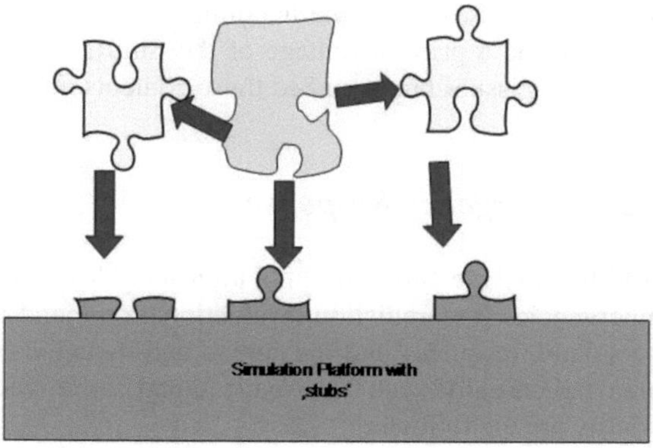

- simulation of (future) events including feedback loops
- exploration, evaluation and verification of technical dependencies (e.g. interplay/interference/compatibility of tools)
- exploration of important influence factors affecting First Responders (e.g. stress, poor visibility, noise, heat)
- experiments with emerging, integrated systems to train and support First Responders.
- combination of tactical and practical issues via plug-ins.

A basis for the selection and further development in future steps of the SimRad framework is a common, standardized interface for all plug-ins. This makes the plug-ins exchangeable and cooperative and simplifies upgrades for tackling evolving dangers etc.

5 Implementation

5.1 Training

Beyond the

- Reality-near training can be conducted without endangering the trainees through hazardous materials or situations by substituting dangerous subprocesses by harmless, simulated equivalents.
- Low running costs allow intensive and regular training sessions.
- The repeatability of training situations enables experts and experienced peers to feedback the performance of the trainee independent of spacious and timely restrictions. The complete recording of all details, facts and data from the training session simplifies the feedback process and also provides the basis for self evaluation – often preferred in vocational training.

- Sub-processes – like the handling of tools – will be supported by simulations. It will be a prior advantage of the SimRad-technology not to replace real life training issues but to embed those contents into simulated reality – or vice versa.

5.2 Emergency Response

Additionally the SimRad simulation modules and tools can also be used in real emergencies for prediction, evaluation, command and control functions. Data on personnel, resources and the spread and dynamics of hazards but also of affected area and casualties can be visualized and communicated. The reasons for this possibility are the following:

- The common, standardized interface and plug-ins of SimRad allows testing the compatibility of current/new (Best?) Practices. This can open the way for a more dynamic development of emergency procedures which was hitherto hampered by tradition and scarce resources.
- Visualization of subtle influences in the course of training and in real missions can support the First Responders but also reveal new models and moderators that could interact and thus affect the response.

5.3 Further Benefits – Unity via Variety

Further development steps of the selection and analysis of SimRad.NBC will result in various benefits:

- high acceptance by end users due to a manageable competitive mix of new technologies and traditional well proven tools
- large scope of potential applications through the adaptability on different levels of abstraction as well as
- great scale of possible scenarios which allows the implementation of new evolving dangers
- possibilities to cover the needs of different response organizations and their specific fields of action, tools and tactical characteristics

6 Summary and Outlook

Up to now no near-reality simulation system existed which enabled the training of larger, heterogeneous groups of First Responders, beyond the available training in individual, regional organizations in specific areas. SimRad provides a solid technological and scientific basis to develop a user centered simulation and communication tool for First Responders in CBRN emergencies. Thus the needs and current Best Practices of different organizations were surveyed and analyzed.

A key in this project is the possibility to model and analyze human and environmental factors concerning CBRN emergencies. Based on the generated models modern ICT can be utilized to create flexible and practical simulations. Most of the simulations will be performed by using Virtual and Mixed Reality environments. As a consequence it will be possible to evaluate the efficiency of existing training and emergency plans and to optimize them. Thus SimRad will contribute to the improvement and structure of First Responders training as well as to missions and will allow quantifiable continuous enhancement of the efficiency and efficacy of First Responders.

While SimRad.NBC provides the foundation, a successor project, SimRad.COMP, starting in November 2009, will subsequently develop further steps to create feasible pre-prototypes for a simulation and communication tool-package for recognizing and reacting to the challenges of the "invisible CBRN dangers". Both projects are supported by the Austrian Federal Ministry for Transport, Innovation and Technology (BMVIT) within "KIRAS", the Austrian security research programme.

Acknowledgements. The work reported is funded by the project "KIRAS PL 2: SimRad.NBC", project no. 813798 from November 2007 under the"Sicherheitsforschungs-Förderprogramm KIRAS" of the Austrian Federal Ministry for Transport, Innovation and Technology (BMVIT).

References

1. Billinghurst, M., Kato, H.: How the virtual inspires the real – collaborative augmented reality. CACM 45(7), 64–70 (2002)
2. Chroust, G., Hoyer, C.: Bridging gaps in cooperative environments. In: Hofer, C., Chroust, G. (eds.) IDIMT 2004, 12th Interdisciplinary Information Management Talks, Budweis, September 2004, pp. 97–110. Verlag Trauner Linz, Linz (2004)
3. Chroust, G., Roth, M., Ziehesberger, P., Rainer, K.: Training for emergency responses – the simrad-project. In: Balog, P., Jokoby, B., Magerl, G., Schoitsch, E. (eds.) Mikroelektroniktagung ME 2008, Vienna, VE, October 2008, pp. 327–334. st. Verband für Elektrotechnik (2008)
4. Fleischmann, M., Strauss, W.: Linking between real and virtual spaces: building the mixed reality stage environment. In: Proc. 2nd Australasian User Interface Conference (AUIC 2001). IEEE Publishing, Los Alamitos (2001)
5. Ramesh, R., Andrews, D.H.: Distributed mission training, teams, virtual reality, and real-time networking. CACM 42(9), 64–67 (1999)
6. Rheingold, H.: Virtuelle Welten – Reisen im Cyberspace. Rowohlt Hamburg (1992)
7. Stone, R.: Virtual reality and telepresence. Robotica 10, 461–467 (1992)
8. Tarumi, H., Morishita, K., Ito, Y., Kambayashi, Y.: Communication through virtual active objects overlaid onto the real world. In: Proceedings of the 3rd International Conference on Collaborative Virtual Environments, pp. 155–164. ACM Press, New York (2000)

Distributed Goal-Oriented Reasoning Engine for Multi-agent Systems: Initial Implementation*

Mihnea Scafeş and Costin Bădică

Abstract. This paper describes the sample implementation of a distributed goal-oriented reasoning engine for multi-agent systems. The paper summarizes part of the design and programming issues that we addressed for providing the initial prototype with customization and self-configuration facilities.

1 Introduction and Background

Complex problem solving, like for example emergency response for disaster management, requires inter-organizational collaboration of multiple and possibly geographically distributed stakeholders, usually assisted by appropriate tools. In such a scenario stakeholders are typically endowed with local processing capabilities and problem solving skills that need to be coherently orchestrated into a cooperative process for efficient problem solving.

In this paper we propose the design and implementation of a distributed goal-oriented reasoning engine that supports flexible dynamic configuration of processes for complex problem solving. Our solution can be described as *agent-based*, i.e. software agents are the main building blocks that: (i) integrate the necessary entities (humans, information sources and specific problem solving tools); and (ii) provide the basic support for autonomous problem solving using goal-oriented reasoning.

When agents cannot successfully achieve goals by themselves, they need to use capabilities of other agents and must ask these others to adopt their goals. We approached this goal delegation/adoption process using dynamic service discovery and negotiation, thus providing our system with self-configuration capabilities, decentralized control and autonomous management of resources.

Mihnea Scafeş · Costin Bădică
University of Craiova, Software Engineering Department, Bvd.Decebal 107, Craiova, 200440, Romania
e-mail: {scafes_mihnea,badica_costin}@software.ucv.ro

* This research is funded by the EC FP7-ICT project DIADEM, ref. no: 224318.

G.A. Papadopoulos and C. Badica (Eds.): Intelligent Distributed Computing III, SCI 237, pp. 305–311.
springerlink.com © Springer-Verlag Berlin Heidelberg 2009

This paper is focused on introducing the design and implementation of a prototype that was initially tested on fragments of utilization scenarios developed within the DIADEM project [5]. In our previous work we considered an utilization scenario from the requirements document [4] and we analyzed this scenario using agent-oriented methodologies [6] to produce initial sets of agents, plans and goals [10].

The paper is structured as follows. We start with the design approach that leads to the definition of goal-oriented agents with goals, plans, tasks and services. We follow with internal agent design. Finally we conclude and point to future works.

2 Goal-Oriented Agent Behaviors

The concept of goal is commonly agreed in organizational and multi-agent modeling. A goal is described as a property of a desired state-of-affair that the system as a whole would like to achieve. Usually goals are captured in the early stages of the system specification [6].

The main steps of our design approach naturally lead to goal-oriented agents [10]: (i) identification of agent types based together with system goals and sub-goals; (ii) assignment of goals to agents; (iii) definition of goals as plans composed of actions, sub-goals and tasks; (iv) detailed design and implementation of agents.

Achievement of agent goals proceeds in top-down fashion by application of plans. A plan consists of: (i) a plan header containing the goal to which the plan is assigned, optionally qualified with a contextual condition, and (ii) a plan body consisting of a sequence of plan elements. A plan element can be a goal (actually a sub-goal of the goal to which the plan is assigned), an action or a task. The contextual condition must evaluate to true before the plan can be executed.

Actions represent agent capabilities to act and change the environment state.

The view of goal delegation on the requester agent side is called *task*, while the view of goal adoption on the provider agent side is called *service*. Tasks model situations when agents cannot successfully achieve goals by themselves and consequently they need to use capabilities of other agents and must ask these others to adopt their goals. Assignment of tasks to services is subject to a one-to-many negotiation process.

The distinction between tasks and services is important for at least two reasons: (i) there might be several agents in the system that provide the same service, i.e. that are able to realize the same task; (ii) we cannot assume that a service provider agent will automatically adopt a task requested by a requester agent; for example the service agent might be overloaded or it might even consider that pursuing the required task is inappropriate given the current context.

A set of goal-oriented agents that interact, negotiate and cooperate to solve a complex problem define a distributed goal-oriented reasoning engine. In the following we provide internal design and implementation details of goal-oriented agents.

An architectural model for goal-oriented agents that supports agent programming using plans and goals is BDI – Belief-Desire-Intention [9]. We are aware of the fact that implementations of the BDI model already exist: Jadex [7] and Jason [8], [2].

Note that we have chosen to implement our own version of goal-oriented agent architecture rather than using Jadex or Jason for the reason of flexibility during the initial stage of the DIADEM project. More precisely, we preferred to have our own design and implementation that we can easily control to fit project's requirements, rather than spending too much time trying to figure out how to configure Jason and/or Jadex to suit our needs. Nevertheless, we plan to spend more effort in the future for comparing our approach with those of Jadex and Jason.

3 Agent Architecture

The internal architecture of our agents is tailored for the implementation using Cognitive Agent Architecture – Cougaar [3] open-source agent platform. We chose Cougaar because of its scalability, robustness and high configurability features.A key aspect of Cougaar agents is that they are plugin-based – they are composed of plugins that communicate efficiently by means of a blackboard architecture.

Our prototype defines two types of agents: (i) *Stakeholder* agents – generic agents that represent various stakeholders involved in the problem solving process; (ii) *Yellow Page* (i.e. *YP*) agents – keep track of services provided by *Stakeholder* agents.

Stakeholder agents are composed of 3 plugins: (i) *ReasoningPlugin* – hosts the plan execution engine; (ii) *TaskManagerPlugin* – deals with management of tasks; (iii) *DirectoryClient* – communicates with *YP* agents. *YP* agents are composed of a single plugin – *Directory Server. Stakeholder* plugins cooperate to achieve goals using plans and task contracting (whenever a goal cannot be achieved by the agent itself and it must be transferred to another agent).

Reasoning Module. A *Stakeholder* agent contains a *reasoning module* responsible with plan execution – *ReasoningPlugin* plugin.

The reasoning module hosts the plan execution engine, an object called *reasoning manager* that can be easily specialized. The current version takes lists of plan elements and sequentially accomplishes them (see the algorithm in Figure 1). In particular, in case of: (i) a goal, it tries to solve that goal; (ii) an action, it executes the action; (iii) a task, it waits until another agent contracts the task (by adopting it as a goal) and solves it. Before starting a goal, the reasoning manager tries to find a plan associated to it. If it does not find a proper plan then it treats the goal as a task that should be delegated to another agent. Note that the current version supports only a single plan element in execution at any time during a reasoning process. We plan to address this restriction in the near future by developing new reasoning managers capable to support more complex forms of execution of plan elements.

A reasoning process is started when a top level goal is triggered (through an event for example). A unique identifier that we call *reasoning instance* is associated with this process. The reasoning instance is shared by agents when they contract tasks and send back task results, therefore creating a distributed reasoning process.

The *ReasoningPlugin* is an interface between the reasoning manager and the *TaskManagerPlugin*. It uses the reasoning manager to trigger goals (execute

```
TRIGGER-GOAL(goalName, reasoningInst)
if (not reasoningInst in execution)
    create new reasoning instance reasoningInst associated to goalName
    execute(reasoningInst)
```

```
EXECUTE(reasoningInst)
currentElem(reasoningInst) ← findNextElement(currentElem(reasoningInst))
while (currentElem(reasoningInst))
        if (currentElem(reasoningInst) is-a goal)
            plan ← findProperPlan(currentElem(reasoningInst))
            if (plan is null)
                delegate(currentElem(reasoningInst)
                return
            else
                for element in plan
                    schedule(element)
        else if(currentElem(reasoningInst) is-a action)
            execute(currentElem(reasoningInst))
        currentElem(reasoningInst) ← findNextElement(currentElem(reasoningInst))
```

Fig. 1 Sequential reasoning algorithm

reasoning instances), creates task contracting requests that will be carried out by the *TaskManagerPlugin* and informs the *TaskManagerPlugin* whenever a reasoning process has been completed. When the reasoning plugin starts, it loads plans, goals and actions descriptions with the help of a *reasoning manager provider* object. In the current prototype there is a particular reasoning manager provider that loads and configures a reasoning manager based on a declarative description known as *process ontology* from an XML file.

Task Manager. The *TaskManagerPlugin* is responsible for delegating tasks to other agents. This process assumes one-to-many negotiation of the agent that is delegating the task with the agents that provide services capable to execute the task. Currently, this plugin: (i) uses Contract Net Protocol (CNET, [11]) to negotiate for tasks, (ii) informs the reasoning plugin to start contracted tasks (on the contractor side) and (iii) returns task results to negotiation initiator agents (on the manager side). We are using original terminology for the description of roles during CNET negotiation: (i) *manager* is the agent that initiates the negotiation, that has a task to offer; (ii) *contractor* is an agent interested to adopt the goal of executing the task.

The *TaskManagerPlugin* uses a *task manager* object to manage active negotiations. The task manager has been gives basic strategy functionalities for CNET, with the help of *Contractor* and *Manager* objects. A *Contractor* contains the logic to decide whether an agent should engage in a negotiation for a task and to create bids. A *Manager* object contains the logic to choose the contractor to which the contract is awarded from all contractors that submitted bids in a given negotiation.

The *TaskManagerPlugin* is configured to use a specific *task contracting provider* object for loading the task manager during the startup. A specific implementation is currently provided to create and set up a task manager based on a *task ontology* from an XML configuration file. The *task ontology* specifies tasks and the implementations of their associated *Contractor* and *Manager* objects.

Plugins Interaction. The complete interaction between the reasoning manager and the task manager is depicted in figure 2. We have taken into account only the current version of *reasoning manager*. While reading the diagram, keep in mind that the reasoning manager is part of the *ReasoningPlugin*.

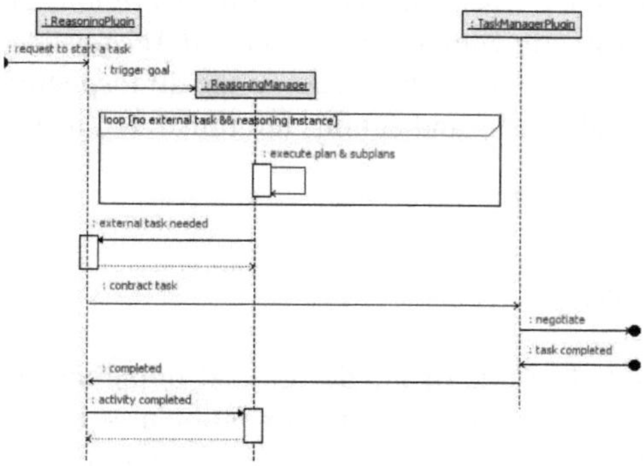

Fig. 2 Interaction between reasoning module and task manager

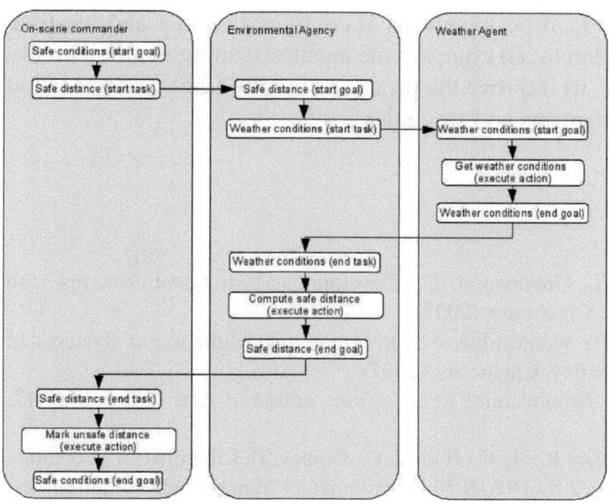

Fig. 3 Example of a reasoning instance

Whenever a request to start a top-level goal arrives at the reasoning plugin (for example when an event handling routine decides that a goal should be triggered, or through a GUI), the reasoning plugin asks the reasoning manager to trigger the goal. Consequently, a new reasoning instance is started and executed until the goal is achieved. Whenever a task is found, the reasoning manager blocks the reasoning instance and transfers control to the *TaskManagerPlugin*, which will award the task to a potential contractor. As soon as the delegated task is completed, *TaskManagerPlugin* informs the *ReasoningPlugin*, which in turn notifies the reasoning manager that a delegated task has been completed. The reasoning manager resumes the reasoning process. This process is repeated until the top-level goal of the reasoning instance is achieved and the reasoning instance is completed. Figure 3 shows an example of a reasoning instance. An *On-scene commander* wishes to achieve *Safe conditions* at the location of an incident and interacts directly with an *Environmental Agency* that has knowledge about chemical substances (e.g. how they react to the atmosphere) and indirectly with a Weather Agent. Goals, tasks and actions are illustrated using rectangles.

4 Conclusions and Future Work

In this paper we have presented our initial results of the design and implementation of a distributed goal-oriented reasoning engine for complex problem solving. The focus of the work was set on application of pragmatic programming practices for producing an initial working prototype. This prototype was tested on part of a sample scenario for solving an emergency response problem. The highly dynamic and exploratory nature of the problem domain was addressed by providing the initial prototype with customization facilities by means of configuration files and run-time classes. As future work we plan to: (i) compare the agent reasoning model with existing BDI implementations; (ii) improve the prototype by providing customization facilities of negotiation mechanisms and strategies.

References

1. Bellifemine, F.L., Caire, G., Greenwood, D.: Developing Multi-Agent Systems with JADE. John Wiley & Sons, Chichester (2007)
2. Bordini, R.H., Hübner, J.F., Wooldridge, M.: Programming Multi-Agent Systems in AgentSpeak Using Jason. Wiley, Chichester (2007)
3. Cougaar: Cognitive Agent Architecture, `http://www.cougaar.org` (Cited May 17, 2009)
4. Damen, D., Pavlin, G., Van Der Kooij, C., Bădică, C., Comes, T., Lilienthal, A., Fontaine, B., Schou-Jensen, L., Jensen, J.S.: DIADEM Environmental Management Requirements Document, Issue 1.12.0 (2009)
5. DIADEM: Distributed information acquisition and decision-making for environmental management, `http://www.ist-diadem.eu` (Cited May 17, 2009)

6. Henderson-Sellers, B., Giorgini, P.: Agent-oriented Methodologies. Idea Group Publishing, USA (2005)
7. Jadex: BDI Agent System, `http://jadex.informatik.uni-hamburg.de` (Cited May 17, 2009)
8. Jason: A Java-based interpreter for an extended version of AgentSpeak, `http://jason.sourceforge.net` (Cited May 17, 2009)
9. Rao, A.S., Georgeff, M.P.: BDI-agents: from theory to practice. In: Proceedings of the First Intl. Conference on Multiagent Systems – ICMAS 1995, USA, pp. 312–319 (1995)
10. Scafeş, M., Bădică, C.: Preliminary design of an agent-based system for human collaboration in chemical incidents response. In: Ultes-Nitsche, U., Moldt, D., Augusto, J.C. (eds.) Proc. of 7th Int.Workshop on Modelling, Simulation, Verification and Validation of Enterprise Information Systems (MSVVEIS 2009), pp. 53–63. INSTICC Press (2009)
11. Smith, R.G.: The contract net protocol: High-level communication and control in a distributed problem solver. IEEE Transactions on Computers 29(12), 1104–1113 (1980)

Global Military Conflict Simulator

Claudiu Tanase and Andreea Urzica

Abstract. The Global Military Conflict Simulator is an application that allows the unfolding of a large virtual war in real time, involving hundreds of thousands of soldiers, with a virtual Earth serving as battlefield. The simulation attempts to achieve unprecedented scale and scope by providing a global map, satellite imagery and elevation coverage through cloud computing, web mapping technologies, providing a flexible real time simulation engine capable of managing a huge number of units and replicating the organization of an actual army through the use of a hierarchical multi-agent system. This paper covers the context, general architecture of the system and the multi-agent system organizational model.

1 Introduction

The advancements in World Wide Web technology, as well as the increased availability of geographic data and images have enabled the development of web mapping, defined as "the process of designing, implementing, generating and delivering maps on the World Wide Web"[10]. Desktop software applications that make use of web mapping, such as virtual globes, have become increasingly popular as a result of their functional aspects (e.g. route planning, local search) and their worldwide addressability. Two examples of freely available virtual globes are NASA WorldWind and Google Earth[1, 7].

It is a known fact that military training comprises of computer simulation in order to test scenarios and put the trainee in a wider array of situations, with much smaller cost than real military exercises. The simulation scenario, as in VBS1 for example, is generally targeted to the trainees needs, and not focused on the entire military as a whole. In order to guarantee the realism of the simulation, artificial

Claudiu Tanase · Andreea Urzica
University Politehnica of Bucharest, Faculty of Automatic Control and Computers, Computer Science and Engineering Department, Splaiul Independentei 313, Bucharest, Romania
e-mail: claudiu.tanase@cs.pub.ro, andreea.urzica@cs.pub.ro

G.A. Papadopoulos and C. Badica (Eds.): Intelligent Distributed Computing III, SCI 237, pp. 313–318.
springerlink.com © Springer-Verlag Berlin Heidelberg 2009

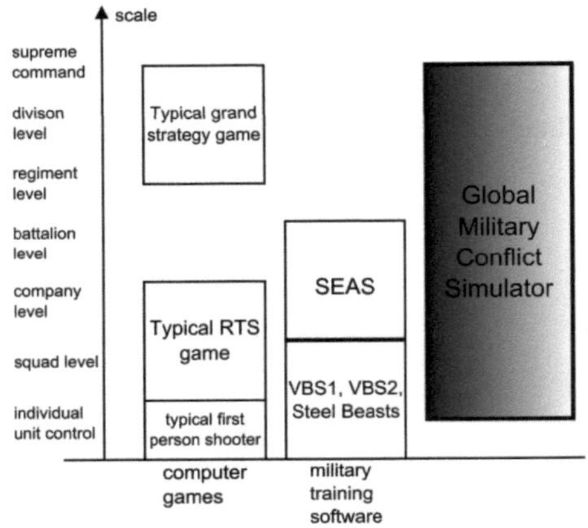

Fig. 1 The control level implemented in the current application compared with other military software simulations. It should be noted that several strategy games work on two disparate levels, one tactical and one grand strategic (e.g. Total War)

intelligence is not used at all, as the other characters involved in the simulation are usually controlled by training staff or other trainees[9].

The entertainment industry has also taken interest in virtual wars. Real Time Strategy (RTS) and Real Time Tactics (RTT) are the two of the genres that deal with computer wargaming. Real Time Tactics games (e.g. World in Conflict) put the player in command of a small force. AI is present on two levels: low-level AI (pathfinding, targeting) and high-level AI which controls the enemy armys commander. One frequent criticism of many titles in this genre is the need of micromanagement as a direct consequence of the large number of subordinates that the player has to manage.

Another genre of computer games related to the concept of the simulator are the Massively multiplayer online role-playing games (MMORPG), in which a large number of players interact with one another within a virtual game world. MMORPGs present a virtual world composed of large stretches of land (called "realms"), each simulated on a special server. The virtual, seamless world is an illusion, because the realm is a relatively small area, linked with other realms through "teleporters" (a.k.a. "portals"). Thus the MMORPG offer a large virtual world, but its geographic accuracy makes the approach unfeasible, because the virtual world is rather a sparse network of small geographic areas. One example of MMORPG game is the very popular World Of Warcraft.

In response to the existing products mentioned above, the Global Military Conflict Simulator attempts to both overcome their shortcomings and offer a new perspective in military simulation, by combining appropriate technologies. The simulator can serve as a military training tool or even an RTS game engine.

2 Motivation

First, this simulator is an attempt to combine existing technology, manifested in the graphical challenges of building a virtual globe and maintaining a suitable level of detail and framerate for a large volume of texture and geometry data, with a proposed application of the multi agent system paradigm that is inspired by the actual structure of an army, namely a hierarchy of units.

Second, the scope of the current military simulation software, training tools, virtual wargames and computer strategy games spans very little on the organizational, hierarchical or global aspect of an army, most of them being focused on a combat on a single level of the command hierarchy (a typical real time strategy game puts the player in command of a small number of units, typically in the tens, all under the direct control of the player, on a small theater of operations; a so-called "grand strategy" game shows the perspective of a high or supreme commander of an army, but handles the underlying levels of command by abstraction). A simulation tool able to work on all the possible levels and provide the largest possible geographic extent would increase the scope, realism and usefulness of computer military simulations and can easily become a game engine for an RTS.

3 System Architecture

Figure 2 shows the main components of the system and the way they are connected. It is a distributed client-server application, with the components on the left (graphics and user interface) on the client side. The components on the upper right are planned to be implemented as distributed servers.

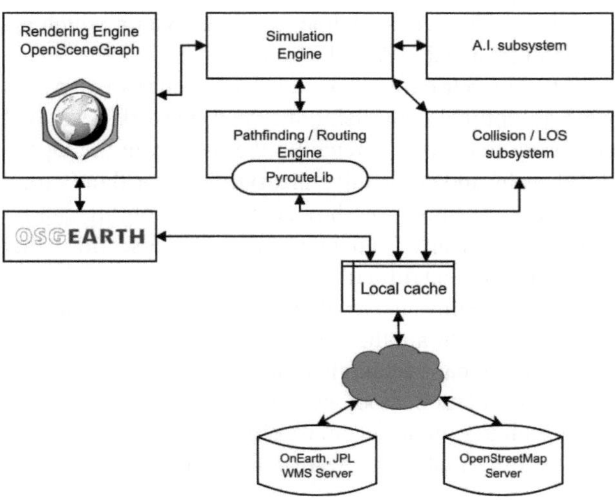

Fig. 2 General view of the application architecture

The most important component of the architecture is the Internet (colloquially called the 'cloud'), which separates the Data Sources from the client side. This is known as cloud computing[8]. The rationale for this architectural choice is that the amount of data is too big to be handled locally (for example, the Earth's surface is $510,072,000km^2$, if this were a single texture sized at 500m/pixel we would obtain a 127,518 MP image).

The Data Sources are servers which transmit geo-referenced data, such as satellite imagery (for the WMS), and a local map portion (OSM). The modules requesting such data also store them in a local cache. osgEarth fetches the satellite images and elevation maps and constructs the 3D model of the Earth. The rendering engine, OpenSceneGraph, takes the 3D model and renders it, along with the other graphical items supplied by simulation engine.

The core elements of the simulator, however, are the Simulation Engine, Routing Engine, Collision Engine and the AI subsystem (the 4 boxes in the upper right of the diagram). The reason why the functionality of the simulator is divided into 4 components is an implementation concern: the type of computations involved in the Routing and Collision "departments" are highly parallel and should be developed and employed as parallel server side components (in the spirit of cloud computing).

The basic element of the real time simulation is the Simulation Engine. This system component keeps track of all the variable (moving) objects in the virtual world: units, messages, models, etc. Its task is to update the simulation objects at each time step and respond to events that require a state change, e.g. adding a new unit to the simulation, modifying the simulation speed, etc. It also communicates with the other modules in order to collect decisive information about the units it manages.

Most modern software that deals with large scale complex AI uses scripting languages to define behaviour (e.g., the RTS Supreme Commander uses the Lua scripting language for almost all functionality). The separation between the Simulation Engine and the AI system is an implementation constraint: the AI system consists of a binding to a scripting language. The scripts define agent behaviour, which allows for easily extensible and interchangeable AI behaviour.

The Routing module ensures that mobile entities can travel between any two points in the virtual world, following an optimal path, using the road network described by the OpenStreetMap. The library in charge of fetching the data from the OpenStreetMap servers into the local cache and performing the routing algorithm (which is a traditional A*) is PyrouteLib. One limitation, however, is that routing can be performed only on roadways (i.e., specifically from intersection to intersection); this will make implementing infantry movement difficult.

The Collision module operates on the physical level of the simulation, and detects two things: whether a unit intersects with another object in the world or scenery (collision detection) and whether a unit can see another unit (Line Of Sight).

As a future option, all the 4 simulation modules should be implemented as services, thus making the distributed simulation more efficient, for example by implementing each module on a more suitable hardware configuration.

3.1 AI Subsystem

The AI subsystem is responsible with determining the actions of each of the simulation actors, known as non playable characters (NPCs). Intelligent entities in the simulation are interconnected agents, thus the AI as a whole is a multi-agent system. Modern militaries are hierarchical, and their organization is reflected in the Order of Battle. In its modern use, the order of battle signifies the identification, command structure, strength, and disposition of personnel, equipment, and units of an armed force during field operations. Carl Philipp Gottlieb von Clausewitz[18] noted that the order of battle depends on the effective span of control by a commander. Clausewitz recommended that armies have no more than eight to ten subunits and subordinate corps four to six subunits. According to this model, each commanding unit in the hierarchy can be assimilated as an acting layer in a subsumption architecture[4] or rather, as described by Spector and Hendler, a supervenience architecture.[16].

Each agent corresponds to a simulated unit. The role of an agent is determined by his place in the chain of command, which means that all but the topmost agent in the hierarchy have a "superior" and all but the lowest ranked agents have one or more "subordinates". This structure is based on, but not entirely faithful to the chain of command in an army. The agent hierarchy model is a simple tree, while the actual command structure of an army is a complex concept involving military ranks, commission of ranks and many types of additional staff[12, 6].

Each agent receives messages representing orders from his superior. The received orders alter the mental state of the agent, thus modifying the agent's behaviour. Based on his perception of the world and his mental state, the agent can formulate orders for his subordinates in order to satisfy his goals. The perception horizon (the range of perceptual events that the agent can interpret) for an agent is the union of its subordinates perception horizons. An order given by a higher command unit which needs to be carried out by a lower command unit will be passed down the chain of command to the relevant unit, and will interrupt whatever activity or goal that unit was trying to achieve. In conjunction with the fact that the lower level units accomplish simple, granular tasks, the multi agent system can be thought of as an example of subsumption architecture[16, 14].

This module will be implemented using a "traditional" agent-based modelling tool, albeit one that can easily support distributed simulation. We have plans to try out Repast, SeSAm and Dex.

4 Conclusions

The paper presented the architecture of a global conflict simulator and presents some aspects of its internal functioning. While the rendering aspect is complete and functional, the rest of the application is under development, and still subject to change. However, the agent-based model and approach is a constant of this project and will comply with the functional description presented in this article, including the adoption of a multi-agent system middleware.

The inclusion of open standards and external data has proved successful and reliable. The use of cloud computing, manifested in web map services and other web technologies, makes for an easy, lightweight client. This client can efficiently obtain data from the web and manage its caching, while keeping the application data away from the local "terminal" (the application itself has no more than 60MB out of the box).

The true scalability of the simulation, as well as the performance will be empirically tested as development continues; this is because the project depends on many libraries (some of which are also under development, e.g., osgEarth), and the software development process is iterative and incremental.

References

1. Beck, A.: Google Earth and World Wind: remote sensing for the masses. Antiquity 80, 308 (2006)
2. Bomford, G.: Geodesy, 855 p. (1980)
3. Brooks, R.: A robust layered control system for a mobile robot. IEEE journal of robotics and automation 2(1), 14–23 (1986)
4. Brooks, R.A.: Intelligence without reason. Artificial intelligence: critical concepts 3 (1991)
5. Dana, P.: The Geographer's Craft Project, Department of Geography, The University of Colorado at Boulder (1999) (accessed February 10, 2005)
6. Gorniak, P., Davis, I.: SquadSmart-Hierarchical Planning and Coordinated Plan Execution for Squads of Characters. In: Proc. of AIIDE 2007, pp. 14–19 (2007)
7. Grossner, K., Clarke, K.: Is Google Earth,Digital Earth?: Defining a vision. In: University Consortium of Geographic Information Science, Summer Assembly, Vancouver, WA (2006)
8. Hayes, B.: Cloud computing (2008)
9. Hill, R., Gratch, J., Marsella, S., Rickel, J., Swartout, W., Traum, D.: Virtual humans in the mission rehearsal exercise system. Künstliche Intelligenz 4(03), 5–10 (2003)
10. Kraak, M.J., Brown, A.: Web cartography: developments and prospects. Taylor & Francis, Abington (2001)
11. Laird, J., Van Lent, M.: Human-level AIs killer application. AI magazine 22(2) (2001)
12. Pechoucek, M., Thompson, S., Voos, H.: Defense Industry Applications of Autonomous Agents and Multi-Agent Systems (Whitestein Series in Software Agent Technologies and Autonomic Computing). Birkhäuser, Basel (2008)
13. Reynolds, C.: Steering behaviors for autonomous characters. In: Game Developers Conference, vol. 1999, pp. 763–782 (1999)
14. Russell, S.J., Norvig, P., Canny, J.F., Malik, J., Edwards, D.D.: Artificial intelligence: a modern approach. Prentice Hall, Englewood Cliffs (1995)
15. Sidran, D.E., Kearney, J.: The Current State of Human-Level Artificial Intelligence in Computer Simulations and Wargames. Computer 22(290), 4 (2004)
16. Spector, L., Hendler, J.: The use of supervenience in dynamic-world planning. In: Proceedings of The Second International Conference on Artificial Intelligence Planning Systems, pp. 158–163 (1994)
17. Thibault, D., Valcartier, D.: Commented APP-6A-Military symbols for land based systems
18. Clausewitz, C.V., Graham, J., Honig, J.: On war. Barnes & Noble Publishing (2004)

Author Index